1972

HEARING MECHANISMS IN VERTEBRATES

HEARING MECHANISMS IN VERTEBRATES

A Ciba Foundation Symposium

Edited by

A. V. S. DE REUCK

and

JULIE KNIGHT

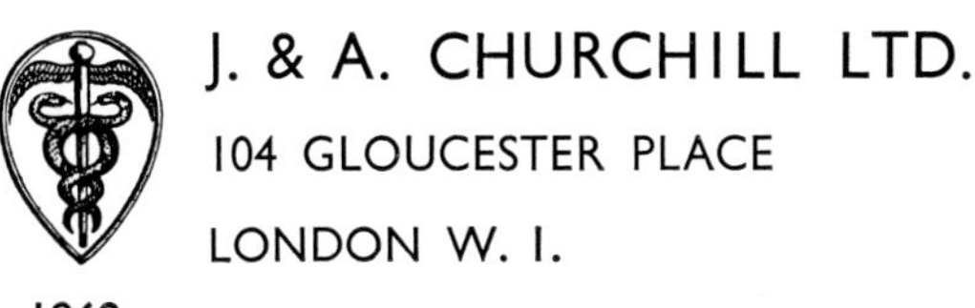

J. & A. CHURCHILL LTD.
104 GLOUCESTER PLACE
LONDON W. 1.

1968

First published 1968

Containing 107 illustrations

Standard Book Number 7000 1360 1

Printed in Great Britain

Contents

O. Lowenstein	Chairman's introduction	1

Section I Structure and function of hearing organs in non-mammalian vertebrates

P. S. Enger	Hearing in fish	4
Discussion	*Batteau, Bosher, Davis, Enger, Engström, Fex, Fraser, Hallpike, Johnstone, Lowenstein, Schwartzkopff, Spoendlin, Tumarkin*	11
A. Tumarkin	Evolution of the auditory conducting apparatus in terrestrial vertebrates	18
Discussion	*Bosher, Davis, Enger, Hood, Lowenstein, J. D. Pye, Schwartzkopff, Tumarkin, Whitfield*	37
J. Schwartzkopff	Structure and function of the ear and adjacent brain area in birds	41
Discussion	*Batteau, Hallpike, Johnstone, Lowenstein, Neff, Schwartzkopff, Whitfield*	59

Section II Structure and function in mammals

J. D. Pye	Hearing in bats	66
Discussion	*Bosher, Davis, Engström, Fex, Johnstone, Lowenstein, Ade Pye, J. D. Pye, Salomon, Schwartzkopff, Tumarkin*	84
H. Spoendlin	Ultrastructure and peripheral innervation pattern of the receptor in relation to the first coding of the acoustic message	89
Discussion	*Batteau, Bosher, Davis, Engström, Fex, Johnstone, Spoendlin*	119
G. Bredberg	Cochlear structure and hearing in man	126
Discussion	*Bosher, Bredberg, Davis, Engström, Fex, Hallpike, Hood, Lowenstein, Ade Pye, Rose, Tumarkin, Whitfield*	139

Section III Afferents and efferents in the auditory nerve

J. E. Rose J. F. Brugge D. J. Anderson J. E. Hind	Patterns of activity in single auditory nerve fibres of the squirrel monkey	144
Discussion	*Davis, Enger, Erulkar, Evans, Fex, Johnstone, Lowenstein, Neff, Rose, Schwartzkopff, Spoendlin, Whitfield*	157
J. Fex	Efferent inhibition in the cochlea by the olivo-cochlear bundle	169
Discussion	*Davies, Erulkar, Fex, Johnstone, Lowenstein, Tumarkin, Whitfield*	181

Section IV Spatial localization

H. Kleerekoper, T. Malar — Orientation through sound in fish — 188

Discussion — *Batteau, Enger, Engström, Johnstone, Kleerekoper, Lowenstein, Neff, J. D. Pye, Schwartzkopff, Tumarkin, Whitfield* — 201

W. D. Neff — Behavioural studies of auditory discrimination: localization of sound in space — 207

Discussion — *Davis, Evans, Neff, Rose, Schwartzkopff* — 231

D. W. Batteau — Role of the pinna in localization: theoretical and physiological consequences — 234

Discussion — *Batteau, Davis, Hallpike, Johnstone, Lowenstein, J. D. Pye, Schwartzkopff, Whitfield* — 239

Section V Central mechanisms

I. C. Whitfield — Centrifugal control mechanisms of the auditory pathway — 246

Discussion — *Bosher, Davis, Erulkar, Evans, Johnstone, Fex, Lowenstein, Salomon, Whitfield* — 254

H. Davis — Auditory responses evoked in the human cortex — 259

Discussion — *Davis, Erulkar, Fex, Hood, Johnstone, Kleerekoper, Lowenstein, Neff, Rose, Salomon, Whitfield* — 269

E. F. Evans — Cortical representation — 272

Discussion — *Davis, Engström, Erulkar, Evans, Fex, Lowenstein, Neff, Rose, Schwartzkopff, Tumarkin, Whitfield* — 287

General discussion — Absolute sensitivity of the hearing mechanism — 296
Possible functions of the cochlear efferent fibres — 298
Bosher, Davis, Engström, Fex, Johnstone, Lowenstein, Schwartzkopff, Spoendlin, Tumarkin, Whitfield

O. Lowenstein — Chairman's closing remarks — 310

Author index — 311

Subject index — 313

Membership

Symposium on Hearing Mechanisms in Vertebrates, held 26th–28th September, 1967

O. Lowenstein (Chairman)	Dept. of Zoology and Comparative Physiology, University of Birmingham
D. W. Batteau*	Dept. of Mechanical Engineering, Tufts University, Medford, Massachusetts
S. K. Bosher	Ferens Institute of Otolaryngology, Middlesex Hospital, London
G. Bredberg	Öronkliniken, Göteborgs Universitet, Göteborg
H. Davis	Central Institute for the Deaf, St. Louis, Missouri
P. S. Enger	Institute of Physiology, University of Oslo
H. Engström	Öronkliniken, Göteborgs Universitet, Göteborg
S. D. Erulkar	Dept. of Biophysics, University College, London
E. F. Evans	Dept. of Communication, University of Keele, Staffordshire
J. Fex	Laboratory of Neurobiology, National Institute of Mental Health, Bethesda, Maryland
F. C. Fraser	78 Hayes Road, Bromley, Kent (formerly British Museum, Natural History, London)
C. S. Hallpike	Ferens Institute of Otolaryngology, Middlesex Hospital, London
D. J. Hood	MRC Otological Research Unit, National Hospital, London
B. M. Johnstone	Dept. of Physiology, University of Western Australia, Perth
H. Kleerekoper	Institute of Life Sciences, Texas A and M University, College Station, Texas
W. D. Neff	Center for Neural Sciences, Indiana University, Bloomington, Indiana
Ade Pye	Institute of Laryngology and Otology, London
J. D. Pye	Dept. of Zoology, King's College, London
J. E. Rose	Laboratory of Neurophysiology, University of Wisconsin Medical School, Madison, Wisconsin
G. Salomon	Øre-naese-halsafdeling, Gentofte Amts Sygehus, Hellerup, Copenhagen
J. Schwartzkopff	Institut für Allgemeine Zoologie der Ruhr-Universität, Bochum
H. Spoendlin	Otorhinolaryngologische Klinik und Poliklinik der Universität, Kantonsspital Zürich
A. Tumarkin	45 Rodney Street, Liverpool (formerly Director, Department of Otorhinolaryngology, The University, Liverpool)
I. C. Whitfield	Neurocommunications Research Unit, University of Birmingham

* Deceased

The Ciba Foundation

The Ciba Foundation was opened in 1949 to promote international cooperation in medical and chemical research. It owes its existence to the generosity of CIBA Ltd, Basle, who, recognizing the obstacles to scientific communication created by war, man's natural secretiveness, disciplinary divisions, academic prejudices, differences of language, or separation by distance, decided to set up a philanthropic institution whose aim would be to overcome such barriers. London was chosen as its site for reasons dictated by the special advantages of English charitable trust law (ensuring the independence of its actions), as well as those of language and geography.

The Foundation's house at 41 Portland Place, London, has become well known to workers in many fields of science. Every year the Foundation organizes six to ten three-day symposia and three to four shorter study groups, all of which are published in book form. Many other scientific meetings are held, organized either by the Foundation or by other groups in need of a meeting place. Accommodation is also provided for scientists visiting London, whether or not they are attending a meeting in the house.

The Foundation's many activities are controlled by a small group of distinguished trustees. Within the general framework of biological science, interpreted in its broadest sense, these activities are well summed up by the motto of the Ciba Foundation: *Consocient Gentes*—let the peoples come together.

Preface

In this, the fourth in a series of symposia on sensory function held at the Ciba Foundation, we turn to vertebrate hearing mechanisms—having previously dealt with *Colour Vision* (1965), *Touch, Heat and Pain* (1966) and *Myotatic, Kinesthetic and Vestibular Mechanisms* (1967). Included in the volume are both formal papers, providing a survey of current research in a range of aspects of hearing in vertebrates, and the informal discussions of the papers by the group of twenty-four specialists who attended the meeting. It is hoped that this combination of the formal and informal, the factual and speculative, may give readers a perspective not only of current developments in the field but also of areas where further investigation may be most rewarding.

The Ciba Foundation is once again much indebted to Professor Otto Lowenstein, for his guidance of the meeting itself and for his continuing help over the preceding months of planning. Our thanks are also due to Dr. I. C. Whitfield for valuable advice during the preliminary stages.

Shortly after the symposium we were shocked to learn of the sudden death of one of the members, Professor Dwight W. Batteau, whose outstanding contribution and personality had made a deep impression on the meeting. We are most grateful to Dr. Sanford J. Freedman, also of Tufts University, Massachusetts, for his help in editing Professor Batteau's often highly condensed comments in discussion, and also for reading his paper in proof.

CHAIRMAN'S INTRODUCTION

Professor O. Lowenstein

This is the fourth in the Ciba Foundation's series of symposia on sensory function, and the third on mechanoreceptive mechanisms, a fact which should assure us of the great importance of mechanoreception. It is curious that whereas there has been hardly any difficulty in defining "vision"—when we talk about vision everybody, quite naturally, knows that what is being discussed is something performed by the eyes and what is being perceived is light—when we come to hearing, unfortunately the history of the subject has turned out differently, and the literature is full of often rather meaningless disputations about what is "hearing". Very often one finds circular definitions, such as "hearing is the perception of sound", and when one wonders what sound is, one is told that sound is what is perceived by the ears. I fervently hope that this sort of scholastic dispute will be avoided during the present meeting. Let us be naïve and take hearing at its face value. I know that this will not be completely possible, because contributors dealing with hearing in fish will have some difficulty in proving that what they are studying is in every instance genuinely hearing. However, the less arguing about the concept of hearing, the better, I feel.

We shall begin quite fittingly with an account of hearing in fish. However much emphasis we may finally place on the mechanism of hearing as found in the mammalian ear, we must never lose sight of the comparative aspect, a thread which has run through all three of the earlier symposia. From hearing in fish we shall follow the great adventure of life in its emergence on to dry land and the adaptations connected with terrestrial existence. Then, with the exception of a paper on hearing in birds and one on orientation through sound in fish, the rest of the symposium will be devoted to mammalian hearing, fittingly culminating in contributions dealing with central nervous phenomena. This in outline is the scope of the symposium which now follows.

SECTION I

STRUCTURE AND FUNCTION OF HEARING ORGANS IN NON-MAMMALIAN VERTEBRATES

HEARING IN FISH

PER S. ENGER

Institute of Physiology, University of Oslo

FOR the purposes of presentation the topic of hearing can conveniently be divided into descriptions of (1) The range of sound frequencies perceived by the animal, and the auditory thresholds; (2) The power of pitch discrimination; and (3) The ability to determine the direction of a sound source. (The last topic will not be discussed here, since Professor Kleerekoper will present a paper on orientation through sound in fish; see p. 188.) A presentation of hearing in aquatic animals, however, should start with a discussion of the acoustic stimulus and the so-called near-field and far-field effects.

NEAR-FIELD AND FAR-FIELD

In reports of sound perception in fish there has been some disagreement and controversy about the auditory stimulus (cf. Harris and van Bergeijk, 1962; Dijkgraaf, 1963*a*), simply because the physical parameters of this stimulus have been unknown or at least not fully understood. Harris and van Bergeijk (1962) clarified some of these difficulties in an important study of the lateral line organ and introduced the terms near-field and far-field. In short, an underwater sound source produces two effects: one is a compression and rarefaction of water particles—that is, a change of pressure; the other is a displacement of water particles around the source. The pressure changes represent the actual propagated sound wave whose pressure amplitude decreases linearly with distance, whereas the displacement amplitude of the near-field decreases with distance to the second or third power. In the case of a spherical, pulsating sound source, the relation between far- and near-fields is given by the equation

$$d = \frac{p}{2\pi f \rho c}\left(i - \frac{\lambda}{2\pi r}\right)$$

where d = displacement amplitude, p = sound pressure, f = frequency, ρ = density, c = sound velocity, λ = wavelength, r = distance from sound source, $i = \sqrt{-1}$ (meaning a 90° phase shift). The propagated sound wave obeys the equation $d=p/2\pi f\rho c$ and the factor $\lambda/2\pi r$ is the correction for near-field displacements. It can be seen that for low frequencies in particular, the near-field effect extends for a considerable distance from the source. For example, for 100 Hz the wavelength is 15 m. and at 2·4 m. distance the displacement amplitudes of the two effects are equal.

Harris and van Bergeijk (1962) did not study sound reception in fish, but pointed out the difficulties one might encounter in such studies. Most experiments on hearing in fish have been performed in relatively small tanks and therefore the acoustic stimulus from an underwater speaker has been a mixture of far- and near-field effects. Inasmuch as the sound pressure is the physical parameter measured with ordinary hydrophones, the distance between the fish and the underwater loudspeaker also becomes a parameter in the auditory stimulus under such circumstances. The total displacement amplitudes at threshold, however, calculated from the formula given, are independent of distance (Olsen, 1965; Enger, 1966).

The near-field effect can stimulate the lateral line receptors and also the auditory receptors directly. The far-field effect can stimulate auditory receptors through the swim bladder, which is caused to vibrate. The net effect in both cases is a displacement of the auditory receptors.

In tanks of a reasonable size there is only one way of avoiding the near-field effect for low frequencies, namely by producing the sound in air (Parvulescu, 1964) in such a way that the aquarium is inside a closed chamber in which the loudspeaker diaphragm constitutes part of the wall. Increased (or decreased) pressure in the air will immediately produce increased (or decreased) pressure in the water in the aquarium.

Enger (1966), using this information from Parvulescu, found considerable differences in pressure thresholds in goldfish exposed to sound from a water loudspeaker and an air loudspeaker, respectively, but the experimental conditions were far from ideal. The system was "open" instead of "closed", so that sound in air was not necessarily transmitted as sound pressure to the water, but could produce displacement on the water surface. In spite of the technical short-comings, there was a clear indication that displacement is a better parameter than pressure in determinations of auditory thresholds.

Jacobs and Tavolga (1967) were closer to creating a pure sound pressure, but the lowest auditory threshold obtained was about the same as that found by Enger (1966), namely about -40 db re 1 μbar at 1,000 Hz. However, for low frequencies (below 800 Hz) they found considerably lower thresholds than did Enger.

In order to avoid using small aquaria, the acoustics of which are little understood, Enger and Andersen (1967) performed some experiments at sea. Reflections from aquarium walls and standing waves can thereby be avoided and, most important, free field acoustic conditions can be obtained. Through implanted electrodes, saccular microphonic potentials were recorded. The fish was held at distances of up to 10 m. from an underwater loudspeaker—that is, in the acoustic far-field. In the codfish (*Gadus morrhua*)—which has a swim bladder—the potential amplitude was clearly a function of sound pressure and not of distance, implying that the swim bladder transforms pressure into displacement. In the sculpin (*Cottus scorpius*)—which lacks a swim bladder—the amplitude varied with distance as well as with pressure, but beyond 1 m. distance no microphonics could be recorded at all. The conclusion drawn from this has been that the swim bladder in teleosts is essential for hearing in the acoustic far-field.

SOUND FREQUENCY RANGE AND AUDITORY THRESHOLDS

The range of sound frequencies perceived by fishes has been determined for a large number of species. As far as hearing was concerned, teleosts were previously put roughly into two major groups: (1) The Ostariophysi, perceiving frequencies up to several thousand Herz; and (2) The non-Ostariophysi, perceiving frequencies up to 1,000 Hz. The former group comprises the Cyprinidae, Characinidae and Siluridae, and the basis for their well-developed sense of hearing is the so-called Weberian ossicles connecting the swim bladder to the labyrinth. Vibration of the swim bladder, due to the passage of sound, will then be directly transmitted to the ear. The extirpation of the malleus of the Weberian ossicles reduces the sensitivity by 30–40 db (Poggendorf, 1952) in the catfish, and mutilation of the swim bladder reduces the sensitivity by 13 db at low frequencies (330–750 Hz) and by 30 db at high frequencies (1,500 Hz) (Kleerekoper and Roggenkamp, 1959). The few audiograms published for ostariophysid species show that their lowest threshold is for frequencies around 1,000 Hz at an intensity of about -40 to -45 db re 1 μbar. These values were determined and found to be the same whether the fish was within the acoustic near-field (Poggendorf, 1952; Enger, 1966) or the far-field (Jacobs and Tavolga, 1967). This value was roughly 20 db higher with the fish in essentially a pure near-field (Weiss, 1966).

Among the non-Ostariophysi the situation is not quite as clear as stated above. Very many species no doubt respond to frequencies below 1 kHz only, but some species with swim bladder extrusions contacting some part of the labyrinth (Sparidae) can perceive frequencies up to 1,250 Hz (Dijkgraaf, 1952). Tavolga and Wodinsky (1963) also obtained responses to higher frequencies than 1 kHz in several species, but found no systematic relation between a close swim bladder–ear connexion and perception of relatively high frequencies, and the converse situation. Thus, the auditory threshold of *Holocentrus ascencionis* which has a swim bladder–ear connexion, is higher than that of *Equetus*, which has no such connexion. The lowest thresholds in most non-ostariophysids are found for frequencies below 500 Hz, although there are species with 600–800 Hz as their best frequency range. The accuracy of the threshold values reported for low frequencies, however, is probably poor. First of all, the near-field effect must have been rather pronounced in most experiments and therefore large variations in the threshold determinations in different reports are to be expected. Secondly, individual differences in auditory sensitivity among several specimens of the same species may be 10–20 db or even more for frequencies below 300 Hz (Tavolga and Wodinsky, 1965). Thirdly, threshold values depend on the duration of the stimulus. In goldfish, the threshold is found to be lowest for continuous tone stimulation (Offutt, 1967), but this does not necessarily have to be so for all fish. (Incidentally, this is in contrast to man, whose threshold for a series of tone pips is lower than for a continuous tone.) Finally, the threshold varies with the background noise. Nelson (1966) reported that an increase in the background noise by 20 db resulted in a corresponding rise in the auditory threshold of the lemon shark (*Negaprion brevirostris*).

Species in which the ear is more or less surrounded by air-filled cavities do perceive considerably higher frequencies than most other non-ostariophysids. Thus, the Mormyridae perceive tones of pitch up to 3,100 Hz (Stipetic, 1939), and the Anabantidae up to 4,700 Hz (Schneider, 1942). There is still another group of fish, the Clupeidae, which on anatomical grounds have long been suspected of possessing a relatively acute sense of hearing. In these fish a thin duct from the swim bladder expands into two air-filled bullae close to each labyrinth (Tracy, 1920; Evans, 1932; Wohlfahrt, 1936). The anterior one, the pro-otic bulla, is aimed directly at the sensory epithelium of the utricle. A membrane in this bulla separates the air from the endolymph, and pressure changes produced by sound would presumably cause the membrane to vibrate. Recently, Enger (1967) published a tentative audiogram for the herring (*Clupea harengus*) based on recordings of the nervous activity in auditory neurons in the medulla oblongata. The threshold is probably low (less than −20 db) for frequencies up to about 1,200 Hz, after which it increases rapidly, but nervous responses were obtained for frequencies up to 4,000 Hz at moderate intensities.

All reports on hearing in teleosts support the notion that some kind of air bubble enclosed within the animal greatly improves hearing, and perhaps is even necessary to make hearing possible in the acoustic far-field.

Although a swim bladder or some other air-filled cavity may be the most effective transducer in the perception of underwater sound, it is not the only way propagated underwater sound can be received. Elasmobranchs have no swim bladder, but there are numerous reports of the perception of sound in these fishes. Thus, Kritzler and Wood (1961) trained bull sharks (*Carcharhinus leucas*) to respond to pure tones and obtained responses in the 100–1,500 Hz frequency range, with the lowest threshold for 400–600 Hz. The distance between fish and loudspeaker in these experiments was approximately 6 m., meaning perhaps that the fish detected far-field sound waves. More convincing in this respect are later publications (Nelson and Gruber, 1963) where sharks are reported not only to hear low frequency pulsed sounds at distances beyond 15–25 m., but also to detect the direction of the source. Observations from aircraft (Nelson, 1966) showed that sharks detected and orientated to sounds in the acoustic far-field from distances as great as 600 feet.

Auditory thresholds obtained are fairly high (above 1 μbar), except in the report of Kritzler and Wood (1961), and there is good evidence that in spite of the sharks being in the acoustic far-field, the displacement rather than pressure is the stimulus for the receptors (Banner, 1967).

Whether the lateral line organ responds to low frequency auditory stimuli in the far-field (which might be suspected, since these fish orientate in the acoustic far-field) is not known. Dijkgraaf (1963*b*) found in the dogfish (*Scyliorhinus caniculus*) that the inner ear was responsible for the detection of a tone of 180 Hz. His experiments were performed in the near-field, but it may well be that the difference in acoustic properties between water and cartilage is sufficient for elasmobranchs to detect low frequency sound in the far-field as well.

PITCH DISCRIMINATION

Studies on the power of pitch discrimination in fish are rather few. Among the Ostariophysi, Stetter (1929) found minnows (*Phoxinus laevis*) able to discriminate two tones an octave apart (100 per cent frequency difference) and in one case even a minor third apart (19 per cent). These values are for absolute pitch, since the fish in these conditioning experiments was not presented with the two tones in immediate succession, but had to "remember" the first tone for some time before the next one sounded. Wohlfahrt (1939) investigated relative pitch discrimination in minnows and found it to be about 6 per cent for tones around 1,000 Hz.

Dijkgraaf and Verheijen (1950), using a better technique, found these fishes able to distinguish two tones differing in pitch by 3 per cent in the 400–800 Hz range. With the same method, Dijkgraaf (1952) found the best pitch discrimination for non-ostariophysids to be 9 per cent. For elasmobranchs, the only report known to the present author is that of Nelson (1966), who found discrimination ability in lemon sharks (*Negaprion brevirostris*) to be half an octave or a little better in the 40–60 Hz range. This refers to absolute pitch, however, since the time-interval between the presentation of the two tones was about one minute.

For animals which have no obvious morphological frequency analyser such as a cochlea, this accuracy in pitch distinction is certainly quite remarkable. One can think of two possible mechanisms for pitch discrimination in these animals; either that a synchronization between sound frequency and the frequency of impulse discharge takes place, or that different sensory units respond to different sound frequency ranges. The first possibility was formulated by Wever (1933) in his so-called volley theory, and in recordings from the elasmobranch nerve the impulse discharge is indeed synchronized with the sound (Lowenstein and Roberts, 1951). The second possibility implies that some kind of frequency analyser is present after all, and Enger (1963) showed that even in fish this possibility cannot be excluded. An analysis of the nervous activity of single auditory neurons in the sculpin (*Cottus scorpius*) revealed four types of neurons, classified according to their spontaneous activity, but at the same time with different properties in respect of their responsiveness to sound: (1) Units with a regular spontaneous discharge did not respond to sound stimulation. This agrees well with the results of Lowenstein and Roberts (1949, 1951), who found that labyrinthine neurons (in the ray, *Raja clavata*) with a regular spontaneous discharge responded to positional changes, but not to vibratory stimuli. (2) Units with no spontaneous discharge responded to low frequencies only (below 200 Hz). (3) Units with an irregular spontaneous discharge pattern responded to all frequencies which the animal presumably can hear. (4) Units with a burst-like spontaneous discharge responded to sound frequencies up to around 300 Hz with a discharge characterized by being synchronized with the stimulating sound and without showing any adaptation.

There is no indication that any frequency analysis is taking place by, for example, the lagena and sacculus covering different sound frequencies. All types of units were found in these two structures as well as in the utriculus. This is in contrast to the situation in rays, in which no sound responses were recorded from the lagena (Lowenstein and Roberts, 1951).

Units of type (4), and to a certain extent also of type (3), support the volley theory insofar as pitch discrimination is concerned. On the other hand, a comparison of the information received by the central nervous system through neurons of type (2) and (3), respectively, should give a rough discrimination of frequency regardless of the eventual information received by the synchronization between sound frequency and discharge rate. From these results one would expect frequency distinction in the sculpin to be at its best for frequencies below 200 Hz, fairly good between 200 and 400 Hz and virtually absent above 400 Hz. Data from conditioning experiments on *Gobius niger* (Dijkgraaf, 1952) compare well with these expectations, inasmuch as the discriminating ability is 9 per cent around 150 Hz and 12 per cent for 300 Hz. In *Sargus annularis*, which is responsive to higher frequencies than *Cottus scorpius*, the values reported are for 150 Hz, better than 9 per cent; for 300–450 Hz, 15 per cent; and for 600 Hz, over 30 per cent.

SUMMARY

An underwater sound source produces a propagated sound pressure wave and a local water displacement. The latter can stimulate auditory receptors directly and will at close range interfere with determinations of sound pressure thresholds. Perception of propagated sound is accomplished by air bladders or vesicles in teleosts. A swim bladder will undergo volume changes in response to pressure changes. These volume changes produce local particle displacements which in turn stimulate the auditory receptors.

Species having a chain of ossicles connecting the swim bladder to the labyrinth (Ostariophysi) perceive frequencies up to at least 4,000 Hz. Non-ostariophysids with air cavities close to the labyrinth (Mormyridae, Clupeidae) have an upper limit of 3,000–4,000 Hz; other non-ostariophysids have upper limits ranging from less than 1,000 Hz to 2,000 Hz. Teleosts without a swim bladder are perhaps unable to perceive propagated sound. Elasmobranchs do hear propagated sound although air-filled body cavities are lacking.

Pitch discrimination in fish may be as good as 3 per cent. The neurological basis for this is in part a frequency analysis taking place in the peripheral auditory system. Not even in fish, therefore, is pitch discrimination based solely on discharges in auditory nerve fibres having the same frequency as the stimulating sound.

REFERENCES

BANNER, A. (1967). In *Lateral Line Detectors*, ed. Cahn, Phyllis H. Bloomington and London: Indiana University Press.
DIJKGRAAF, S. (1952). *Z. vergl. Physiol.*, **34**, 104–122.
DIJKGRAAF, S. (1963*a*). *Biol. Rev.*, **38**, 51–105.
DIJKGRAAF, S. (1963*b*). *Nature, Lond.*, **197**, 93–94.
DIJKGRAAF, S., and VERHEIJEN, F. J. (1950). *Z. vergl. Physiol.*, **32**, 248–256.
ENGER, P. S. (1963). *Acta physiol. scand.*, **59**, Suppl. 210.
ENGER, P. S. (1966). *Comp. Biochem. Physiol.*, **18**, 859–868.
ENGER, P. S. (1967). *Comp. Biochem. Physiol.*, **22**, 527–538.
ENGER, P. S., and ANDERSEN, R. (1967). *Comp. Biochem. Physiol.*, **22**, 517–525.
EVANS, H. M. (1932). *Proc. R. Soc. B*, **111**, 247–280.
HARRIS, G. G., and BERGEIJK, W. A. VAN (1962). *J. acoust. Soc. Am.*, **34**, 1831–1841.
JACOBS, D. W., and TAVOLGA, W. N. (1967). *Anim. Behav.*, **15**, 324–335.
KLEEREKOPER, H., and ROGGENKAMP, P. A. (1959). *Can. J. Zool.*, **37**, 1–8.
KRITZLER, H., and WOOD, L. (1961). *Science*, **133**, 1480–1482.
LOWENSTEIN, O., and ROBERTS, T. D. M. (1949). *J. Physiol., Lond.*, **110**, 392–415.
LOWENSTEIN, O., and ROBERTS, T. D. M. (1951). *J. Physiol., Lond.*, **114**, 471–489.
NELSON, D. R. (1966). *Diss. Abstr.*, **27**, no. 1.
NELSON, D. R., and GRUBER, S. H. (1963). *Science*, **142**, 975–977.
OFFUTT, G. C. (1967). *J. acoust. Soc. Am.*, **41**, 13–19.
OLSEN, K. (1965). Thesis, University of Oslo, unpublished.
PARVULESCU, A. (1964). In *Marine Bio-Acoustics*, pp. 87–100, ed. Tavolga, W. N. Oxford: Pergamon.
POGGENDORF, D. (1952). *Z. vergl. Physiol.*, **34**, 222–257.
SCHNEIDER, HILTRUDE (1942). *Z. vergl. Physiol.*, **29**, 172–194.
STETTER, H. (1929). *Z. vergl. Physiol.*, **9**, 339–477.
STIPETIC, ELISABETH (1939). *Z. vergl. Physiol.*, **26**, 740–752.
TAVOLGA, W. N., and WODINSKY, J. (1963). *Bull. Am. Mus. nat. Hist.*, **126**, 177–240.
TAVOLGA, W. N., and WODINSKY, J. (1965). *Anim. Behav.*, **13**, 301–311.
TRACY, H. C. (1920). *J. comp. Neurol.*, **31**, 219–257.
WEISS, B. A. (1966). *J. aud. Res.*, **6**, 321–335.
WEVER, E. G. (1933). *Physiol. Rev.*, **13**, 400–425.
WOHLFAHRT, T. A. (1936). *Z. Morph. Ökol. Tiere*, **31**, 371–410.
WOHLFAHRT, T. A. (1939). *Z. vergl. Physiol.*, **26**, 570–604.

DISCUSSION

Fraser: Dr. Enger, you have shown that the Ostariophysi which have a swim bladder seem to have a better perception of sound than those non-Ostariophysi which do not have a swim bladder. Is it not a question of the opportunity of transferring a molecular vibration into a molar vibration on going from water into air that gives the better performance in the Ostariophysi?

Enger: The difference between the Ostariophysi and non-Ostariophysi is that the former have a swim bladder *and* a chain of ossicles connecting it to the inner ear, which compare functionally to our middle-ear ossicles. The non-Ostariophysi have no such connexion but they may have swim bladders.

An ostariophysid of course has a good mechanism for transferring vibrations of the swim bladder to vibrations of the endolymph of the inner ear. A non-ostariophysid with a swim bladder is still not at a complete loss, because vibrations of the swim bladder produce a secondary near-field effect which goes through the tissues of the fish and also to the auditory receptors. That is how I see it. Without a swim bladder you may be in difficulty, and from the data presented here I think you will agree that sound perception in the far-field is impaired, to say the least. On the other hand, sharks have been shown to perceive sound in the far-field, but it was concluded that in these fish also the actual stimulus is displacement or shearing and not pressure (Nelson, D. R. [1966]. *Diss. Abstr.*, **27**, no. 1); there is no pressure receptor.

Fraser: Among the non-Ostariophysi you mentioned the Mormyridae which have air spaces in the labyrinth, which again indicates that it is better to have an air space than not to have one.

Enger: Yes. And of course the closer this air space is to the labyrinth, the stronger are the near-field effects, so that when you get right next to the sensory cells, the displacement will produce large effects on them.

Lowenstein: When you recorded the saccular microphonics, Dr. Enger, where exactly were the electrodes implanted?

Enger: The electrodes were rather thick (0·1 mm. diameter) and they were placed between the saccule and the brain, but not so far ventrally as to destroy the sensory cells.

Lowenstein: Secondly, did you try tilting the fish to see whether the units which did not respond to sound were sensitive to changes of position?

Enger: No: my set-up did not allow that.

Schwartzkopff: The microphonics that you recorded showed the double wave very clearly, which has also been seen in the lateral line system. My co-worker Miss B. Grözinger and I ([1967]. *Naturwissenschaften*, **54**, 446) found the same to be true occasionally even at midbrain level, which means that this double response goes up through the neural pathways and is not only a microphonic effect from the periphery. We have no explanation for this but we would like to know more about it. We have considered the possibility that one could influence the double wave by tilting the fish, but we have not done experiments to this effect. We thought that by changing the position of the fish, vestibular effects might come in and one or other part of the auditory events might become stronger. Have you any ideas about that?

Enger: Not really. But one can change the relative amplitudes of the two components of the microphonic response very much, for example by touching the receptors. The microphonic potentials from fish can be completely sinusoidal, yet the frequency is double that of the stimulating sound. On the other hand you can also get different distorted wave-forms with a fundamental frequency equal to that of the stimulus. I am surprised that you can record these double potentials in the brain, but of course with sinusoidal stimulation, in fish as in other animals, you can have two spikes on each wave, if the sound intensity is high enough.

Schwartzkopff: We certainly found the double spikes on these waves, though not at very high intensity, and they change with the relative amplitude of the wave. However, this is not a microphonic in the sense of being produced by the hair cells, but some potential of the nervous brain structures which is not exactly sinusoidal, but shows two peaks and spikes may appear on top of both peaks, or sometimes only on one.

Tumarkin: Å. Flock ([1964]. *Acta oto-lar.*, Suppl. 199) also found in the lateral line organ that the microphonic had double the frequency of the stimulus. He related this to the fact that alternate hair cells are orientated in opposite directions, so that a displacement that hyperpolarized one would depolarize the other, and *vice versa.* He confirmed this by recording the effect on the microphonic of a static displacement of the cupula. This would appear to be exactly the experiment that Professor Schwartzkopff suggests. A. Zalin ([1967]. *J. Lar. Otol.*, **81**, 119–135), discussing the function of the hair cell, has suggested that the kinocilium is a true motile organ exactly as it is elsewhere and that its function is to bias the stereocilia so as to set up a resting discharge. This would explain the double microphonic and also the effects produced by static displacement of the cupula. I would go further and suggest that the hair cells may turn out to be coupled in pairs so that they act in push-pull. This might considerably improve not only sensitivity but also directionality.

Lowenstein: In the sacculus of all the forms I have seen (cyclostomes and elasmobranchs), the hair cells point in two opposite directions (see also Wersäll, J., Gleisner, L., and Lundquist, P.-G. [1967]. *Ciba Fdn Symp. Myotatic, Kinesthetic and Vestibular Mechanisms*, pp. 105–116. London: Churchill). There are always two populations, divided by a midline, and it is quite likely that the two peaks are created by the two populations responding alternately.

Batteau: Dr. Enger, I was interested to learn that you feel that displacement is the dominant stimulus, rather than pressure, because my colleagues and I at Tufts University have looked at the problem of sensing pressure and have found that biochemical reactions are very insensitive to pressure—as opposed to temperature, for example. We suggested that if there was shear, produced by pressure gradients or motion, the strong forces produced by the shearing of surfaces would be sufficient to disturb the biochemical rates, whether they be shifts in energy levels or actual reactions. I did not then know anything about the actual structure, and I would be very interested if it is so constructed that shearing is produced in the sensory regions.

Lowenstein: Shearing is apparently the *only* stimulus that affects the hair cells. My slight bewilderment has always been in connexion with this question of pressure *versus* displacement as the stimulus, because I do not know of a pure pressure receptor. There is always displacement in the last instance.

Fex: Since we do not actually know what the mechanical transducer is, at what level do we speak about the shearing force as the *critical* force? We do not know where it is critical that something changes.

Lowenstein: It is at the hair cell itself, which has processes—one kinocilium with a root, and a group of stereocilia. There are ancillary structures over the hair cell, either otolith organs or cupulae, which undergo displacement, and the shearing force then acts somewhere at the base of the hair processes. So the shearing force acts right at the point where transduction must begin.

Fex: We do not really know that transduction does begin at the base of the hair processes.

Lowenstein: This is surmise, but the mechanical events certainly involve the bases of the hairs and therefore it is plausible that mechano-electric transduction begins there. In the organs of dynamic and static equilibrium shearing forces are bound to act on the bases of the cilia.

Hallpike: Dr. Enger, have you any idea to which half-cycles of the sinusoidal sound stimulus the evoked action potentials correspond?

Enger: I do not know.

Schwartzkopff: They respond to both; we found a double response.

Enger: In your case, yes, but I found a discharge rate corresponding to the frequency of the sound. The spikes were phase-locked to the sound waves, but I do not know to which half-cycle.

Johnstone: Recent work of Furukawa (Furukawa, T., and Ishii, Y. [1967]. *Jap. J. Physiol.*, **17**, 572–588) in the goldfish might throw light on this. He was able to record intracellularly from giant primary nerve fibres to the macula which multiply innervate hair cells. The macula is divided into hair cells which have different directional sensitivity; their kinocilia point in opposite directions. A certain number of the nerve fibres branch and innervate cells pointing in both directions. He records intracellularly from them; hence he can pick out the microphonic and postsynaptic potential. It is also possible to bias the whole mechanical system by putting a collar around the animal's head and increasing the pressure on one side or the other.

He obtains microphonics of three kinds, depending upon where the nerve that he is recording from terminates. Some nerves give responses for the positive half-cycles of pressure, some for the negative half-cycles, and some give responses for both—that is, twice per cycle. Hence it is possible to obtain action potentials at double the input frequency.

A nerve that gives single spikes goes wholly to one side or the other. Those that give double the input frequency go to both sides. For such a nerve, the response can sometimes be changed to a single-frequency type by compressing or decompressing part of the fish.

Engström: In the maculae of most mammals there are at least three separate regions, one in the striola region and one on each side of this, which all seem to have different morphology (Lindeman, H. [1968]. *Acta oto-lar.*, in press). It would be most interesting to know what can be recorded from the striola region; whether a small population of exceptionally sensitive cells can be found there, or of cells with a different kind of response. Also on the cristae ampullares, at the top of the crest there is a population of exceptionally richly innervated sensory cells.

Davis: I am glad that the concept of the "sensory unit"—the single nerve fibre and the receptor transducer mechanisms which excite it—has already been introduced, because this is the fundamental concept for analysing the activity of the ear and the laws of its sensitivity to differences of frequency and intensity. For example, the overall threshold is the envelope of the response areas of the different sensory units.

I want to call attention to the fact that for a single sensory unit there is a relation between frequency and intensity that describes its sensitivity; the curves that bound the response area are sloping curves. It becomes an interesting question whether the important change in the acoustic stimulus should be considered as a change of frequency or a change of intensity. The important point is that it crosses a contour. I would like to know whether the effective percentage difference in intensity at a single frequency has been investigated in fish. If we want

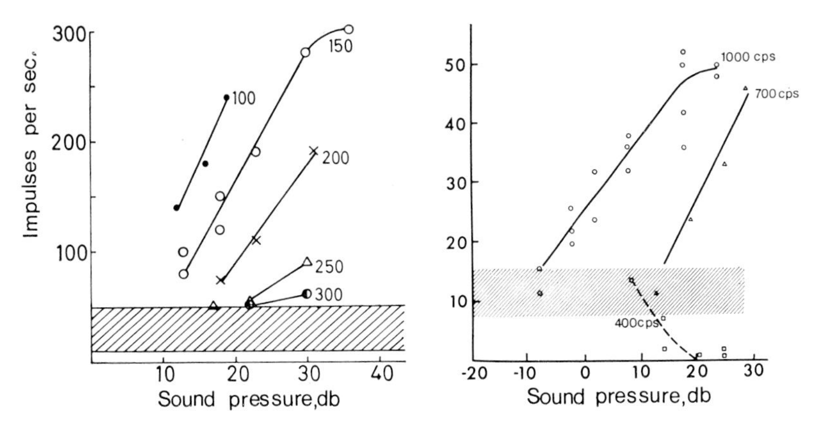

FIG. 1 (Enger). Plots of intensity (sound pressure) against number of impulses per second, for single units from the sculpin (*Cottus*; left) and the herring (*Clupea harengus*; right).

to be anthropocentric, we can ask how the fish hears a change in sound. Does it perceive the change as a difference of frequency, meaning some different *kind* of source, or simply as a difference of intensity, meaning a changing intensity from the *same* source?

Enger: In my experiments the thresholds of single units were estimated for different sound frequencies by plotting the intensity (sound pressure) against the number of impulses. In Fig. 1 (to the left) are shown data from *Cottus*. For a given sound pressure there was a certain discharge rate for a given frequency, and this discharge rate falls as the sound pressure is decreased.

The crossing points between the curves and the level of spontaneous activity are what I estimated as thresholds. For 300 Hz this is some 20–25 db; for 150 Hz it is around 10 db. The right half of the figure gives data from the herring (*Clupea harengus*) (Enger, P. S. [1967]. *Comp. Biochem. Physiol.*, **22**, 527–538), probably from the second-order auditory neurons, and again the threshold depends on the frequency. For 1,000 Hz the estimated threshold is −8 db, for 700 Hz, +14 db. For 400 Hz it is also +14 db, but estimated as a threshold for inhibition. What I am pointing out is that even in this primitive ear, the fish ear, there is a possibility of peripheral frequency discrimination, but I also think that more work should be done in fish on the spectrum or critical bandwidth, as we know really nothing about that. We know only the sinusoidal frequency ranges, and nothing about the envelopes.

Lowenstein: Dr. Davis, would you find it easy to imagine that absolute pitch discrimination—remembered pitch on consecutive occasions (the minor third is the best a minnow can do in this way)—is possible on a memory of intensities?

Davis: Definitely not; this is very good evidence that there is frequency selection. The response depends on which unit or group of units is being activated. My intention was to bring out just this point.

Schwartzkopff: I could add a little information in answer to the question posed by Dr. Davis. Miss B. Grözinger has studied the intensity function at different levels of the fish's brain, and in some cases, in the midbrain, the intensity slope was less pronounced than at the level of the lower centres, which is opposite to what Katsuki reported in higher vertebrates. We therefore do not feel that intensity becomes better evaluated at higher levels of the brain. If any difference exists, it is worse at higher levels because there is less discrimination. Certainly we have never found better transformation. (See Grözinger, B., and Schwartzkopff, J. [1967]. *Naturwissenschaften*, **54**, 446; Grözinger, B. [1967]. *Z. vergl. Physiol.*, **57**, 44–76.)

On the question of frequency differentiation, we failed to find evidence of this, but perhaps this was because, working in the tench, we could only occasionally hold units. In about a dozen different places in the brain we found auditory activation but only one of these, in the medulla, showed different frequency contours with a maximum at 300 cyc./sec. Generally, for a certain auditory area or auditory fibre bundle, we found almost identical envelopes, which in the tench had a maximum at about 600 cyc./sec. We supposed that this one maximum at 300 cyc./sec. might have come from the lateral line, or from one of the parts of the peripheral organ which we cannot discriminate, but according to your experiments, it would not be expected to come from the peripheral organ—that is to say, not from the lagena.

Spoendlin: Dr. Enger, if you find different central units responding to different frequency ranges in fish, do you find them mixed throughout the whole nerve or are they arranged according to topographical principles, as for instance in the acoustic nerve of mammals?

Enger: I cannot answer that, because I cannot say whether particular units are on the top of the nerve or deep in it.

Bosher: Some work at Princeton University (Weiss, B.A. [1966]. *J. aud. Res.*, **6**, 321–335) seems to show that 100 cyc./sec. stimulation may be of importance for fish, and I was interested to note that in the present investigation such low frequency sounds appear to have low thresholds in a number of animals. Weiss found in goldfish (*Carassius auratus*) that the responses behaviourally were different below this stimulation frequency, so-called "startle" responses being produced which were more primitive in nature than the conditioned responses. These he attributed to the lateral line organ. I wonder whether you have noticed any similar responses and whether, in the biological environment of the fish, sounds of such low frequencies have some warning significance?

Enger: In behavioural studies there is the problem in the low frequency range that one does not know which sensory system the fish is using for detecting the signal. When one records from the auditory nerve, one avoids this problem. One cannot yet assess the sensitivity of the fish at low frequencies, because practically all reports so far have been contaminated with the near-field effect and so we do not know how much is due to pressure being transformed into displacement and how much is direct displacement from the sound source.

EVOLUTION OF THE AUDITORY CONDUCTING APPARATUS IN TERRESTRIAL VERTEBRATES

A. TUMARKIN

Former Director, Department of Otorhinolaryngology, The University, Liverpool

THE auditory conducting mechanisms of the fish of the Devonian period must have been useless on land, and the emerging amphibians had therefore to evolve entirely new hearing mechanisms. The first amphibians were completely prostrate but as the invasion of dry land proceeded, more erect postures evolved. First the head and later the whole body was

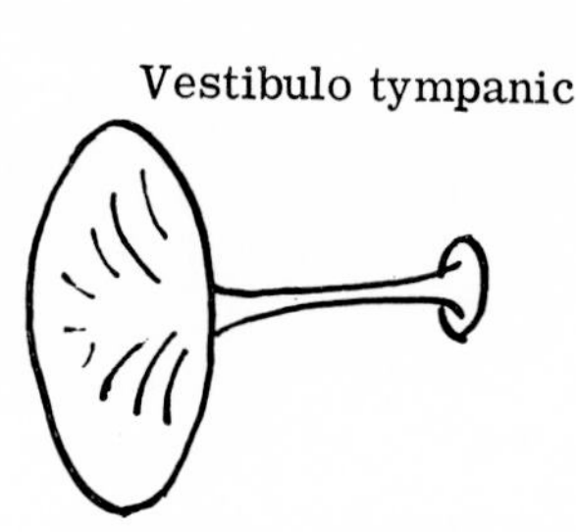

FIG. 1. Air sensitive middle ears.

According to text-book theory, the vestibulo-tympanic and vestibulo-ossicular mechanisms are the only functioning middle ears, and all other forms are degenerate. According to the functional theory presented here, there are—or have been—at least five non-homologous air sensitive ears and at least four non-homologous ears sensitive to substrate-borne sound.

elevated. In due course quadrupedal, bipedal and flying reptiles appeared. Every change of posture made fresh demands on the auditory apparatus. In some cases these were met by modification of an existing structure, but in other cases a totally different ear had to be created.

Modern ears may be classified* as follows:

Air conducting (Fig. 1)

(1) Vestibulo-tympanic: found in frogs, birds and many reptiles.

(2) Vestibulo-ossicular: the triple ossicle mechanism of the mammal.

* The nomenclature used is based on the attachments of the stapes. Vestibulo-scapular is exceptional. Its ossicle, the operculum, and the associated muscle have no relations whatsoever with the stapes and the stapedius.

Bone conducting

(1) Vestibulo-quadrate: found in snakes and many lizards. (Fig. 7)
(2) Vestibulo-hyoid: found in amphisbaenid reptiles. (Fig. 7)
(3) Vestibulo-squamosal: found in urodele amphibians. (Fig. 7)
(4) Vestibulo-scapular: found in both urodele and anuran amphibians (Fig. 8).

Transitional forms

(1) Between vestibulo-tympanic and vestibulo-quadrate. Within the lizards, genera such as *Gekko* have a perfect vestibulo-tympanic mechanism. At the other extreme, *Chamaeleo*, *Holbrookia* and many others have no drum or middle ear cavity, and the stapes articulates with the quadrate—thus the mechanism is vestibulo-quadrate. In between the extremes, transitional forms abound. In some the drum is covered in scales; in others it is sunk in muscle. The middle ear cavity in some cases is comparatively roomy; in others it is a mere chink. The delicacy and mobility of the stapes vary in the same way.

(2) Between vestibulo-hyoid and vestibulo-tympanic. *Sphenodon*, the rare New Zealand rhynchocephalian lizard, is the link in this chain. The connexion between the stapes and the hyoid is more tenuous than it is in the amphisbaenid lizards and a rudimentary tympanic membrane appears as a fibrous condensation within the muscles of mastication.

(3) Between vestibulo-squamosal and vestibulo-tympanic. While many frogs have a perfect vestibulo-tympanic ear, many toads such as *Bombinator* have no such air sensitive ear, nor have any of the aquatic Pipidae. Intermediate forms abound.

According to the accepted text-book theory, an air sensitive ear of the vestibulo-tympanic type appeared in the very earliest amphibians such as *Eogyrinus*, as a result of the closure of the first gill cleft, the spiracle, by a tympanic membrane. At that time the upper jaw, which in fish is slung from the cranium by the hyomandibula, was beginning itself to fuse with the cranium. Thus the hyomandibula lost its suspensory function and became the stapes or columella (Fig. 2). Much later the mammals jettisoned the reptilian articulo-quadrate jaw joint and formed a new one between the dentary and the squamosal bones. Being thus left with two bones (articular and quadrate) surplus to requirement, they interpolated them between the tympanic membrane and the stapes, the quadrate becoming the incus and the articular, the malleus. This theory was reinforced when the early amphibians were found to have a large notch in the skull bones between the tabular and the squamosal, directly over the hyomandibula and ideally placed to carry a tympanic membrane. Nevertheless, on closer scrutiny, this theory is found to be unacceptable, for the following reasons.

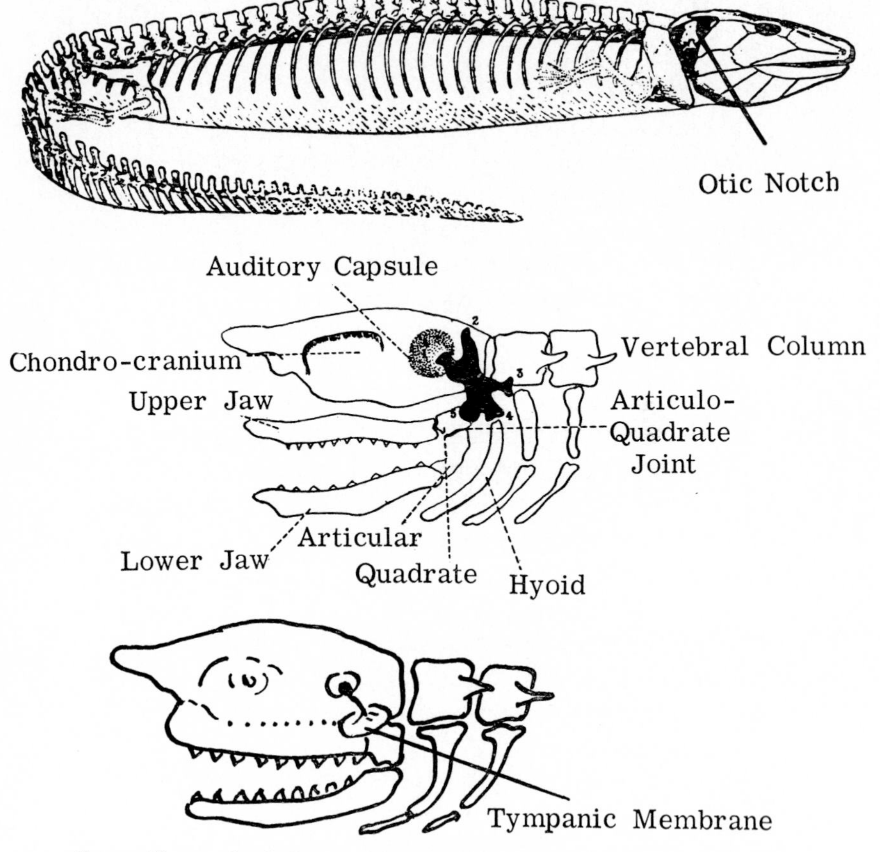

FIG. 2. The text-book theory of the evolution of the monossicular middle ear.
Above: Eogyrinus, one of the earliest known amphibians.
Centre: The fish skull, showing the various processes of the hyomandibula.
Below: The upper jaw has fused with the chondrocranium in the amphibian skull. The hyomandibula therefore loses its suspensory function and becomes the stapes or columella.

(1) No account or explanation is offered for the bone conducting and transitional forms. They are usually dismissed as degenerate. The idea that a premier sense organ should undergo such widespread degeneration in so many different families and orders is repugnant and entirely fails to explain the vestibulo-scapular mechanism which exists side by side with

a perfect vestibulo-tympanic mechanism in the frog. It has been suggested that the bone conducting ears may not be actually degenerate but are functional adaptations to the prostrate habitus. Most, if not all, prostrate animals do indeed use one or other of these mechanisms. If, however, the prostrate habitus calls for a bone conducting ear, it is difficult to believe that a completely prostrate creature such as *Eogyrinus* could have evolved an air sensitive ear, especially as it lived most of its life in water. Moreover, *Eogyrinus* did not possess any sign of an oval window (or for that matter

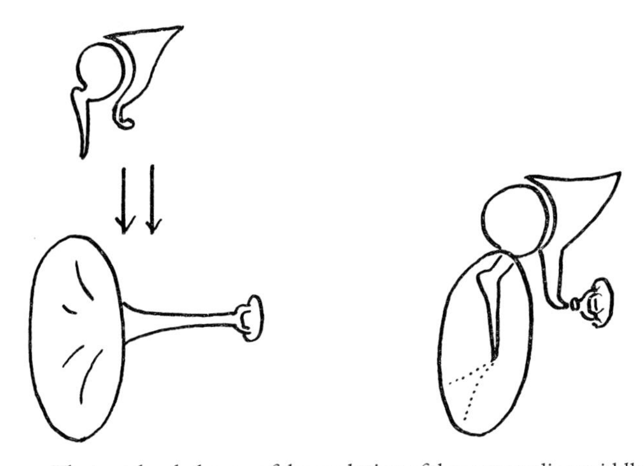

FIG. 3. The text-book theory of the evolution of the mammalian middle ear. If the Devonian amphibians already had air sensitive hearing, the much later pre-mammalian Therapsida must surely have improved it. In that case, the interpolation of an obsolescent jaw joint (the articular and quadrate, becoming the malleus and incus) would have gravely impaired the efficiency of hearing for a time.

of a round window), so that its tympanic membrane would have been quite useless.

(2) The interpolation of two extra bones into the mammalian middle ear is quite inexplicable on this basis (Fig. 3). Békésy has pointed out (1936) that this elaborate triple ossicle mechanism has no known mechanical advantage over the monossicular system of the lower animals. The mammals emerged about two hundred million years after *Eogyrinus*. If the latter really had air sensitive hearing then we can be confident that under the influence of evolutionary functional adaptation the middle ear would have shown progressive improvement. If, however, the emerging mammals had inherited a refined vestibulo-tympanic mechanism after such a prolonged period of improvement, it is impossible to believe that they would have interpolated the massive bones of an obsolescent jaw

joint into it. Such an alteration could not have been achieved without disruption of the existing ear with great reduction in its efficiency.

(3) Above all, however, the theory fails to account for the accepted facts of palaeontology. As noted above, the tympanic membrane of *Eogyrinus* would have been useless in the absence of an oval window. In the stem reptiles, that succeeded the amphibia, an oval window does indeed appear, but this cannot be interpreted as evidence of air sensitive hearing. As will appear later, it merely formed part of a bone conducting apparatus. On the other hand the otic notch, which is the very foundation of the air sensitive theory, completely disappears. The hypothetical drum, instead of being held, as it should be, in a ring of bone or cartilage, is believed to have floated precariously somewhere within the muscles of mastication, as indeed it still does in many of the transitional forms described above. It is difficult to believe that functional adaptation could deal with a sense organ of such supreme importance in such a cavalier fashion.

If we follow the fate of the middle ear in the later reptiles we encounter two further inexplicable facts. In the Permian period a fantastic proliferation of reptilian orders took place, but for our purpose it will suffice to distinguish the two main lines:—

(1) The Synapsida or premammalia, giving off first the Pelycosauria, next the Therapsida and finally leading up to the mammals.
(2) The Archosauria, starting with the Thecodontia, leading to the enormous radiation of the dinosaurs and represented to this day by the crocodiles and the birds.

These two great super-orders have many distinguishing features but one of the most striking is the size of the stapes. It is unnecessary to point out that in the air sensitive ear the ossicular chain is characteristically light, mobile and balanced, yet within the premammalia the stapes shows no sign whatsoever of those features (Fig. 9). It seems to become ever more clumsy and locked to adjoining bones. In some cases it is positively gigantic, weighing as much as twenty or thirty grammes. The archosaurian stapes is even more mysterious, since is it practically unknown. Up to 1960 only a single dinosaur stapes had ever been discovered, despite the fact that thousands of skulls in an excellent state of preservation, many of them several feet in length, had been described.

The paradoxes described above all stem from the one erroneous assumption that the primitive amphibian *Eogyrinus* possessed an air sensitive ear. They will be found to disappear if the alternative hypothesis is adopted, namely that initially hearing in terrestrial vertebrates was entirely by bone conduction. The following arguments are adduced in favour of that hypothesis.

The earliest amphibians spent most of their lives in water. They sallied on to land only when the lagoons dried up (Fig. 4). On land they would be interested in two things: water on which their lives depended, and similar creatures that might turn out to be predators or prey. In either case any sound generated would be more intense in the substrate than in air. Moreover, because of impedance matching, substrate vibrations would enter the body easily, whereas airborne vibrations would not. It follows that these creatures must have evolved detectors for substrate vibrations.

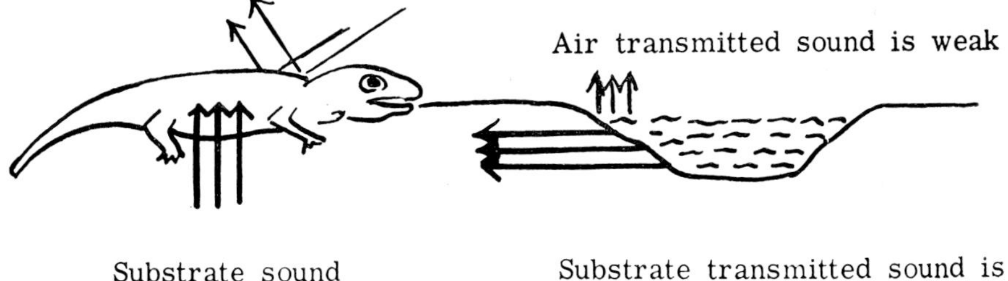

FIG. 4. The auditory interests of *Eogyrinus*.
Substrate sound is powerful and significant; airborne sound is weak and meaningless.

Moreover, because they were, audiologically speaking, completely unspecialized, and because the detection of sound must have been of immense survival value, every route that could possible carry vibrations was pressed into service by some group or groups. It is common ground that the hyomandibula, by virtue of its attachment to the otic capsule, played a key role in all these primary ears, but while text-books declare that an air sensitive ear formed immediately, I suggest that no less than three bone conductors evolved. Fig. 5 shows the fish hyomandibula with its internal process attached to the otic capsule and its external processes leading to the squamosal, hyoid and quadrate respectively. Each of these external processes in combination with the internal process could have given rise to a conducting route producing, in fact, the bone conductors already described as vestibulo-squamosal, vestibulo-hyoid and vestibulo-quadrate respectively. These three ears, therefore, are seen to be valid functional adaptations to the prostrate habitus, and it is wrong to regard them as degenerate versions of an air sensitive ear which clearly could not have existed.

The actual mode of action of these ears is perhaps not quite so simple as appears at first sight. It is natural to assume that vibrations would be picked up from the substrate by the contiguous body tissues and canalized on to the oval window by the individual hyomandibular processes. There is, however, an alternative route. Vibrations might stream into the inner ear via the body structures and the cerebrospinal fluid and emerge on to

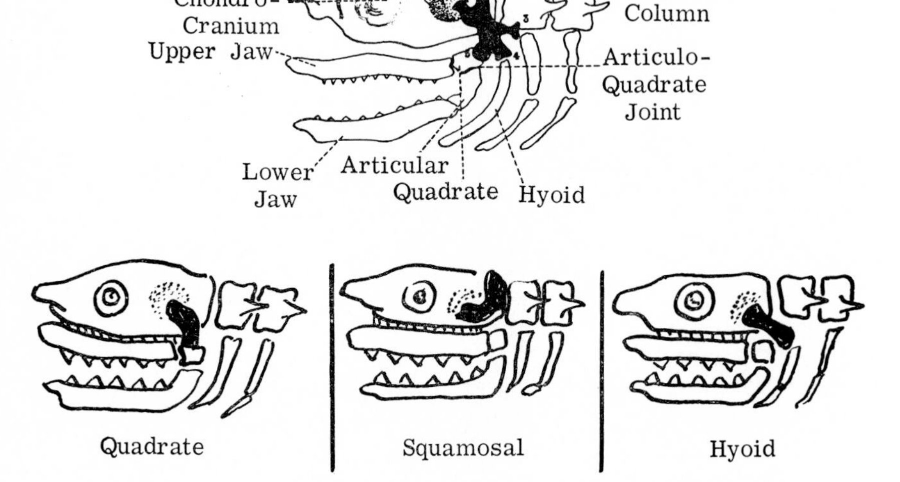

FIG. 5. The metamorphosis of the hyomandibula according to the functional theory.
Above: The fish skull, showing the various processes of the hyomandibula.
Below: Three bone conducting mechanisms produced by different combinations of four of the hyomandibular processes—1-5, 1-2, 1-4—giving vestibulo-quadrate, vestibulo-squamosal and vestibulo-hyoid mechanisms respectively.

the stapes at the oval window. This idea is based on the familiar dynamic principle that in order to detect movement it is necessary to have two structures, one to respond to the movement and the other to remain stationary (or at any rate out of phase) to act as a frame of reference. This principle is embodied in the standard tuning fork tests of hearing. When air conduction is being tested the fork, held near the pinna, sets the ossicles vibrating while the head stands still; but when bone conduction is tested, the base of the fork, being placed on the mastoid process, sets the skull in vibration while the ossicles stand still (or at any rate vibrate out of phase) (Fig. 6). A striking feature of this latter test, and one that is highly relevant to the present argument, is that this type of bone conducted hearing can be substantially improved by increasing the inertia of the ossicular chain as, for example, by loading the tympanic membrane with mercury. It follows, therefore, that bone conducted hearing can arise either by a direct route, in which vibrations are conducted by the stapes to the inner ear, or by a reverse route in which vibrations are conducted to the inner ear by the general body structures and the cerebrospinal fluid and *emerge* via the stapes. It is, of course, possible for a given mechanism to act in both ways; nevertheless, two important points of difference may be noted between

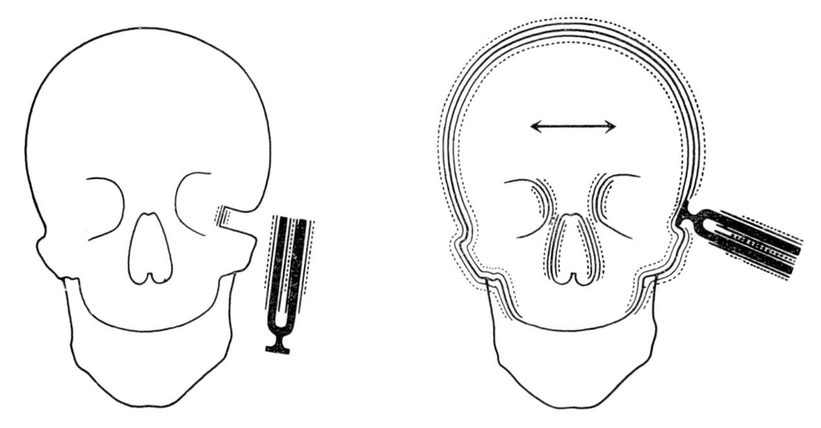

FIG. 6. The relativity of the detection of movement, as in tuning-fork tests of hearing. A stationary reference point is necessary for movement to be detected.
Left: Test of air conduction. The ossicles move and the head stands still.
Right: Test of bone conduction. The head moves and the ossicles stand still.

the direct and the reverse routes. As noted above, in the reverse route, it is advantageous to make the stapes massive, whereas in the direct route that feature is not required. Secondly, in picking up substrate vibrations by a direct route it is clearly necessary for the particular stapes to lie on the *incoming* route, whereas in the reverse route is is essential that it should lie away from the incoming path.

With these considerations in mind let us now return to the primordial three. In their day they represented an evolutionary triumph; nevertheless, their limitations are obvious. The direct conductors all went out of action as soon as the head was lifted, since the stapes lost its contact with the ground (Fig. 7). On the other hand the reverse-route conductors could still act after a fashion since vibrations could ascend the limbs to the pectoral or even the pelvic girdle and so enter the cerebrospinal fluid. Both these solutions to the dilemma of the raised head were adopted by different groups. The direct-route users, having lost contact with the substrate, re-established it by a different route. The nearest alternative was via the forelimb. This was a totally new creation and is found to-day in the

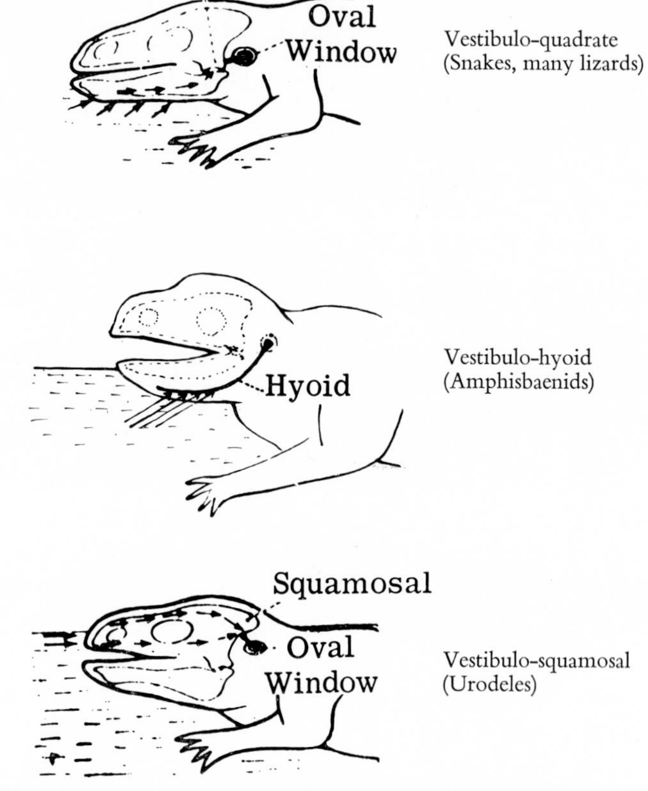

FIG. 7. These three bone conducting mechanisms are not degenerate, but are different solutions to the problem of detecting substrate vibrations. They were developed by the earliest amphibians long before air sensitive hearing appeared.

vestibulo-scapular mechanism of the amphibia (Fig. 8). The reverse-route users also used the forelimb, or indeed any other part of the body that was in contact with the substrate; but they "perfected" their ear by *increasing* the mass of the stapes. This, in my opinion, is the explanation of the enormous stapes of the premammalia. Also, in a way, it was an evolutionary triumph, since it made possible the quadrupedal stance, while all others were limited to the prostrate or semi-prostrate habitus. Thus the reverse vestibulo-quadrate of the premammalia in its day was the best bone conductor on earth.

Nevertheless, this ear, like its predecessors, suffered from serious limitations. First, it was tied to the body dimensions; the bigger the animal the more remote the vibrations. Secondly, incoming sound would be masked by the noise of the creature's own movement. Substrate detectors are at their best in sluggish postrate creatures when the biologically significant sound is substrate-borne, as in the case of *Eogyrinus*. They become of diminishing value as the habitus becomes more erect and as significant

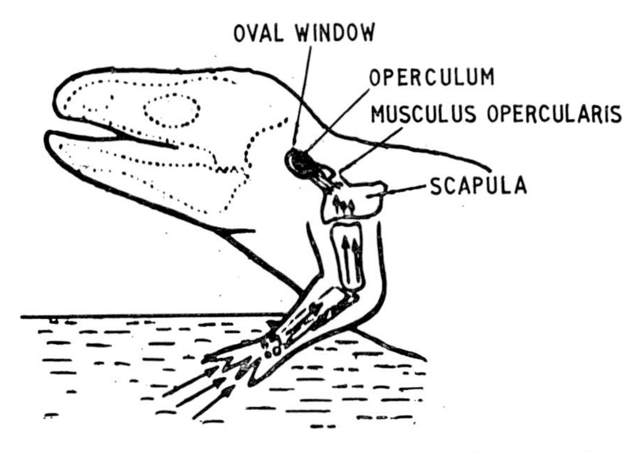

FIG. 8. Vestibulo-scapular mechanism, making use of the forelimb.

This mechanism, found in all modern amphibians, enabled primitive amphibians to detect substrate vibrations even when the head was raised. It has no relation to the hyoid anlage.

sound comes increasingly over the air. But this is exactly what was happening as time went on. Possibly the reptiles were themselves uttering biologically significant sound.

And so at long last, late in the Permian period, the stage was set for the emergence of the air sensitive ear. This new mechanism was obviously of enormous survival value and so, as in the case of the primordial bone conductors, every route that could possibly transmit airborne sound, was utilized. In some cases an existing bone conductor was modified but in other cases a totally new ear had to be created. The various amphibian and reptilian orders all struggled with this problem with varying success. The modern prostrate Urodela (salamanders and newts) and Squamata (snakes and lizards) represent groups that entirely failed to solve the problem and so have remained condemned to the prostrate habitus. The various transition groups comprise creatures that are still struggling with the problem with varying degrees of success.

It remains to consider the means whereby the more progressive members

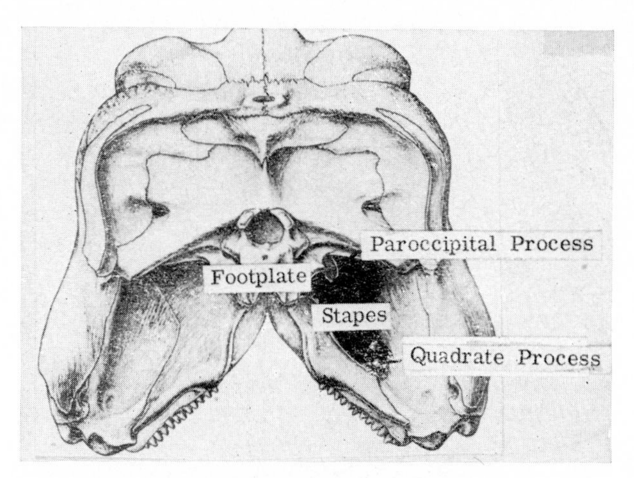

FIG. 9. Rear view of the skull of a pelycosaur.

Note the colossal stapes. There was no room within the skull bones for a tympanic membrane big enough to energize this monstrosity. (But compare Fig. 10.)

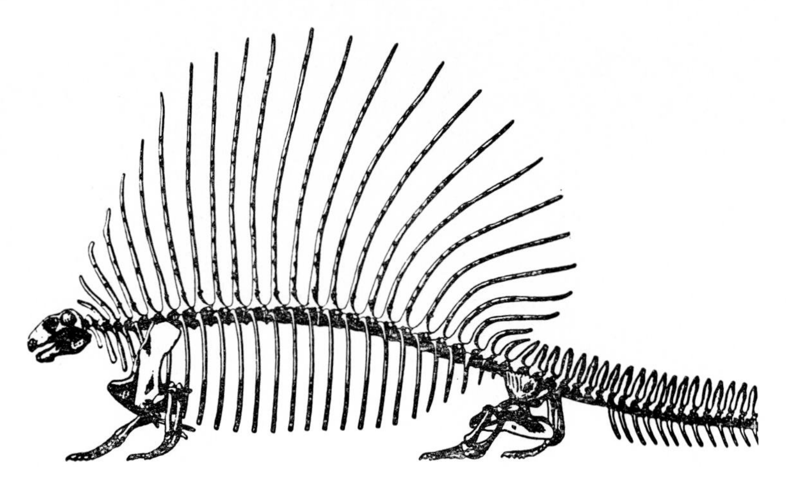

FIG. 10. The skeleton of the dorsal sail of the pelycosaur.

It is suggested that this was the first air sensitive ear on earth. Despite its absurd appearance (but hardly more absurd than that of a bat) it was superior to all the contemporary bone conducting mechanisms (vestibulo-quadrate, vestibulo-squamosal and vestibulo-scapular).

of the different orders solved the problem. Five entirely different solutions call for consideration.

(1) *The Premammalia* in this competition were gravely handicapped by their massive stapes. To change completely to air conduction they would have had to replace it by a light ossicle. But in the transition they would have passed through a stage in which the ear was neither one thing nor the other. Instead they made the all too common error of trying to keep the best of both worlds. They retained the massive stapes for substrate detection (Fig. 9), but they also contrived to drive it by airborne sound by erecting an enormous tympanic membrane across the specially elongated spines of the vertebrae (Fig. 10). Pelycosaurs, such as *Dimetrodon*, therefore had two reverse-route ears, one for substrate and the other for air sensitivity. Both these mechanisms appear crude in the extreme, nevertheless they

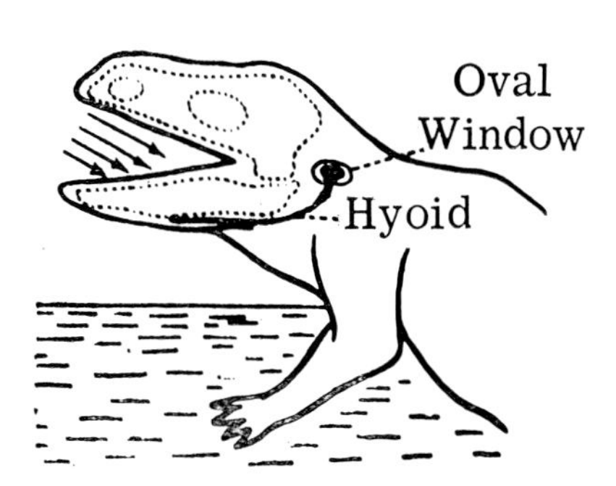

FIG. 11. The vestibulo-hyoid mechanism as an air sensitive ear.
Although primarily a bone conductor, the vestibulo-hyoid mechanism can act as a rudimentary air sensitive device.

should not be underrated, if only because the dorsal sail was the first air sensitive ear on earth. All other creatures, both amphibians and reptiles, up till then, heard by bone conduction. Not surprisingly, *Dimetrodon* was one of the most successful vertebrates of its day. Nevertheless the dorsal sail had the obvious limitations of being linked to the body dimensions. As the animal grew larger the conducting route became longer and less efficient and the dorsal sail must have been an increasing liability.

(2) The *Archosaurs* were far more fortunate. The thecodonts of the Permian were already bipedal and it is impossible to believe that they could have achieved that posture unless they had efficient air sensitive hearing, since no known bone conductor would have been of real value. We can deduce how and why this happened by examining the ear of modern descendants of the Archosauria—the birds. In them a persistent hyo-columella ligament betrays the origin of the archosaurian ear from the primordial amphibian vestibulo-hyoid ear. If we look at the primitive

trio it is apparent that the vestibulo-hyoid mechanism was superior to the others in possessing a rudimentary sensitivity to airborne sound (Fig. 11). The soft tissues of the floor of the mouth could respond, albeit crudely, and provided that the ossicle was a direct conductor and not a reverse one, little modification would have been necessary to transform and perfect it. *Sphenodon* still shows the transition stage through which it must have passed. Two further points should be made. First, this ear was immensely superior to the dorsal sail insofar as it was independent of body size, so that it could reach its own optimum dimensions regardless of any other considerations. Secondly, the mystery of the missing dinosaur stapes is solved. Precisely because it was sensitive to airborne sound the ossicle

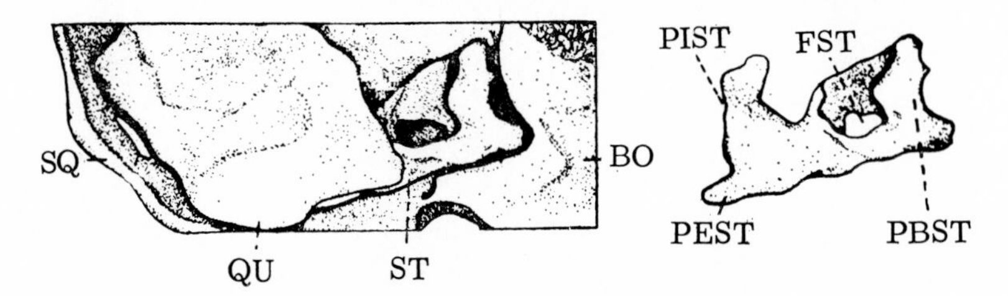

Fig. 12. The therapsid stapes.

On the left the stapes (ST) is seen locked between the basi-occiput (BO) and the quadrate (QU). The isolated stapes is shown on the right. PIST, PEST, PBST—processes of the stapes. Observe the close attachment to adjacent bones and the lack of space for a tympanic membrane. SQ, squamosal. FST, stapedial foramen.

was so delicate and mobile that it could not withstand the crushing forces of fossilization.

(3) The *Lacertilia* (lizards) had a rather more difficult problem insofar as the vestibulo-quadrate that they seem to have started with did not possess the same secondary air sensitivity as the vestibulo-hyoid. Just like the snakes, many Lacertilia have failed to progress and so have remained prostrate. Other, such as *Gekko*, have triumphantly solved the problem, while many others are still in various stages of evolution.

(4) The *Anura* possess the vestibulo-tympanic ear side by side with the vestibulo-scapular hearing mechanism, while the *Urodela* possess vestibulo-squamosal side by side with vestibulo-scapular, and it is natural to assume that the anuran vestibulo-tympanic mechanism evolved out of the primordial vestibulo-squamosal. This, however, turns out to be incorrect. Ontogenetically it is quite easy to demonstrate that the stapes of the reptile, the bird, and even the mammal, all derive from the hyoid anlage, but it has been repeatedly shown that the anuran columella never has any

association with the hyoid anlage. The anuran columella is a totally new creation derived partly from the chondrocranium and partly as a local condensation of mesoderm (Gaupp, 1904; and see p. 35).

(5) The *Mammalian* vestibulo-ossicular mechanism evolved as a result of the interpolation of the obsolescent reptilian jaw joint into the middle ear, and we have already commented that this would have been little short of a disaster if that ear already acted as an air conductor. It now appears that that was not the case. The ear in fact was vestibulo-quadrate. Fig. 12 shows the therapsid stapes and we recognize the massive bone stretching across the whole base of the skull. The therapsids lived in the early Mesozoic. By that time the pelycosaurs had become extinct and the therapsids, with their vestibulo-quadrate ear, had to face the terrible onslaught of the dinosaurs with their perfect vestibulo-tympanic mechanism. Not surprisingly the therapsids were utterly defeated and went extinct. It seems that the larger forms were destroyed first and as they approached their final doom they became progressively smaller until at the crossing of the mammalian horizon they were no bigger than a mouse. Ironically this very miniaturization in defeat gave them the air sensitivity that had eluded them in their prime, because of the concomitant increase in the area to mass ratio. The essence of an air conductor is a large collecting area and a light ossicle, and in shrinking from as much as 10 feet to as little as three inches they would increase that ratio by a factor of as much as forty. The proto-mammals, therefore, inherited this potential sensitivity and it would be natural to assume that, starting as they did with the vestibulo-quadrate ear, they would proceed to transform it into a vestibulo-tympanic ear exactly as the Lacertilia are doing at this very moment. The fact that they did not do so suggests that the stapes alone could not reach the surface. That this was indeed the case can be deduced from an examination of the therapsid lower jaw. Fig. 13 shows that there was a progressive enlargement of the muscular process of the dentary, indicative of a corresponding hypertrophy of the masticatory muscles. The stapes must have been ever more deeply overlain by them. At the same time the articular and its joint surface with the quadrate became progressively smaller. The transition to the mammal took place when the dentary finally made contact with the superjacent squamosal to form a new joint. The articular and quadrate, being no longer needed for mastication, presumably acted for a time as a subsidiary suspension of the lower jaw, but finally fell into atrophy through disuse. Meanwhile the deep-sunk stapes, unable to contact the surface unaided, enjoyed a tenuous indirect contact via these two bones, and in due course the very disuse atrophy into which they had fallen made

them more and more suitable for auditory conduction. These factors, together with the stimulus of the increased area to mass ratio, finally resulted in the emergence of the mammalian triple ossicle mechanism (Fig. 14).

It is customary to regard this whole process as determined solely by the mechanical requirements of the masticatory function, and certainly the resulting shortening provided a more efficient masticatory mill. In my opinion, however, the auditory mechanism was by no means passive in

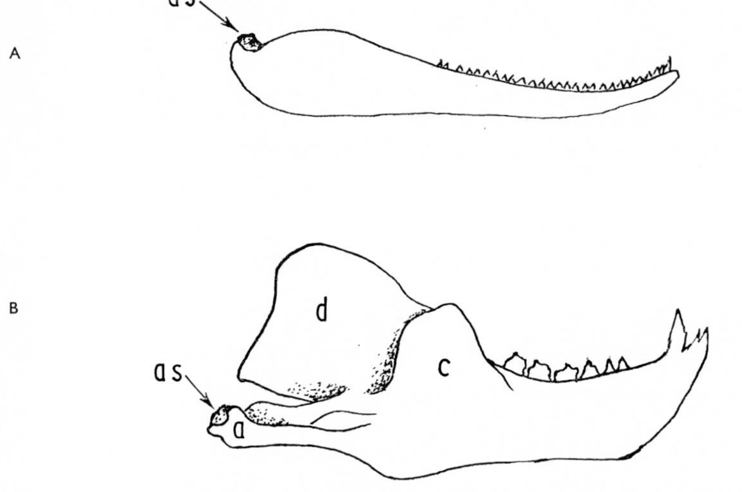

FIG. 13. The evolution of the pre-mammalian jaw.

A. *Ophiacodon*, an early pelycosaur; B. *Cynognathus*, a late therapsid. *as*, articular surface of the articular (*a*); *c*, coronoid; *d*, dentary. The lower jaw of the pelycosaur was an engulfing maw; the lower jaw of the therapsid was a masticatory mill. Observe the enormous muscular processes of dentary and coronoid in B. The former eventually articulated with the superjacent squamosal to form the mammalian joint. Observe the slender articular, which seems almost ready to abandon its masticatory function. It ultimately did so in the mammal and became the malleus.

this transaction. It takes no great stretch of imagination to envisage the articular and quadrate torn, so to speak, between the conflicting demands of the masticatory and the auditory functions—the former calling for massive, stable muscle-clad bones, the latter for delicate airborne mobile ossicles. If this dilemma of Solomon really existed it follows that the mammals owe their very existence to some unknown therapsid that hit on the inspired compromise of placating the masticatory function with a brand new joint in order to dedicate the bones of contention to the overriding needs of air sensitive hearing.

The functional theory may perhaps best be evaluated in terms of the various paradoxes listed on pp. 20–2. All modern ears are shown to have emerged as functional adaptations to specific environmental niches under the prevailing biological conditions. The concept of degeneration is rejected entirely. The existence in the various orders, sub-orders and even families of members in all stages of hearing, from primordial bone conduction to perfect air sensitivity, is interpreted simply as evidence that creatures can move at widely different speeds along the evolutionary trail. That their progress is in considerable measure determined by functional considerations is indubitable. Yet it would be naïve to imagine that we fully understand the complex interplay of biophysical factors that has left, for example, *Chamaeleo* at the starting line of vestibulo-quadrate

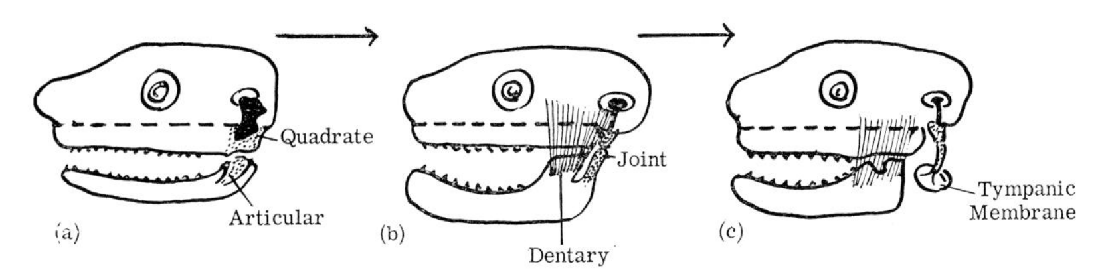

FIG. 14. The evolution of the mammalian middle ear according to the functional theory.

(*a*) Vestibulo-quadrate mechanism in an early therapsid. Note the massive interlocked stapes.

(*b*) The condition in a late therapsid, which was a tiny creature with powerful masticatory muscles overlying the stapes. Note the massive dentary and the fragile articulo-quadrate joint (compare Fig. 13).

(*c*) The mammalian squamosal-dentary joint takes over as the jaw joint. The articulo-quadrate joint, deprived of its primary masticatory function, develops its secondary function as part of the middle ear. The articular, being subcutaneous, is able to pick up airborne sound and transmit it via the quadrate to the deep-sunk stapes.

while *Gekko* has reached the finishing post of vestibulo-tympanic. Puzzling though this may be, it is surely more plausible than the alternative that *Chamaeleo* and countless other species should be bent on a suicidal abandonment of air sensitive hearing bequeathed to them three hundred million years ago.

Eogyrinus lacked oval and round windows because it was still essentially a fish and had no vestige of terrestrial hearing. The so-called otic notch never carried a tympanic membrane. The degradation of the tympanic membrane from the security of the otic notch in *Eogyrinus* to the muscles of mastication in the stem reptiles did not in fact take place. Neither *Eogyrinus* nor the stem reptiles had a tympanic membrane.

The twin mysteries of the gigantic premammalian stapes and the missing archosaurian stapes are explained by the fact that the former was a reverse bone conductor, while the latter was the first direct air sensitive ear on earth. An interesting pendant to that hypothesis occurred in 1958 when Colbert stumbled almost by accident on the stapes of the dinosaur *Dromeosaurus*. Fig. 15 shows this side by side with the stapes of a modern lizard. The resemblance is indeed startling.

The interpolation of articular and quadrate into the middle ear, so meaninglessly disastrous on the assumption of a pre-existing air sensitive ear, is seen to be logical and indeed necessary once it is recognized that the therapsid ear was vestibulo-quadrate in type and not vestibulo-tympanic.

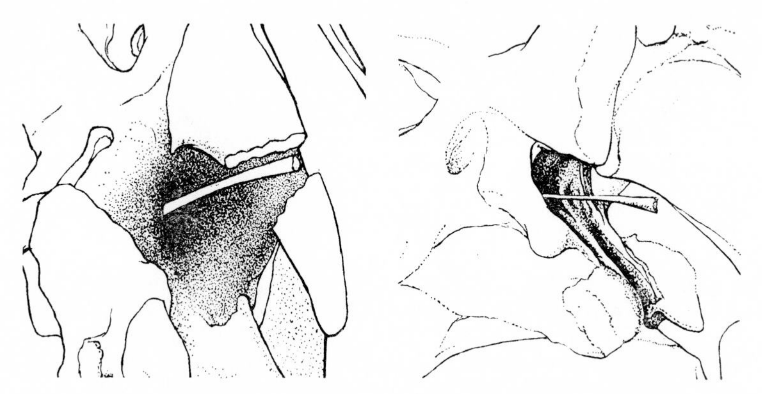

FIG. 15. The air sensitive ear of the dinosaur (left) compared with that of a modern lizard (right).

The resemblance between these structures is quite startling. Even more startling is the contrast between these ossicles and the "landlocked" massive ossicles of the pelycosaurs and the therapsids (compare Figs. 9 and 12).

Certain broad conclusions may be drawn. The central issue might be crystallized as homology *versus* analogy. The idea that all hearing mechanisms might be variations on a single theme has by its very simplicity dominated the thinking of generations of morphologists. It is, however, utterly contrary to modern concepts of evolution in which far more importance is attached to the ideas of analogy and convergence.

Sound is ubiquitous. Auditory detectors, as shown in this paper, have played a major role in the fluctuating fortunes of the vertebrates and there is no reason to believe that the bewildering variety of ears described could possibly be subsumed under a single mechanism rigidly derived from the hyoid anlage. Yet this view is still vehemently maintained. Hotton (1959) for example says:

> "The combination of tympanum, stapes and tympanic cavity that first appeared in the amphibians was a functional unit. The tympanum

was set in motion by vibrations of a fluid medium, at first water and later air. The stapes communicated these vibrations to the inner ear, at first via the wall of the otic capsule and later via the perforate fenestra ovalis. Changes in the ear were merely reflections of changes occurring in other structures, the ear making do at any time, so to speak, with what it had at the time."

Recently Bergeijk (1966) has adopted an even more extreme viewpoint:

"Until now the ear drum and stapes were assumed to have made their first appearance in the primitive amphibia. The evidence presented here pushes the origin of the eardrum stapes complex back to the ancestral fish (Rhipidistia) where it appears to have been entirely functional for air hearing as well as under-water hearing. Thus continuity in both structural and functional evolution is demonstrated."

In this context "structural continuity" is tantamount to "homologous" while "functional continuity" is synonymous with "ear drum plus stapes". Yet within the amphibians alone there is ample evidence of a multiplicity of non-homologous structures carrying out a multiplicity of functions. Three have already been mentioned—namely the vestibulo-squamosal, the vestibulo-scapular and the vestibulo-tympanic mechanisms. To these may be added the lateral line of the tadpole. Witschi (1955) has also described in the tadpole a pulmonary mechanism reminiscent of the swim bladder ear of the fish. A sixth possibility may be mentioned because of its relation to the vertebral ear of *Dimetrodon*. The endolymphatic system of the frog sends down the vertebral column a prolongation lined by the glands of Swammerdam. Nothing is known of the function either of the duct or of the glands, but analogy with the dorsal sail suggests that they might be vestiges of a former vertebral ear. Be that as it may, it is clear that the amphibians in the course of evolution have utilized five and possibly more middle ears. Of these the lateral line, the pulmonary, the scapular and the vertebral ears clearly have no relation to the hyoid anlage. The vestibulo-squamosal ear may be a hyoid derivative, but the vestibulo-tympanic certainly is not. Although the hyoid derivation of the stapes in reptiles, birds and even in mammals is demonstrable with the utmost ease in the embryo, in the tadpole there is never at any time any vestige of a hyoid connexion. I have personally confirmed this but a far more authoritative statement can be found in Gaupp (1904):

"Recently, Villy, Killian and I myself have investigated the development of the sound conducting apparatus in the anura. We have reached results which in all essentials agree amongst themselves and with Huxley's theory; namely that ontogenetically neither the operculum nor the columella has anything to do with the hyoid arch, and that the auditory capsule is the foundation out of which their structure develops."

It is true that the frog's skull has undergone considerable specialization; nevertheless there is no reason to believe that this could have entirely wiped out all evidence of a hyoid origin if, as postulated by Hotton, Bergeijk and others, the vestibulo-tympanic mechanism had been directly handed down in continuity from *Eogyrinus*.

We are compelled to conclude what is indeed obvious from functional considerations, that the anuran vestibulo-tympanic ear came quite late on the evolutionary scene, having been preceded by various bone conducting mechanisms. The hyoid was only one of many possible embryonic anlagen that contributed to this complex. Indeed its role seems to have been quite minor, since at most it provided the vestibulo-squamosal bone conductor. Presumably it was no longer available when, very much later, the air sensitive ear emerged and so, as Gaupp says, the columella had to be carved out of the auditory capsule and the adjacent mesoderm.

It thus appears that the concept of structural continuity (homology) has no foundation in fact, while the concept of functional continuity (the archetypal ear drum plus stapes) can only be regarded as an anthropocentric prejudice.

This contribution has been confined almost exclusively to the morphology and palaeontology of the conducting ossicles. Parallel information is available from other characteristic features of the middle ear. The intratympanic muscles, the round window, the Eustachian tube and even the perilymphatic duct all show bewildering variations within the different vertebrate orders. These have by no means been adequately codified but enough is now known about them to justify the same conclusion, namely that the "biblical" concept of a single ear "created in the beginning" is without foundation.

SUMMARY

The emerging amphibia, being completely prostrate, evolved detectors for substrate vibrations. Three bone conducting devices evolved out of the various processes of the hyomandibula; subsequently a fourth route via the fore limb appeared. They can all be recognized to-day in prostrate vertebrates.

The direct route in which vibrations enter the inner ear via the stapes is contrasted with the reverse route in which they reach the inner ear via the cerebrospinal fluid and emerge via the stapes. The efficiency of the reverse mechanism is increased by augmenting the mass of the stapes, and this is offered as the explanation of the gigantic premammalian stapes.

By contrast the direct-route air sensitive ear, as encountered in birds, calls for a delicate free-floating ossicle and this is offered as the explanation of the missing dinosaur stapes. The stapes, being delicate, did not survive the trauma of fossilization.

An account is given of the dorsal sail vertebral air sensitive ear of the pelycosaurs.

The interpolation of the reptilian jaw joint into the mammalian middle ear is discussed.

Altogether four non-homologous bone conducting ears and five non-homologous air conducting ears are described. All are shown to be functional adaptations to different enviromental niches.

Many so-called degenerate ears are shown to be transitional forms between bone conduction and air conduction.

REFERENCES

BÉKÉSY, G. VON (1936). *Akust. Z.*, **1**, 13–23.
BERGEIJK, W. A. VAN (1966). In *Contributions to Sensory Physiology*, vol. 2, ed. Neff, W. D. New York and London: Academic Press.
COLBERT, E. H. (1958). *Am. Mus. Novit.*, no. 1900.
GAUPP, E. (1904). *Ergebn. Anat. EntwGesch.*, **8**, 1,058.
HOTTON, N. (1959). *Evolution*, **13**, 93–121.
TUMARKIN, I. A. (1949). *J. Lar. Otol.*, **63**, 119, 193.
WITSCHI, E. (1955). *J. Morph.*, **96**, 497–512.

DISCUSSION

Schwartzkopff: Mr. Tumarkin said that ground-conducted sound, which I would call vibration, becomes useless in erect animals such as the archosaurs. I suggest that the archosaurs probably did use ground-conducted sound, since their descendants the birds use it perfectly, but that this sound no longer reached the ear. Modern birds have a well-developed sense organ of vibration in the tibia which has a sensitivity of about 40 db, better than the sensitivity of human finger-tips to vibration. This is the bundle of 70–80 Herbst corpuscles in a row under the tibia, connected by muscles with the toes. So the perception of ground-conducted sound is sustained in erect vertebrates, but while air-conducted sound goes to the head, ground-conducted sound is perceived by some mechanism close to the feet. This is known behaviourally too; there are records of birds having signalled earthquakes in advance, and much more common ecological observations show that birds do depend on ground-conducted sound (see Schwartzkopff, J. [1949]. *Z. vergl. Physiol.*, **31**, 527–608). Probably the upright archosaurs did the same.

Enger: I would like to question Mr. Tumarkin's suggestion that the dorsal sail of the pelycosaurs was an effective hearing device. The big sail would not be sufficient to detect sound as such. If you compare it to the middle ear, sound detection is possible only when there is a pressure difference on the two sides of a membrane. The middle-ear cavity is at constant pressure, whereas the sound pressure bounces on the eardrum, so there is a pressure difference and therefore a displacement of the tympanic membrane. With a big sail there is no reference point—no pressure difference from one side of the sail to the other. As to its actual function, it might have been a means of temperature regulation; we know that some modern reptiles can keep a constant body temperature.

Whitfield: Surely this criticism is not true. A vibrating membrane can set the air in motion, and in reverse the membrane could act as a receptor. The range of "hearing" would depend on the relation between the wavelengths concerned and the size of the membrane.

Enger: A near-field effect in air could set up vibrations, but with a sound source at a considerable distance, the sail would not vibrate unless there were a reference point on the further side of it.

Tumarkin: Why not? You speak of a pressure wave, but there is no such thing as a pure pressure wave. There is always pressure plus displacement. The sound particles have a direction of movement, and that is why, for example, a ribbon microphone can detect the direction of a sound source. The dorsal sail could to some extent act like a ribbon microphone, the amplitude of its vibration depending on the direction of the incoming sound. The spines would act as levers, or impedance matching transformers, so that airborne energy could enter the vertebrae and so pass into the cerebrospinal fluid.

Enger: If vibrations are going down the spines of the sail, your mechanism could work.

Davis: Man has made use of just such a device as a hearing aid. Before the days of electrical aids, one device used was the acoustic fan, which was essentially a sheet of metal with a prolongation that was placed in contact either with a tooth or with the skull. It was the size of a ladies' fan of the conventional type. This was effective enough to be helpful in marginal cases. There was a plane pressure wave on one side relative to the other, a movement of air particles perpendicular to the surface that faced the source of the sound.

Enger: At close-range, this would work.

J. D. Pye: Surely the point is that there can be a difference in pressure on two sides of a free vane in air, provided that it is very big compared with the sound wavelengths, and therefore effective only at high frequencies; but at high frequencies it becomes very directional, because if the sound is not coming at right-angles to the plane of the vane it can be out of phase in different areas. The animal would be sensitive only to sounds coming from the sides and would not be able to tell from which side.

Tumarkin: It would be all the better for being a directional device.

J. D. Pye: One point puzzles me; if the stapes is to act as an inertial reference source "for sound to come out", in the detection of substrate sound, the end that is not fitted into the window should surely be free? In many cases it articulated firmly with another part of the skull in the premammalia.

Tumarkin: The stapes of the pelycosaurs certainly had a very considerable articulation with the para-occipital. I don't claim to understand the physics entirely, but I cannot think of any reason, apart from substrate hearing, why the stapes should be so colossal. It surely could not have responded to any ordinary tympanic membrane.

Hood: So far as I know, bone conduction in man does not appear to serve any useful purpose; in fact the conductive part of the mechanism seems to be optimally designed to reduce sensitivity to bone conduction to a minimum. Any interference with the conductive mechanism results in an enhanced sensitivity to bone conduction. I know of no scientific experiments in which the process can be shown to be reversed—namely reduction of sensitivity to bone conduction—apart of course from interference with the cochlea. This is a remarkable tribute to the almost ultimate completeness of the evolutionary process so far as reducing bone conduction in man is concerned.

Tumarkin: Mammals certainly have been amazingly successful in abandoning bone conduction, I agree; obviously it is of no real value to them, in comparison with the supreme value of air conduction. I cannot agree that interference with the conduction mechanism improves bone conduction. In my experience a true increase of bone conduction is comparatively rare and in most of the cases operated on for middle-ear deafness, bone-conducted hearing is impaired.

Lowenstein: Would it be wrong to suggest that bone conduction in man might serve in the feedback control of voice production?

Hood: It probably does, but I would have thought that this was incidental, and more in the nature of a by-product usefully employed.

Schwartzkopff: It also plays a role in the control of the jaw movements: if you bite on something hard with your teeth, you hear it immediately and stop masticating.

Bosher: Certain pathological conditions occur in man which produce persistent patency of the Eustachian tube. Such patients complain that their voices sound unpleasantly loud and that the noise generated by the respiratory air movements deafens them. I should have thought that in one sense bone conduction was probably essential for making us aware of what is happening while we are speaking, chewing and so on, since, being relatively insensitive, it does not allow these effects to predominate over what is happening in the external world. One is, in effect, deliberately using under normal circumstances the less sensitive mechanism for the presumably essential monitoring of such activities.

Davis: Von Békésy gave a very illuminating discussion of this question of hearing the sound of one's own voice, treated from the point of view of an acoustic physicist (Békésy, G. von [1949]. *J. acoust. Soc. Am.*, **21**, 217–232; also in Békésy, G. von [1960]. *Experiments in Hearing*, ed. Wever, E. G. New York: McGraw-Hill).

Lowenstein: He attributed great importance to bone conduction.

Davis: He did not take much of a stand on its importance, but emphasized the way in which the structure and location of the voice mechanism and the ear in man seem to minimize the probability of overwhelming the auditory apparatus by the sound of one's own voice.

Whitfield: It would surely be very difficult to get rid of bone conduction. It must exist whether it is made use of or not. It is an epiphenomenon in the sense that unless you went to enormous lengths you could not eliminate it.

Schwartzkopff: Mr. Tumarkin has shown ([1965]. *J. Lar. Otol.*, **99**, 667–694) that in the whale ear, tissue conduction is eliminated to a high degree by the isolation of the petrous bone from the rest of the body by air spaces. There is a very similar mechanism in some birds, where most of the hearing organ, especially in song birds, is isolated from the bone by air cavities. Bone conduction can be reduced, but not completely eliminated.

Whitfield: But in the whale it is more or less essential that the ear should be isolated, whereas in terrestrial mammals conduction through the tissues seems to be harmless, and so it has not been worth the "expense" of eliminating it.

STRUCTURE AND FUNCTION OF THE EAR AND OF THE AUDITORY BRAIN AREAS IN BIRDS

J. SCHWARTZKOPFF

Institut für Allgemeine Zoologie der Ruhr-Universität, Bochum

PROBLEM

THE comparative physiology of hearing can profit in many ways from human physiology, which has already successfully begun to investigate the functions of the ear. This rather young branch of science may, however, also run the risk of unjustifiably transferring particular findings from the mammalian ear to the hearing organ of other vertebrate classes, and so making false generalizations. The full benefit of comparative studies can therefore be derived only through critically reconsidering whether generalizations are justified. This is even more important in the approach to the physiology of hearing in the highest vertebrates. This critical approach is one of the objects of my contribution.

Research on the hearing of one group of vertebrates is faced from the outset with phylogenetic problems which were recognized only recently in human physiology. Morphologically, the bird's ear is closer to the reptilian than to the mammalian one, corresponding to the phylogenetic status of this group; but its performance closely resembles that of the ear of mammals. Both classes share homoeothermia, a high level of metabolism, and an all-round increase of sensory and nervous functions. But it is by no means true that similar achievements are brought about by identical mechanisms, as has been pointed out especially by Pumphrey (1961).

Besides the vast problems of functional evolution, zoologists have to face special questions arising from ecological adaptations within the class of birds. These may, for instance, be concerned with the nocturnal life of owls, with the song of the song birds, or with orientation in some cave-dwelling species.

A theoretical problem, characteristic of comparative physiology, appears when data have to be correlated which have been gathered by very different techniques. Such data may concern the anatomy of the bird's ear as studied by classical zoology, ecology, ethology and general ornithology, and finally physiology, comprising quantitative conditioning experiments, experimental anatomy and electrophysiology.

PERFORMANCE OF THE EAR ACCORDING TO BEHAVIOURAL STUDIES

The major part of our present knowledge about hearing in birds has been contributed by occasional ethological observations, systematic bio-acoustic studies and conditioning experiments.

Acoustic behaviour

Nearly all avian species produce sound signals, and many of them a considerable number of different ones which can be answered by specific actions—that is, which are understood. No other vertebrate class equals the birds in this respect.

At least two orders of birds with numerous species—the song birds and the parrots—have generally acquired the capacity to reproduce sound signals by imitation. This can otherwise be done only in man and his closest relatives. Thorpe (1963) and Thorpe and North (1965) have emphasized the significance of imitation as a very highly developed form of behaviour. Since sound signals acquired by learning are sometimes associated with a special "meaning", a fundamental analogy seems to exist to human language.

Finally, all species of birds appear to share the ability to localize a calling fellow bird, though this type of hearing performance does not reach the perfection it has in many mammals (see Schwartzkopff, 1962*a*). While most mammals are adapted to a nocturnal way of life, birds are primarily diurnal "eye animals", depending less upon auditory localization than nocturnal animals.

The prerequisite of all bioacoustic behaviour is the analysis of sound signals by the ear, in spite of the experimentally established fact that songs or other sound signals can be produced without further auditory control, once they have been acquired; it is even possible that certain calls are innately produced, without ever having been heard by the singer (Schwartzkopff, 1949; Hüchtker and Schwartzkopff, 1958; Konishi, 1963, 1965).

Since man apparently differentiates songs and other sounds of birds in a similar way to birds themselves, many authors have concluded that they have more or less equal hearing abilities. This deduction is of course not valid, as indicated by the following observations. Firstly, human hearing is surpassed in certain cases—for example, when a bird, imitating a song, reproduces quick changes of frequency which are too fast for human perception (Borror and Reese, 1956; in *Mimus polyglossus*). Secondly, the antiphonal calls and songs studied by Thorpe (1963) and Thorpe and North (1965) are indicative of a very efficient temporal resolution of acoustic signals.

Quantitative determination of efficiency

Direct measurements of the hearing capacities of birds by means of conditioning experiments confirm the general impression derived from behavioural studies. However, no unequivocal statement can be based upon these so far, regarding possible divergencies from the mammalian ear. There is only one reservation to this restriction: birds do not show significant hearing abilities within the true ultrasonic range. Though bird voices and flight sounds contain measurable frequency components between 20 and 40 kcyc./sec. (Thorpe and Griffin, 1961), virtually no hearing has been detected above the upper frequency limit of the human ear, in spite of repeated attempts (see Frings and Cook, 1964; for older literature, see Schwartzkopff, 1949). The lower limits of hearing is of minor physiological interest and can be defined only with difficulty. It coincides roughly with the human one.

The frequency function of threshold sensitivity has been studied by Trainer (1946) in *Anas*, *Columba*, *Sturnus*, *Parus*, *Corvus* and *Bubo*; by Schwartzkopff (1949) in *Pyrrhula*; by Heise (1953) in *Columba*; and by Stewart (1955) in *Phasianus*. All the threshold curves have in common the coincidence of optimal sensitivity with the frequency centre of the particular voice. It is always situated between 1,000 and 4,000 cyc/sec., similarly to man (Fig. 1), but sensitivity decreases more steeply towards higher pitch than in man. In about half the birds studied this is true for lower tones as well. Maximum sensitivity matches almost that of man or surpasses it slightly (e.g. in *Bubo*). On the other hand, some species, like the pigeon, remain behind by about 20 db. It can be said that in general the threshold efficiency in birds and mammals coincides. Both groups reach the limits imposed by thermal noise. However, the area of main sensitivity is somewhat narrower in birds than in man and other mammals. This appears to correspond to the conclusions drawn from acoustic behaviour. Only the ultrasonic overtones of the sound production are apparently not perceived and are supposed to be a by-product, as has also been concluded by Thorpe and Griffin (1961).

The capacity for analysing sound signals has been insufficiently investigated, except for the general sensitivity just dealt with. Frequency discrimination in birds was the subject of a few rather old studies. According to these, the pigeon is left far behind in frequency discrimination by man, while the parakeet and the crossbill with difference limens of 0·3–0·7 per cent are doing almost as well as he does (0·1 per cent) (according to Knecht; see Schwartzkopff, 1949). But the figures mentioned require confirmation by modern methods.

I do not know of any direct investigations of the differentiation of intensity steps and the temporal resolution of acoustic signals. Both abilities are likely to be linked complementarily with frequency analysis. Merit for repeatedly pointing out this gap is due to Professor Pumphrey. He has studied (1961) song records published by Thorpe in which chaffinches reproduce heard tunes by imitation. Pumphrey has calculated from the accuracy of the reproduction that temporal resolution in birds must be ten times better than in man.

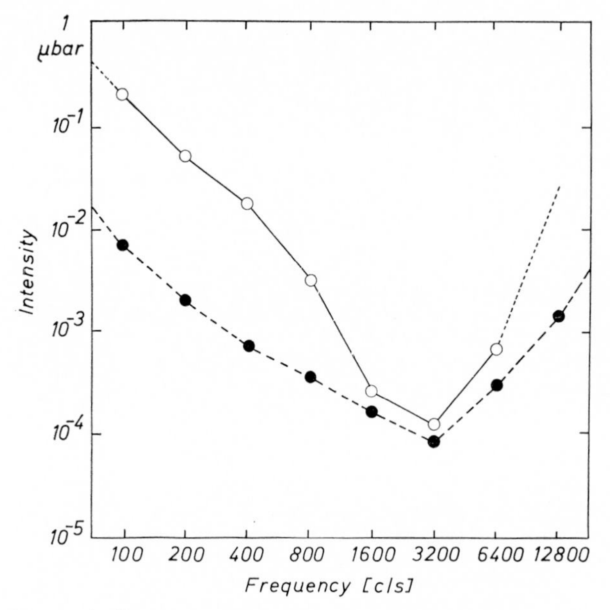

FIG. 1. Auditory threshold curves of bullfinch (*Pyrrhula pyrrhula*) ——o——, and man --●--; modified after Schwartzkopff (1949).

ANATOMY AND FUNCTION OF THE EAR

External and middle ear

Acoustic stimuli pass through the sound-collecting structures of the external ear and through the transforming middle ear on their way from the carrier medium of the air to the actual sense organ. The sound-conducting apparatus in birds has developed in an essentially different way from mammals, but both groups are at about the same level of special adaptive variability and are distinguished in this form from the lower vertebrates.

Russian scientists have compared the shape of the ear opening and the peculiarities of the surrounding feather structures with ecological performance in a large and systematically arranged number of birds. There are interesting correlations: the intensity of social behaviour (acoustic communication), for example, interferes with feeding habits (hunting by diving, with protective mechanisms for the external ear), as shown by Kartaschew and Iljitschow (1964) in various auks. Or the width of the ear feathers is related to auditory localization, when different birds of prey and owls are compared (Dementiew and Iljitschew, 1963). Furthermore the fine structure of the feathers surrounding the external meatus varies in accordance with hearing efficiency (Ilytchev and Izvekova, 1961; Ilytchev, 1962). These studies of ecological and morphological correlations require, of course, to be supplemented by physiological testing of the actual auditory capacities. They are a useful foundation for such tests.

The difficulties of combining morphological and ecological data with physiological findings are illustrated most clearly by the almost historical problem of explaining the asymmetry of the external ear in some owls. Pumphrey in 1949 clarified the physical preconditions for a three-dimensional analysis of sound direction in these owls. The asymmetric ears differ in the frequency function of their directional characteristics. Later, Payne showed how each ear of *Tyto alba* receives sound fractions of different pitch depending upon the direction of the sound source. Unfortunately his paper, though read before the XIIIth International Ornithological Congress held at Ithaca in 1962, was not published in the proceedings. Payne's experiments were made using stuffed heads of barn owls, the middle ear of which was replaced by a microphone. To his conclusions it can be objected that auditory cues large enough to be utilized appear only at very high frequencies (above 10 kcyc./sec.). In this range, the ear of the owl, like that of other birds, shows poor sensitivity. Hence, it is doubtful whether different pitches can be discriminated at all, the prerequisite for this kind of orientation. Schwartzkopff (1962*b*) has studied in a fundamentally similar way the cochlear potentials as a function of sound direction. Living but anaesthetized *Asio otus* were used in these experiments. Very little difference between the asymmetric ears were found when frequencies were used to which the owl is highly sensitive (3–5 kcyc./sec.). Since this does not agree with its ecological performance, the assumption is made that the asymmetry becomes fully efficient only in the living and flying bird, which operates the ear flaps voluntarily (and asymmetrically). According to this hypothesis, the asymmetry is primarily a dynamic process which only secondarily became visible as a static, morphological adaptation. Therefore only a small fraction of the functional capacity of the asymmetric ears could be determined in the experiments of Payne and Schwartzkopff.

Pumphrey's theory of acoustic localization through asymmetric ear openings requires excellent monaural frequency discrimination. This is met at least morphologically by the extraordinary length of the basilar membrane in owls (Winter and Schwartzkopff, 1961; Winter, 1963). Some difficulties arise from the assumption that the very long sense organ is specially adapted for discriminating the highest tones. Unfortunately, behavioural confirmation is still lacking.

The middle ear of birds has been the subject of several reports (Schwartzkopff, 1955; 1957*a*), and therefore only a short review is given here. Great similarity is found in the matching of impedance between the air and the fluids of the inner ear when mammals and birds are compared. This is a clear case of functional convergence, since the underlying structures are of quite different morphology. The ratio of the eardrum to the size of the columella footplate varies in mammals and birds within the same range, from 10 to 40. Birds which can hear very well, such as owls and song birds, have the highest ratios. Diving birds, on the other hand, have coarse ear drums, and relatively sturdy columellae with larger footplates. Furthermore, in owls the membrana tympani secundaria, covering the round window, is extremely tender and large.

Anatomy of the inner ear

The actual sense organ of hearing in birds differs considerably from that in mammals. This is expressed in its general anatomical appearance. Firstly, in birds it is stretched out in almost a straight line with only slight bends and turns. Secondly, the length of the cochlea—which, according to traditional views, corresponds to the capacity for sound analysis—is considerably less than that of mammals, as Schwartzkopff (1957*a*) has shown by allometric measurements in 45 species of birds (Fig. 2). A swan has an inner ear of 6 · 8 mm. length, which does not even reach the length of the basilar membrane of the mouse (7 mm.), though the swan weighs 250 times as much. The mouse has a cochlea of more than twice the length of that of a bird of the same size.

Regarding a correlation between the size of the cochlea and auditory performance, only owls are clearly distinguished from the majority of the other birds by much longer cochleas.

Another peculiarity of the avian inner ear is the well-developed lagena at its distal end. This underlines the relationship with the lower vertebrates. The possible perception of sound through this sensory end-organ was discussed by Pumphrey (1961). This author emphasized the analogy with the fish ear, in which the lagena (and the sacculus) serves as sound receptor.

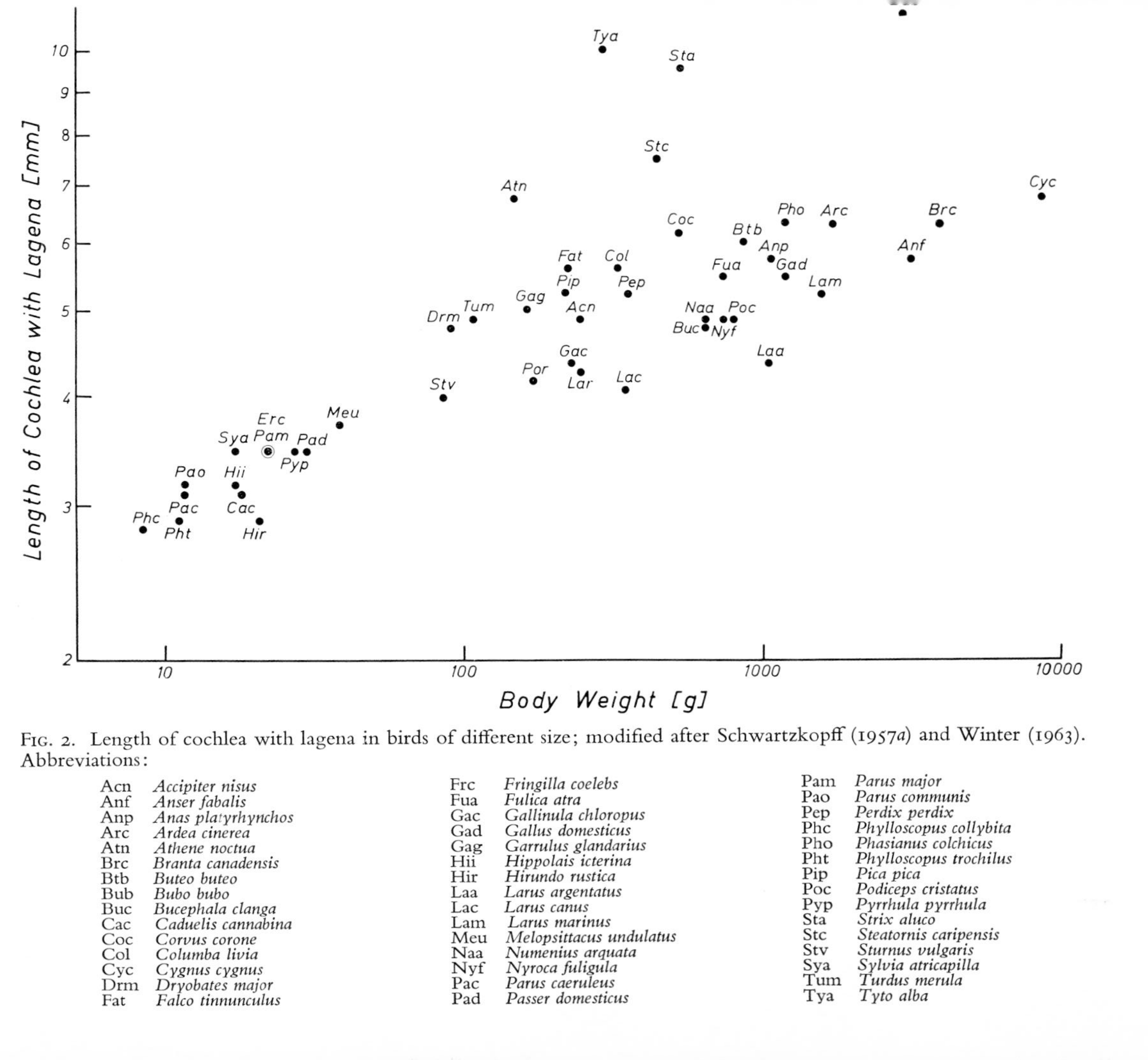

FIG. 2. Length of cochlea with lagena in birds of different size; modified after Schwartzkopff (1957*a*) and Winter (1963). Abbreviations:

Acn	*Accipiter nisus*	Frc	*Fringilla coelebs*	Pam	*Parus major*
Anf	*Anser fabalis*	Fua	*Fulica atra*	Pao	*Parus communis*
Anp	*Anas platyrhynchos*	Gac	*Gallinula chloropus*	Pep	*Perdix perdix*
Arc	*Ardea cinerea*	Gad	*Gallus domesticus*	Phc	*Phylloscopus collybita*
Atn	*Athene noctua*	Gag	*Garrulus glandarius*	Pho	*Phasianus colchicus*
Brc	*Branta canadensis*	Hii	*Hippolais icterina*	Pht	*Phylloscopus trochilus*
Btb	*Buteo buteo*	Hir	*Hirundo rustica*	Pip	*Pica pica*
Bub	*Bubo bubo*	Laa	*Larus argentatus*	Poc	*Podiceps cristatus*
Buc	*Bucephala clanga*	Lac	*Larus canus*	Pyp	*Pyrrhula pyrrhula*
Cac	*Caduelis cannabina*	Lam	*Larus marinus*	Sta	*Strix aluco*
Coc	*Corvus corone*	Meu	*Melopsittacus undulatus*	Stc	*Steatornis caripensis*
Col	*Columba livia*	Naa	*Numenius arquata*	Stv	*Sturnus vulgaris*
Cyc	*Cygnus cygnus*	Nyf	*Nyroca fuligula*	Sya	*Sylvia atricapilla*
Drm	*Dryobates major*	Pac	*Parus caeruleus*	Tum	*Turdus merula*
Fat	*Falco tinnunculus*	Pad	*Passer domesticus*	Tya	*Tyto alba*

In addition, the course of the lagenar nerve fibres hints at this, since they end partially together with the cochlear fibres in the secondary auditory centres of the medulla (see Boord and Rasmussen, 1963, for the older literature). Pumphrey's further assumption, that the lagena receives low-pitched sound through bone conduction, does not appear to be quite unequivocal. It has been established in man that bone conduction plays a role in the transmission of higher frequencies only. Furthermore, the cochlea of birds and the lagena in particular are poorly qualified for picking up

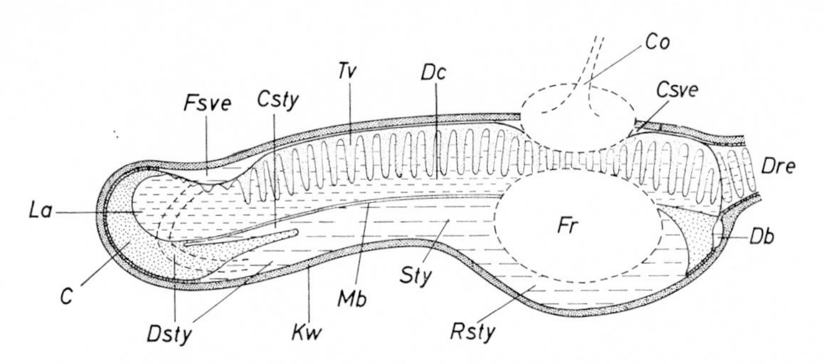

FIG. 3. Diagrammatic longitudinal section of the cochlea and lagena of the bird after Schwartzkopff and Winter (1960).

Abbreviations:

C	Cartilage	Fr	Fenestra rotunda
Co	Columella	Fsve	Fossa scalae vestibuli
Csty	Cavum scalae tympani	Kw	Bony wall
Csve	Cisterna scalae vestibuli	La	Lagena
Db	Ductus brevis	Mb	Membrana basilaris
Dc	Ductus cochlearis	Rsty	Recessus scalae tympani
Dcsty	Ductus scalae tympani	Sty	Scala tympani
Dre	Ductus reuniens	Tv	Tegmentum vasculosum

bone-conducted vibrations, since they protrude rather freely into the air-filled cavities of the avian skull (Schwartzkopff,1949; 1955). In this they are quite different from the mammalian labyrinth, which is solidly embedded in the os petrosum. Clearly, an experimental test of lagenar function is badly needed.

Finally, the peculiar orientation of the round and oval windows with regard to the basilar membrane should be emphasized. Since Satoh (see Schwartzkopff, 1949) published a misleading schematic drawing, the spatial relations have been described wrongly again and again (e.g. Pumphrey, 1961). In birds, both windows are situated about one-quarter of the way along the basilar membrane, not at the end of it. The axis of the columella is roughly at right-angles to the cochlea. Only in owls does the arrangement come closer to that in mammals (Schwartzkopff, 1955; Schwartzkopff and Winter, 1960).

Inside the cochlea, the gelatinous arch of the tegmentum vasculosum almost completely fills the spaces of the scala vestibuli and is in contact with the bony walls (Fig. 3). This can be observed only in freshly dissected organs. Von Békésy (see Schwartzkopff and Winter, 1960) first noticed the absence of a true scala vestibuli in birds. But Retzius (see Pumphrey, 1961) had already depicted the fresh membranous labyrinth of a duck, in which the arching tegmentum can be seen, in contrast to the shrunken appearance after histological treatment, presented by many later authors (see Schwartzkopff, 1955).

More meticulous study shows perilymphatic fissures to exist between the tegmentum vasculosum and the endosteum which become larger just beneath the columella footplate to form the "cisterna scalae vestibuli". These fissures correspond to the scala vestibuli of the mammalian ear, but they are also connected with the scala tympani at the basal end through the ductus brevis, described by de Burlet (see Schwartzkopff and Winter, 1960). Apically a helicotrema has developed between lagena and cochlea. It is located at the end of a channel (ductus scalae tympani) of some length which penetrates the supporting cartilage of the inner ear. The apical part of the basilar membrane is isolated from the helicotrema, lying above a blind sac of the ample scala tympani.

There is no way of extending conclusions from the well-studied hydrodynamics of the mammalian ear to the bird cochlea as long as only the highly divergent anatomy is known. Only one general assumption appears to be sufficiently founded, namely that the tegmentum has an overall damping function, filling most of the cochlear duct.

The older suggestion that the glandular folds of the tegmentum which correspond to the supplying blood vessels ("auditory teeth" of birds) have something to do with frequency analysis is obviously disproved by the fact that the number of tones to be discriminated is much higher than the number of folds or "auditory teeth". In the song bird, for instance, there are 30–40 (Winter, 1963).

The actual sense organ of the inner ear is formed by a stretched epithelial layer of hair cells embedded between suppporting cells and through these connected with the basilar membrane (Vinnikov *et al.*,1965). The hair processes are anchored within the tectorial membrane. The sensory cells of the bird's cochlea have a polarized kinocilium and 40–50 stereocilia (Vinnikov *et al.*, 1965), like the hair cells of other labyrinthine organs in vertebrates (Wersäll, Flock and Lundquist, 1965). In the mammalian cochlea, kinocilia are present only during foetal life. Since birds lack almost all the criteria by which the organ of Corti in mammals is distinguished from the homologous papilla basilaris in lower vertebrates, the use of the latter more general term must be recommended.

The papilla basilaris of birds, while being shorter, is at the same time broader than the organ of Corti in mammals. This means that representatives of equal size from the two groups may have rather similar numbers of hair cells. It is possible that birds have even more. Therefore, the assumption is justified that comparable amounts of evaluated information are received by the sense organ in both cases. But the anatomy does not reveal which parameters of a complex acoustic signal can be utilized preferentially by the bird, and which are of less importance. Pumphrey (1961) assumes that the comparatively broader basilar membrane of birds leads to an improved analysis of sound events. This ought to be confirmed by behavioural studies. The damping of the whole system, combined with the reduced length of the basilar membrane, suggests in any case a limitation of frequency analysis, while temporal resolution should be enhanced. Correspondingly the behavioural studies of Knecht (see Schwartzkopff, 1949) show good frequency discrimination to be present within a narrower band in birds than in mammals. The indication, from bioacoustic findings, of better temporal analysis, has already been mentioned. Furthermore, Schwartzkopff (1957*b*; 1958*a*) has emphasized a particular phase sensitivity of the cochlear potentials (microphonics as well as nervous components) and of higher auditory neurons. This confirms the conclusions just made and underlines the differences between the hearing mechanism of birds and mammals.

As to the synapses of the papilla basilaris, no essential difference seems to exist from the organ of Corti and other labyrinthine sense organs (Wersäll, Flock and Lundquist, 1965; Vinnikov *et al.*, 1965). The nerve endings form several small buttons at each hair cell, thus resembling the type II synaptic connexions of Wersäll (1956), and are in contrast with the large nerve chalices of type I found in the utriculus of birds and mammals. Afferent and efferent connexions can also be distinguished in the papilla basilaris of birds; according to Cordier (1964), a presynaptic body is found at the side of the hair cell in afferent connexions and various mitochondria within the nerve ending. The efferent fibre endings contain numerous synaptic vesicles and the adjacent hair cell area has additional membranes.

Biochemistry and electrophysiology of the inner ear

The biochemistry of excitation and its transmission has been particularly studied by Russian scientists (Vinnikov and Titova, 1964; Vinnikov, Titova and Aronova, 1965). According to their investigations, the birds are to be placed on the one hand between the reptiles and the mammals: cholinesterase is richly present in the synaptic region of the papilla basilaris, as in mammals, and in much larger amounts than in reptiles. It is also found at the hair-bearing pole of the cells, as in mammals but unlike reptiles. On the other hand, the Russian scientists think in terms of a third type of cholinesterase, found exclusively in the birds' hearing organ and giving them a particular position among the other vertebrates.

The electrical phenomena of the avian inner ear correspond in principle with what is known from mammals and reptiles. The registration of microphonics and action potentials from the round window of birds belongs to the routine methods of sensory physiology (Schwartzkopff and Bremond, 1963).

The microphonic potentials differ from the mammalian microphonics in so far as the amplitude increases considerably less than linearly with intensity, even at very low sound pressure (Schwartzkopff, 1960). The same is true for the reptile ear. The non-linearity is likely to produce distortions of the stimulating sound signal, about which nothing is known, however. The particular properties of the microphonics might be explained by nonlinear damping through the tegmentum vasculosum.

The nerve component of the cochlear potentials represents the activity of the auditory nerve and its synapses. Applying suitable techniques, Schwartzkopff and Bremond (1963) could observe stimulus-synchronous discharges up to 3 kcyc./sec. The action potentials further show adaptation and latency changes as a function of stimulus intensity and are influenced by the metabolic state, altogether very much as in mammals (Schwartzkopff, 1960).

Stopp and Whitfield (1964) have shown that summating potentials also occur in the bird's ear. These cannot be understood as an indication of inner hair cell activity, as is assumed for the mammalian ear. Current studies by R. Necker in our laboratory at Bochum reveal the complex composition of the summating potentials in birds. There are two positive components and one negative, which depend differentially upon oxygen supply. In part the components behave similarly to the microphonics, but in part they do not. The kind of metabolic interference appears to indicate that the full identification of summating potentials in birds with those in mammals is questionable.

The endocochlear d.c. potentials, which have been studied in birds by Schmidt (1963), are characteristic of warm-blooded animals with a high level of metabolic activity. My co-worker Necker was able to confirm and further elaborate on the statements of earlier studies indicating an improved resistance of the potentials from the avian ductus cochlearis to metabolic stress. The normal voltage is at +20 mv and goes down to −30 or −40 mv during short-time nitrogen respiration. Whereas in the mammalian ear the action potentials depend upon the endocochlear potential, in the avian ear the auditory nerve fibres can discharge some time after the endocochlear potential has turned negative. We hope that a more detailed conception will result from the experiments now under way, explaining the biochemical events behind the electrical phenomena. One special goal is to clarify the metabolic role of the tegmentum vasculosum, developed so exuberantly in birds.

CENTRAL AUDITORY PATHWAY AND PROCESSING OF INFORMATION

Eighth nerve

The auditory nerve (ramus cochlearis and lagenaris) of an average bird weighing 100 g. contains about 7,500 fibres, according to Winter (1963). A little more than 10 per cent of these elements belong to the lagena. The amount of fibres changes with an exponential constant of 0·15 of the body weight. It cannot be compared satisfactorily with the figures known for some mammals (see Winter, 1963), but it is safe to assume that mammals of the same size have about twice as many auditory nerve elements.

The nerve fibres, originating from cell bodies of the cochlear and lagenar ganglia, are distributed in an orderly way in the secondary centres in the medulla, as Boord and Rasmussen (1963) have found through experimental anatomical studies. The fibres of the ramus lagenaris end partially in the vestibular and cerebellar nuclei, as well as within the reticular formation. The other portion enters both the secondary nuclei of the auditory system —that is, from ventro-laterally the nucleus magnocellularis and from ventrally the nucleus angularis. The cochlear elements are distributed, proceeding from the apex to the base, in the nucleus magnocellularis from caudo-laterally to rostro-medially, and in the nucleus angularis from rostro-ventrally to caudo-ventrally. An organization is apparently present which corresponds mainly with the tonotopic order known from the mammalian auditory pathways. Unfortunately, neurophysiological confirmation of the morphological findings is not yet available.

As found in all vertebrates studied so far, the auditory nerve of birds contains efferent fibres which enter through the vestibulo-cochlear anastomosis and end within the papilla basilaris and the lagena (Boord, 1961). The efferent fibres originate contralaterally but the exact position of the respective neurons is not yet known. The efferent bundle can be activated in birds, as in mammals, from the floor of the fourth ventricle through electrodes introduced stereotactically (Desmedt and Delwaide, 1965). Stimulation produces changes in the inner ear of birds which are similar to those in mammals: the microphonics of the hair cells are slightly enhanced while the action potentials of the auditory nerve fibres are inhibited. The former phenomenon can be related to the theory that the microphonics are the generator potential of nerve excitation; it increases when current is prevented from flowing through the synapse. The latter indicates the control of afferent sensory input through negative feedback from the central nervous system. These authors also report some differences in the details of the dynamics of the efferent activity between birds and mammals, which I shall not deal with here.

Medullary centres

The secondary auditory nuclei of birds—the nucleus magnocellularis and nucleus angularis—can be compared without difficulty with the parts of the nucleus cochlearis in mammals, the nucleus angularis corresponding to the dorsal part. The differentiation of ganglionic cell types and synaptic connexions, however, does not approach the level of complication found in mammals. Nevertheless, Winter (1963) as well as Boord and Rasmussen (1963) could distinguish several types of neurons.

At the tertiary and higher neuronal stations comparison with the homologous centres in mammals becomes increasingly difficult, regarding position, size and nervous connexions. The superior olive does not show up very distinctly in birds, while it is the most important tertiary station of the auditory pathway in mammals. It is of considerable size only in the owls (Winter and Schwartzkopff, 1961). This nucleus receives fibres mostly from the nucleus angularis, while the ascending fibres from the nucleus laminaris pass through it, mainly without making synaptic connexion. The fibres leaving the olive decussate through the ventral stria and continue their way through the lateral lemniscus to the midbrain. The ventral stria is considerably less developed in birds than in mammals, where most of the decussating fibres run ventrally, forming the corpus trapezoideum. In birds, the dorsal commissure is more prominent. The deccussating part of the fibres originating from the nucleus magnocellularis passes through it and enters the ventral surface of the contralateral nucleus laminaris. The remaining magnocellular fibres enter the nucleus laminaris ipsilaterally by the shortest route, through its dorso-lateral surface.

The neurons of the nucleus laminaris thus receive sensory information from both ears through corresponding secondary nerve fibres. Since the contralateral fibres enter from below and the ipsilateral ones from above, some kind of polarization results. This shows up in electrophysiological experiments, indicating that the orientation of the fibres and synapses participates in the processing of binaural information.

The laminary nucleus in particular is correlated with the ecologically determined efficiency of hearing, in its shape as well as in its size. Commonly, the nucleus appears in transverse sections as a simple layer of neurons, extending roughly in one plane. The shape becomes undulatory and the cells are subdivided in those owls which are remarkable for the asymmetry of their external ears. These birds, hunting at night, are able to catch moving prey by acoustic localization exclusively, an outstanding performance among birds. The number of neurons in the laminary nucleus behaves rather conservatively in most birds. In the "asymmetric" owls, however, it shows an extraordinary increase—for example, to more than 10,000 in *Tyto alba*, while *Falco tinnunculus* of the same size has only about 2,000 auditory neurons within this binaural relay station. The augmentation of cell number, but not the change of the shape of the nucleus, is true also for other medullary centres of "asymmetric" owls (Winter and Schwartzkopff, 1961). Therefore the "total cell number" (auditory neurons of one half of the medulla) varies between 28,700 (*Strix aluco*) and 46,600 (*Tyto alba*), compared with 12,000–14,000 in magpies and crows of similar size. Owls with symmetric ears behave more like other species of birds than like their specialized relatives. The little owl (*Athene noctua*) has a total of 11,700 auditory neurons in the medulla and the great horned owl (*Bubo bubo*) only 18,000, though it weighs ten times as much as *Tyto*.

These relations are of special interest because the correlation between ecological performance and information-processing structures is, as far as we know, most prominent in the medulla of birds. Thus the length of the basilar membrane and the number of auditory nerve fibres increase only twofold, even in extreme cases, while the number of secondary and tertiary neurons of the medulla reaches 4–6 times the value (Winter, 1963).

The auditory centres of the medulla show some histological and anatomical peculiarities in other birds as well, by which families, and even species within one genus, can be distinguished (Winter, 1963; Ilytchev and Dubinskaja, 1966). In these cases no sufficient physiological or ecological data for comparison are available, so that further conclusions cannot be drawn.

General neurophysiology of the medulla

The functions of medullary auditory neurons have been studied in parakeets and pigeons by Schwartzkopff (1957*b*) and by Stopp and Whitfield (1961), using microelectrodes. It was found that single neurons respond synchronously to the stimulus vibrations at lower and medium frequencies. When tested by different tones, a cell responds to only a limited proportion. The width of the response area—that is, the accuracy of frequency discrimination through single neurons—in general corresponds with the findings from the mammalian medulla. The shape of the response area shows certain differences, since its borders with the lower and higher frequencies are mostly not symmetrical in mammals, while in birds they are. This asymmetry is said to originate from the special oscillating movements of the basilar membrane in mammals and from the distribution of nerve endings upon it.

Altogether, about 180 auditory neurons have been tested for frequency sensitivity in the two investigations mentioned. A remarkable degree of conformity is shown when the whole spectrum of frequencies "heard" by these neurons is plotted (Fig. 4). But it must be assumed that both results are incomplete, in spite of the satisfactory agreement. The neurophysiological experiments appear to prove that the upper limit of hearing in birds

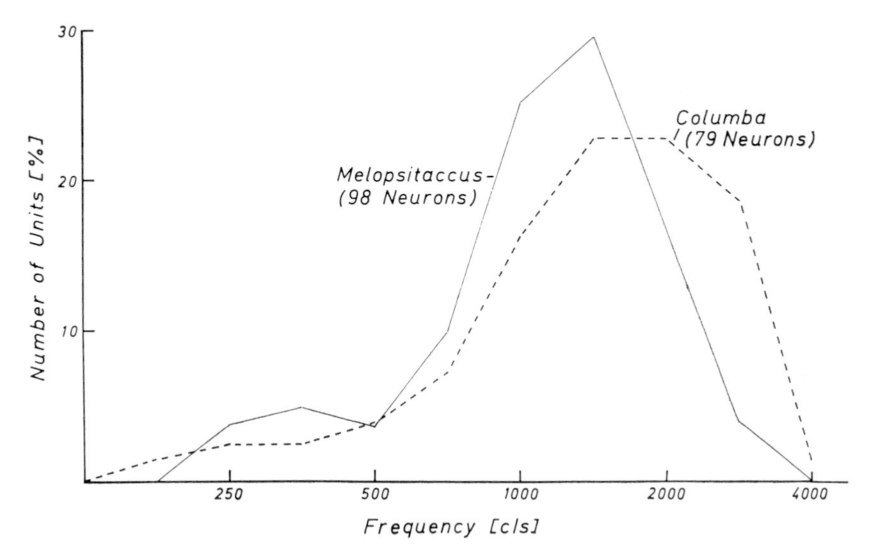

Fig. 4. Number of auditory neurons representing characteristic frequencies in the medulla of the parakeet (*Melopsitaccus undulatus*) and the pigeon (*Columba livia*); modified after Schwartzkopff (1957*b*) and Stopp and Whitfield (1961) respectively.

is at 4 kcyc./sec. This does not conform with the results of behavioural studies. It is well established that birds can hear tones ranging two or three octaves higher than 4 kcyc./sec., and this hearing must be processed by corresponding neurons. Representation up to 40 kcyc./sec. has been found for instance in the brain of cats, in similar experiments. Therefore I suggest that some as yet undiscovered methodological difficulties have prevented the microelectrodes from entering the region of "high frequency cells" in the bird's brain. This assumption is confirmed by a sensitivity curve of the forebrain published by Ilytchev (1966*c*), showing an additional maximum at 3–4 kcyc./sec.

The British scientists have further studied the question of neuronal inhibition. Sound stimuli can reduce or suppress the activity of about two-thirds of all units studied. This indicates contrast formation and in general a high level of neuronal processing of sensory information fed into the medulla in birds, similar to that found in mammals.

Schwartzkopff (1958*b*) has found nervous elements in the medulla (and possibly already in the auditory nerve) of birds which respond to a non-periodic (click) stimulus with a periodic discharge pattern. The same was found in mammals. When a periodically responding element is stimulated by continuous tones and one determines the highest sensitivity, its "best frequency" corresponds with the period of response elicited by the non-periodic stimulus. This phenomenon was called "neurophysiological resonance" and is said to indicate a general mechanism of central frequency analysis common to all vertebrates but specially used by lower forms.

Higher brain centres

Anatomists have only occasionally studied the further course of the auditory pathways above the medulla of birds by classical methods, and it has also not yet been established sufficiently by physiological research. The medullary fibres ascending through the lateral lemniscus end in the nuclei isthmi, including the nucleus semilunaris of the midbrain, according to Craigie (see Erulkar, 1955). Erulkar succeeded in eliciting evoked responses here by sound stimulation. The identification of the torus semicircularis with the colliculus inferior of mammals (which correspond morphologically; see Cobb, 1962) is controversial. Erulkar was unable to elicit evoked potentials from this region. Recently Brown (1965) has established that the torus semicircularis in birds has become a motivation centre for sound production (alarm call). This may indicate a fascinating change of function from sensory-acoustic towards motor-acoustic activity in the bird's midbrain.

Information on the auditory activity of the forebrain of birds has essentially been produced by recent research by Russian authors (see Ilytchev, 1966*a,b,c*). For a long time it was known to physiologists that certain areas of the hemispheres could produce acoustic behaviour when stimulated artificially, as Kalischer reported in his classical paper (see Brown 1965), but nothing more. Erulkar (1955) and with more detail Ilytchev, have shown click stimuli to produce evoked potentials within the nucleus basalis, part of the neostriatum frontale and of the ectostriatum, after surprisingly short latency (4–6 msec.). The very short time required for neuronal transmission of the signal from ear to forebrain may indicate that fewer relay stations are intercalated than in mammals. This fits in with the statement of Erulkar (1955), who found no acoustic activation of the dorsal thalamus in birds, which corresponds to the corpus geniculatum mediale in mammals.

The click-evoked potentials from the forebrain of birds cannot be masked by sinusoidal tones even of very high intensity. This conflicts with the findings in mammals. Furthermore, almost no influence of the direction of the sound could be detected in the bird's hemispheres, though it is present in the midbrain. It seems, therefore, that several functions of the auditory cortex or the dorsal thalamus in mammals are not localized in the forebrain of the birds.

The question of which part of the bird's central nervous system contains the higher stations of frequency discrimination, directional hearing and even more complex auditory functions cannot yet be answered. At present it is possible only to say that already within the medulla the processing of auditory information plays a considerable role. But this does not appear to be more important than in the mammal.

SUMMARY

The findings on the hearing of birds, gathered by different authors by various methods, are critically reviewed. The performance of the bird's ear almost equals that of man as far as absolute sensitivity and frequency discrimination are concerned; the temporal analysis of birds is better.

The structures of the external ear are correlated with ethology and specific auditory performances. The difficulties of functional interpretation of morphological findings are demonstrated by the example of the asymmetric ears of owls.

The inner ear of birds differs considerably from that of mammals because of the simple structure of the papilla basilaris and the absence of an open scala vestibuli. The dynamics of sound propagation through the inner ear are not clear.

The nerve supply and biochemical and electrical phenomena of the inner ear and the auditory nerve resemble those in mammals. The central auditory pathway can be compared with that of mammals only at the first station. The anatomical development of the (tertiary) nucleus laminaris, the centre of binaural interaction, is correlated with ethology for the owls.

The auditory neurons in the medulla of birds show response areas similar to those in mammals. Stimulus-synchronized responses are evoked by medium and lower frequencies. Neuronal inhibition is commonly found. The "representatives" of certain frequency areas do not cover the range of hearing known from behavioural studies. The activity of the forebrain (nucleus basalis) is more closely correlated to behaviour.

REFERENCES

Boord, R. L. (1961). *Expl Neurol.*, **3**, 225–239.
Boord, R. L., and Rasmussen, G. L. (1963). *J. comp. Neurol.*, **120**, 463–475.
Borror, O. J., and Reese, C. R. (1956). *Ohio J. Sci.*, **56**, 177–182.
Brown, J. L. (1965). *Science*, **149**, 1002–1003.
Cobb, S. (1962). *Archs Neurol., Chicago*, **6**, 43–48.
Cordier, R. (1964). *C. r. hebd. Séanc. Acad. Sci., Paris*, **258**, 6238–6240.
Dementiew, G. P., and Iljitschew, W. D. (1963). *Falke*, **10**, 123–125, 158–164, 187–191.
Desmedt, J. E., and Delwaide, P. J. (1965). *Expl Neurol.*, **11**, 1–26.
Erulkar, S. D. (1955). *J. comp. Neurol.*, **103**, 421–452.
Frings, H., and Cook, B. (1964). *Condor*, **66**, 56–60.
Heise, G. A. (1953). *Am. J. Psychol.*, **66**, 1–19.
Hüchtker, R., and Schwartzkopff, J. (1958). *Experientia*, **14**, 106–107.
Ilytchev, V. D. (1962). *Dokl. Akad. Nauk SSSR*, **144**, 1185–1188.
Ilytchev, V. D. (1966*a*). *Zh. vȳssh. nerv. Deyat. I.P.Pavlova*, **16**, 480–488.
Ilytchev, V. D. (1966*b*). *Vest. mosk. gos-Univ., Biol.*, **4**, 32–36.
Ilytchev, V. D. (1966*c*). *Dokl. Akad. Nauk SSSR*, **166**, 246–249.
Ilytchev, V. D., and Dubinskaja, G. R. (1966). *Zool. Zh.*, **45**, 1580–1582.
Ilytchev, V. D., and Izvekova, L. M. (1961). *Zool. Zh.*, **11**, 1704–1714.
Kartaschew, N. N., and Iljitschow, W. D. (1964). *J. Orn., Lpz.*, **105**, 113–136.
Konishi, M. (1963). *Z. Tierpsychol.*, **20**, 349–367.
Konishi, M. (1965). *Z. Tierpsychol.*, **22**, 584–599.
Pumphrey, R. J. (1949). *Rep. Smithson. Instn* 1948, pp. 305–330. Publn. no. 3954.
Pumphrey, R. J. (1961). In *Biology and Comparative Physiology of Birds*, vol. II, pp. 69-86, ed. Marshall, A. J. New York and London: Academic Press.
Schmidt, R. S. (1963). *Comp. Biochem. Physiol.*, **10**, 83–87.
Schwartzkopff, J. (1949). *Z. vergl. Physiol.*, **31**, 527–608.
Schwarztkopff, J. (1955). *Acta XIth Int. orn. Congr., Basle*, 1954, pp. 189–208.
Schwartzkopff, J. (1957*a*). *Z. Morph. Ökol. Tiere*, **45**, 365–378.
Schwartzkopff, J. (1957*b*). *Verh. dt. zool. Ges., Graz*, 374–379.
Schwartzkopff, J. (1958*a*). *Z. vergl. Physiol.*, **41**, 35–48.
Schwartzkopff, J. (1958*b*). *Z. Naturf.*, **13B**, 205–208.
Schwartzkopff, J. (1960). *Verh. dt. zool. Ges., Bonn/Rhein*, 416–424.
Schwartzkopff, J. (1962*a*). *Ergebn. Biol.*, **25**, 136–176.
Schwartzkopff, J. (1962*b*). *Z. vergl. Physiol.*, **45**, 570–580.
Schwartzkopff, J., and Bremond, J. C. (1963). *J. Physiol., Paris*, **55**, 495–518.
Schwartzkopff, J., and Winter, P. (1960). *Biol. Zbl.*, **79**, 607–625.
Stewart, P. A. (1955). *Ohio J. Sci.*, **55**, 122–125.
Stopp, P. E., and Whitfield, I. C. (1961). *J. Physiol., Lond.*, **158**, 165–177.
Stopp, P. E., and Whitfield, I. C. (1964). *J. Physiol., Lond.*, **175**, 45.
Thorpe, W. H. (1963). *Nature, Lond.*, **197**, 774–776.
Thorpe, W. H., and Griffin, D. R. (1961). *Ibis*, **104**, 220–227.
Thorpe, W. H., and North, E. W. (1965). *Nature, Lond.*, **208**, 219–222.
Trainer, J. E. (1946). Thesis, Cornell University, pp. 246–251.
Vinnikov, J. A., and Titova, L. K. (1964). *The Organ of Corti: Its Histophysiology and Histochemistry*. New York: Consultants Bureau.
Vinnikov, J. A., Titova, L. K., and Aronova, M. Z. (1965). *Acta histochem.*, **22**, 120–154.
Vinnikov, J. A., Osipova, I. V., Titova, L. K., and Govardovskij, V. I. (1965). *Zh. obshch. Biol.*, **26**, 138–159.
Wersäll, J. (1956). *Acta oto-lar.*, Suppl. **126**.

WERSÄLL, J., FLOCK, A., and LUNDQUIST, P. G. (1965). *Cold Spring Harb. Symp. quant. Biol.*, **30**, 115–132.
WINTER, P. (1963). *Z. Morph. Ökol. Tiere*, **52**, 365–400.
WINTER, P., and SCHWARTZKOPFF, J. (1961). *Experientia*, **17**, 515–516.

DISCUSSION

Whitfield: I agree with Professor Schwartzkopff that one does not see tuning curves with characteristic frequencies above about 4,000 cyc./sec., when recording from the pigeon brainstem. I don't know what the evidence is that these birds can respond to higher frequency tones, but are we quite sure, if they can, that there is not some other sensory route involved? It is true that we did see some units in the pigeon nucleus magnocellularis which were phase-locked at lower frequencies, but equally we saw others which were not and whose responses were quite irregular.

One of the big problems is the question of the innervation of the hair cells, because there are a large number of cells across the width of the membrane. As far as we can see, the pigeon has about the same number of hair cells as the cat; yet it has many fewer auditory nerve fibres and apparently, at least electrophysiologically determined, a much shorter frequency range. There seems to be no information about how the nerve fibres are distributed to these hair cells or why there are so many. We have speculated that the arrangement of the organ of Corti may be some kind of economy device which enables the same dynamic intensity range to be encompassed with fewer cells, but there is so far no evidence for this.

Schwartzkopff: Conditioning experiments by G. A. Heise ([1953]. *Am. J. Psychol.*, **66**, 1–19) showed that pigeons perceive sound over a range of frequency extending up to 11,000 cyc./sec.

Whitfield: Heise certainly *tested* his pigeons at 11,000 cyc./sec., but whether he got a response is not certain. He merely records the observation that the threshold was "above 88 db" (re 0·0002 dynes/cm.^{2}). His threshold response curve, which does not change by more than about 10 db over the range 400–4,000 cyc./sec., suddenly starts to shoot up abruptly above 4,000 cyc./sec. at a rate of over 70 db/octave. This suggests that some "break" occurs at around the 4,000–5,000 cyc./sec. mark. Our highest unit had a characteristic frequency of 4,130 cyc./sec. and of course its cut-off on the high frequency side was not infinitely steep, so that the correspondence does not seem to be too bad. It is true that Heise got responses at 8 kcyc./sec., but these were at threshold intensities around 80 db. Without controls in deafened animals, can we be sure that these relatively intense sounds are not providing some other clue? At what level for instance is the "threshold of feeling" in pigeons?

Schwartzkopff: It seems then that I am partially mistaken in quoting Heise, who does not give clear figures of upper limits. However, J. E. Trainer ([1946]. Thesis, Cornell University) shows threshold curves up to 7 and 8 kcyc./sec.

(pigeon), M. P. Wassiljew ([1933]. *Z. vergl. Physiol.*, **19**, 424–438) has found the upper limit at 12 kcyc./sec. and E. G. Wever and C. W. Bray ([1936]. *J. comp. Psychol.*, **22**, 353–363) could produce cochlear microphonics up to 11·5 kcyc./sec. in the pigeon. Further figures of upper limits in different birds are to be found in Schwartzkopff ([1955]. *Acta XIth Int. orn. Congr., Basle, 1954*, pp. 189–208). While the pigeon does not hear very much above 10 kcyc./sec., other birds do.

My point was that the neurophysiological findings of P. E. Stopp and I. C. Whitfield ([1961]. *J. Physiol., Lond.*, **158**, 165–170) and of Schwartzkopff ([1957]. *Verh. dt. zool. Ges., Graz*, 374–379) in the medulla of the pigeon or the parakeet do not correspond with the sensitivity curve from the forebrain obtained by Ilytchev. He could show that the pigeon can hear (neurophysiologically) at least up to 5 kcyc./sec. This is in agreement with the older studies but not with our recent results. And I believe that Ilytchev is right ([1966]. *Dokl. Akad. Nauk SSSR*, **166**, 246–249).

The question of "feeling" in birds was studied by Schwartzkopff ([1949]. *Z. vergl. Physiol.*, **31**, 527–608). Certainly, the vibration threshold is better than in man but not for the higher frequencies which are in discussion here. The whole problem of "extra-auditory hearing" was discussed thoroughly in my 1949 paper and seems to have been settled since then. There is no chance that deafened birds can hear high frequency sound, since the experiments Dr. Whitfield is asking for have been performed.

On the question of the peripheral distribution of nerve fibres, we are now engaged on counting the hair cells, and it is clear that in the bird ear there are fewer fibres than and at least as many hair cells as in the mammalian ear, which means that there is at least a ratio of 1:2 and mainly a ratio of 1:4, which is not so good as in the mammal. This can be stated so far, but the question of whether it is intensity or frequency which is transmitted is the essential point.

Batteau: I have been interested in the temporal relationships of sounds, and it seems to me that pure tones are extremely rare in nature; in fact, I know of no natural pure tones. In evolution, perception-processing was developed to handle structural form in sound, which comprised time-relationships and intensity, and the perception of tonality is a byproduct of the system's ability to perceive many kinds of structures, so that using it as a primary indicator can be misleading. But I cannot be much help here, partly because I do not know what kind of structural temporal form, except that for localization, could be useful for determining structural perception. However, two pulses separated by 50 μsec., with one changing its position by 2 μsec., can be perceived by man.

Whitfield: This is the real problem which we have been trying to tackle in the mammal. Dr. Evans and I (Whitfield, I. C., and Evans, E. F. [1965]. *J. Neurophysiol.*, **28**, 655–677) started investigating this problem using frequency modulated tones to give some simple temporal structure to the stimuli, and we found units at the cortical level unresponsive to steady tones which are readily activated by such modulated tones. Beyond this there are plenty of units which respond to the kind of complex noises you can make with your mouth, yet which will not respond even to frequency modulated tones. We have been experimenting with combinations of simple stimuli to try to stimulate these latter units, but it is extraordinarily difficult to conceive what *is* a simple auditory structured pattern. It is easy for us to think of simple visual patterns, but we have found it very difficult to think analogously in auditory terms. We have hit upon one or two simple patterns which will excite particular units but we have not so far been able to find any general set of rules for generating such patterns.

Schwartzkopff: One approach that has been tried, so far without success, by J. C. Bremond ([1959]. *C.r. hebd. Séanc. Acad. Sci., Paris*, **275**, 1643) is to play back the song of a bird (the author used the robin) and to find out by changing the pauses or the frequency what the bird is actually interested in. This work was without final conclusions but the approach was right; the method needs to be improved. Another way would be to use conditioning experiments on the basis you have mentioned for man—what can the bird differentiate, how short can the sound wave be (which comes to the same thing), or if you offer two clicks, at how short an interval does the bird hear them as two events?

There is also the result from Pumphrey that I mentioned (p. 44). Thorpe had collected figures on the songs of chaffinches which had been learned and in these cases the learner was reproducing the time-relationships with astonishing accuracy, and Pumphrey has been able to show that the resolving power of one bird which learned from another must have been at least ten times better than that of man.

Batteau: If the interval between two pulses fed into opposite ears very close together in time (within 80 µsec.) is varied, the subjective effect is a change not in quality but in (apparent) location, so the difference is of frequency response or the ability to correlate in time. A note of very high frequency which is "invisible" at short duration can become highly visible if correlated over a sufficient number of cycles.

You may ask what kinds of sensory mechanism will perform these tasks. Some years ago when we were interested in the perception of sound in the presence of noise, a test was run at the Naval Ordnance Station at Pasadena to discover whether or not a correlated, integrated computated process was involved. Pure tones, which had been identified to the people taking the test, were then immersed in noise. The subjects were asked to indicate by pushing a button when the tone came on and when the tone went off. There was a delay between the start of the tone and the time when the button was pushed, a decision lag, roughly corresponding to the amount of computation time necessary to average out the noise used in order to compute a signal-to-noise ratio of one. The higher the noise level, the longer the time for the decision. The lag also appeared when the tone was turned off, and the time required to decide that the tone was off was just about the same as that taken to decide that the tone was on.

Therefore if you have no evidence of the perception of high frequency sounds from the electrical results, it might be that there is a small signal-to-noise ratio but that neural processing would build up over a large number of cycles to a point where the sound becomes perceptible.

Whitfield: On Professor Schwartzkopff's suggestion that playing around with the signal is the way to tackle it, we have tried (in cats) with units responsive only to "noises", putting these noises through bandpass filters. You reach the stage where you find that when you take out everything above, say, 20 kilocycles there is no response at all, so you then take out everything *except* frequencies above 20 kilocycles and there is still no response; but if you restore the whole 0–70 kilocycles bandwidth you get the response. Clearly, such analysis is not going to give a quick answer.

Hallpike: Professor Schwartzkopff, I recall a suggestion of Pumphrey's that a certain asymmetry of the owl's ears enables it to catch mice in the dark. Is that so, and how does it work?

Schwartzkopff: The asymmetry provides an additional cue to time and to the intensity difference which can work on only one plane. The ear openings are characteristically arranged so that frequency differentiation occurs along one plane for one ear and along another plane for the other. If the sound comes from one point, a different fraction of the sound frequencies is heard by each ear, so that for a complex source, the owl can with some efficiency differentiate the three-dimensionality of the sound, which cannot be done by time or by intensity difference.

Lowenstein: I believe it is true that mammals such as the dog cannot localize sound coming from above. A dog does not react to his master's voice coming from a balcony above.

Neff: Has it been shown experimentally that the owl can localize sounds coming from above or beneath it?

Schwartzkopff: We can show that this localization occurs experimentally, using high frequencies, but at the normal frequencies the effect of the difference between ears is so small that our method is not sufficient to demonstrate it. Evidently the bird must be awake and opening and closing its earflaps and changing the shape of the openings; then the ears become really efficient. What we see in the dead bird is just the structures which enable it to do this. But I think Pumphrey is right, although it is not easy to do this experiment with flying owls, and one must use the anaesthetized bird.

Lowenstein: Turning to the lagena, there is evidence that it is a statoreceptor, because its removal interferes with the counter-rolling of the eyes; so the lagena may be just a statoreceptor that has been carried forward to the tip of the cochlea (Benjamins, C. E., and Huizinga, E. [1927]. *Pflügers Arch. ges. Physiol.*, **217**, 105–123; [1928]. *Ibid.*, **220**, 565–582; [1928]. *Ibid.*, **221**, 104–118).

Schwartzkopff: This would be in conformity with my older findings. However, Boord and Rasmussen showed that about half the fibres from the lagena end in the auditory nuclei and the other half end in the vestibular nuclei. Therefore the lagena may do both.

Lowenstein: This does not surprise me, because I found the same in the sacculus in the elasmobranchs (Lowenstein, O., and Roberts, T. D. M. [1951]. *J. Physiol., Lond.*, **114**, 471–489).

Johnstone: Is there an anomaly in the flightless birds, particularly in regard to what Mr. Tumarkin was saying about substrate hearing? Here we have birds that do not fly and who are in contact with the ground. Have any physiological investigations been made? And where would such birds fall on your graph of cochlear length *versus* body weight?

Schwartzkopff: There have been some studies on the middle-ear apparatus of flightless birds but nothing on the hearing apparatus. They would belong to the group with less good hearing, with a shorter cochlea than the song birds and the owls.

SECTION II
STRUCTURE AND FUNCTION IN MAMMALS

HEARING IN BATS

J. D. Pye

Department of Zoology, King's College, London

The upper frequency limit of human hearing is generally about 17–20 kHz in young people and, although for technical reasons audiometry is seldom continued above 12 kHz, it is fairly easy to show that the limit tends to decline with age as an expression of presbycusis. It has been claimed, however (Pumphrey, 1950) that the limitation is entirely due to conduction loss and that sounds presented at sufficient intensity by bone conduction are audible, without further sensation of pitch change, to more than 100 kHz. It appears that the human cochlea is capable of responding to high "ultrasonic" frequencies but does not normally experience them and is unable to perform frequency analysis upon them.

These limitations of the human ear are not shared by many smaller mammals, for there is good evidence that many species not only hear and analyse ultrasound, but use these frequencies for communication. Ralls (1967) has shown that mice and *Peromyscus* show posterior collicular responses to at least 100 kHz, with greatest sensitivity in the range 10–30 kHz. Noirot (1966) found that baby mice communicate with their mother at 60–90 kHz. Sewell (1967) has found similar behaviour in ten other species of Myomorph rodents and shown that the adults of some species indicate aggressive and submissive intentions by ultrasonic signals.

The hearing of bats is especially interesting, because the Microchiroptera show a highly developed form of ultrasonic echolocation. High frequencies are used because, despite their high attenuation in air, the shorter wavelengths permit better angular resolution and the possibility of wide bandwidth allows better range resolution. Since Pierce and Griffin (1938) first detected the emission of ultrasound by Vespertilionid bats, echolocation signals have been recorded from 11 of the other 15 families of Microchiroptera and it is probable that ultrasonic guidance is used by all the 700 or so species. Echolocation has been demonstrated also in Cetacea (Kellogg, 1959, 1961; Norris *et al.*, 1961), in shrews (Gould, Negus and Novick, 1964), in the birds *Steatornis* (Griffin, 1953) and *Collocalia* (Novick, 1959; Medway, 1959) and in the Megachiropteran genus *Rousettus* (Möhres and Kulzer, 1956; Novick, 1958; Kulzer, 1960), although the last three forms use signals which in part at least are audible to man. The hearing abilities of all these forms are obviously of interest but little information is available apart from the Vespertilionidae of the Microchiroptera.

EXTERNAL EAR

The external ears of bats are very variable in form. The pinna is generally large and strengthened by ribs of cartilage (Fig. 1) and the two may be

FIG. 1. *Taphozous mauritiana* (Emballonuridae). The ears, like those of most Vespertilionids and several other families, are relatively simple and immobile. They are well separated and face forwards, with a small tragus in front of the meatus.

joined together above the forehead as in *Lavia* (Fig. 2). At the other extreme they may be deeply folded with the aperture facing sideways as in *Tadarida* (Fig. 3). Muscles may be present in the pinna (Schneider, 1961) to change its shape by a small but maybe important degree. In *Plecotus*, which folds its enormous ears away under its wings when asleep, it seems that blood pressure may play a part in their erection. Most bats have a well developed tragus which often stands erect as a separate "earlet" in front of the pinna and is especially well seen in those forms with very large pinnae (Fig. 2).

In most families the ears are more or less immobile upon the head. But the pinnae of the Rhinolophidae and Hipposideridae, while lacking a tragus (Fig. 4), are remarkable for their mobility. Schneider and Möhres (1960) have shown that in *Rhinolophus ferrumequinum* the ears change shape considerably, can be turned independently through wide angles and also

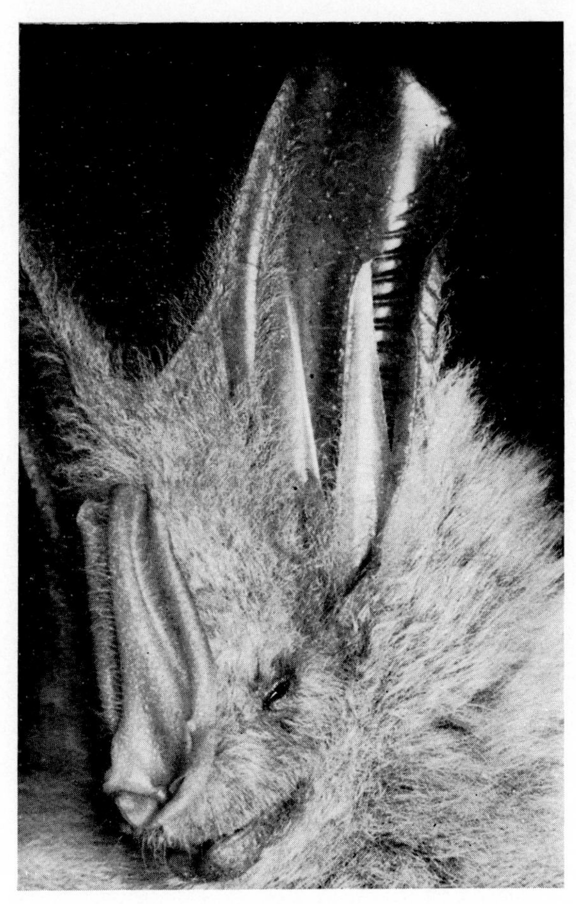

FIG. 2. *Lavia frons* (Megadermatidae). Here the ears are relatively enormous, are joined together well above the head and possess large, pointed tragi.

vibrate rapidly and alternately with a peak-to-peak amplitude at the tip of over 1 cm. These movements are essential to accurate echolocation, for their abolition by surgical means led to disorientation, although some recovery was shown if the bat learned to nod its whole head rapidly. Griffin and co-workers (1962) and Pye, Flinn and Pye (1962) showed that the alternating movements are synchronized with the emission of ultrasonic pulses, a forward movement of one ear and a backward movement of the other accompanying each pulse up to the highest rates of 60–80 pulses per second. Observation suggests, although it has not been shown conclusively, that other members of both families behave in an identical way, and this may be related to their use of very constant frequency "Doppler" signals.

FIG. 3. *Tadarida leonis* (Molossidae). The ears are again joined together in the midline but are deeply folded from behind, giving a heavy "fleshy-lobed" appearance and causing them to face almost laterally. The tragus is very small and cannot be seen in this view.

Great mobility of the pinnae is also shown by the Pteropidae (Megachiroptera) and by many Phyllostomatids, and the latter also produce vibrations of the ears. But the Pteropidae (except for *Rousettus*) have no echolocation and in Phyllostomatids the movements are not synchronized to the production of their rapid trains of very short pulses. In these animals, therefore, ear movements appear to play a part in auditory localization of external sources rather than in echolocation and they may be elicited by any sound, preferably high-pitched, which disturbs the animals and attracts their attention. In this respect, however, it is of interest that Henson (1967) stated that *Chilonycteris*, a Phyllostomatid with relatively immobile ears but long pulses with constant frequency sections, nods it heads vigorously in a way reminiscent of a *Rhinolophus* whose ears have been immobilized surgically.

Wever and Vernon (1961*a*) and Henson (1967) used the amplitude of cochlear microphonic potentials to measure the directional sensitivity of the ear in *Myotis* and *Tadarida* respectively. More elaborate measurements have been made by Grinnell and Grinnell (1965) by observing the sound intensities needed to produce collicular evoked potentials (N_4) of a standard

FIG. 4. *Rhinolophus fumigatus* (Rhinolophidae). Here there is no tragus at all. The separated ears can be turned independently through wide angles and show rapid alternating movements synchronously with pulse production.

magnitude in *Myotis* and *Plecotus*. The stimulating sound source could be varied in frequency and placed anywhere on a spherical locus around the head. Reception was found to be strongly directional, especially at higher frequencies, and was greater in *Plecotus*, owing to the larger ratio of pinna size to wavelength. The direction of greatest sensitivity was different at different frequencies. This permits the possibility of locating the direction of broad-band sources by intensity clues alone, as suggested for the passive hearing of owls by Pumphrey (1948). Grinnell and Grinnell also found that slight distortion of the pinnae, for instance by attaching them together by their medial edges, produced "dramatic" changes in the response patterns of *Plecotus*. Griffin (1958) had earlier shown that this distortion leads to considerable disorientation in the same species. Removal of the tragus produced less pronounced effects and its function is still somewhat problematical.

Wever and Vernon (1961*b*) have shown that in *Myotis* the tip of the pinna can fold forwards under conditions of intense ultrasonic stimulation. By imitating this movement while measuring cochlear microphonic potentials they produced an attenuation at the cochlea of about 20 db for sound sources directly in front of the head. In addition they found a small cartilaginous fold in the external auditory meatus which could occlude its lumen as a second protective mechanism during loud sounds. When moved artificially in an anaesthetized bat this "meatal valve" produced a further attenuation of 20–40 db.

MIDDLE EAR

The middle ear of a number of bats has been described in detail, and compared with that of some insectivores, by Henson (1961). There was a marked correlation between the ability to hear high frequencies and the small size of the tympanic membrane. The ossicles were also smaller in bats, with deep sulci to reduce their mass. They appeared to be tightly coupled, especially at the incus–malleus articulation, presumably to reduce transmission losses. The most remarkable feature, however, was the relatively enormous size of the middle-ear muscles (Fig. 5) which was also remarked by Wever and Vernon (1961*b*). The tensor tympani is very long and generally spindle-shaped, with a low tendinous content. Wassif (1946) had earlier stated that in many bats this muscle is divided and has a double insertion upon the malleus. The stapedius is broadly conical and has a mass comparable with, or even exceeding, that of the tensor tympani, in striking contrast to most other mammals.

Two other unusual features are that Paaw's cartilage is present within the stapedius and that the stapedial artery, which degenerates in most adult mammals, is retained throughout life in bats. The presence of a functional artery passing through the arch of the stapes suggests that low frequency pulse movements cause little disturbance in ears which are specialized for high frequencies. These features have been investigated by A. Pye and Hinchcliffe (1968) in a large variety of bats and other mammals.

Activity of the middle-ear muscles of anaesthetized *Myotis lucifugus* has been studied by Wever and Vernon (1961*b*) by observation of cochlear microphonic potentials. They found that sounds of 3 kHz and exceeding 1 μbar in one ear produced a tympanic reflex which attenuated cochlear microphonics produced by an independent test source in the other ear.

Between 0 and 23 db re 1 μbar the attenuation was a linear function of stimulating intensity, indicating almost perfect regulation at the cochlea. This formed a third protective mechanism which was effective for all test frequencies up to 100 kHz.

Henson (1965) obtained similar results in *Tadarida* and measured latencies of about 10 msec. for the reflex action. But he felt that more interesting

FIG. 5. Horizontal section through the middle ear and cochlea of *Pipistrellus pipistrellus* (Vespertilionidae), showing the spindle-shaped tensor tympani muscle alongside the Eustachian tube and the broadly conical stapedius muscle. The stapes foot-plate is inserted between cochlea and vestibule, and the stapedial artery is seen in cross-section. ×21

and meaningful results would be obtained from unanaesthetized bats during the self-stimulation of normal echolocation activity. He succeeded in chronically implanting electrodes to record both cochlear microphonic potentials and electromyographic activity of the stapedius muscle in *Tadarida brasiliensis*. After recovery the bats showed spontaneous contraction of the stapedius beginning about 4–10 msec. before the emission of each orientation pulse. The stapedius was therefore shown to be under central command, linked to that of sound production by the larynx. By observing the cochlear microphonic potentials caused by a constant intensity external sound, Henson showed that attenuation reached a maximum by the time the vocal pulse was produced and then decayed again over the next 10 msec.

Hartridge (1945) had already suggested that such a contraction, if synchronized with the larynx, would attenuate the initial call and restore sensitivity for echoes by analogy with the transmit–receive (T.R.) switch which protects the receiver during pulse transmission in many radar and sonar sets. This speculation has now been shown to be entirely correct.

In later experiments (1967) Henson allowed *Chilonycteris parnelli* with chronically implanted electrodes to fly across a 3–m. cube recording chamber and to land on a small screen. Recordings made under these conditions showed that as the pulse repetition rate increased prior to landing, the middle-ear muscles were maintained in a state of smooth tetanic contraction.

From these results Henson argued that at lower pulse repetition rates the twitch contraction is useful in attenuating the emitted pulse at the cochlea and that during relaxation it progressively restores sensitivity as echoes from greater distances arrive at the ear. In this way the ear can compensate for the reduction in intensity (due to attenuation and the inverse fourth or other power law) of the more delayed echoes. At close range and high pulse repetition rates the bat is presumably interested only in detecting short-range echoes and in regulating their intensity. This could be achieved by the tetanic contraction, and relaxation between pulses might be unnecessary.

THE COCHLEA

Despite the small size of the head in most bats, the cochleas are relatively very large. In some forms such as *Rhinolophus* and *Chilonycteris* they may occupy a large part of the posterior skull and almost meet in the mid-line. They are very loosely attached to the rest of the skull by layers of fat and loose connective tissue, so that they often drop out during dissection or the preparation of museum skulls. Henson (1961, 1967) has suggested that they thereby obtain some acoustic isolation from the head to attenuate the tissue conduction of sound from the larynx.

The structure of the cochlea has been investigated by means of histological sectioning by A. Pye (1966*a*,*b*, 1967 and personal communication), who also reviewed the remarkably small previous literature. She has described 31 species of ten families to date. The number of turns varies from two in

Pteropus to three and a half in some Microchiroptera and the height of the cone from 1·7 mm. to 2·8 mm. The size of the cochlea, however, bears little relation to the size of the bat and tends to be greater in those forms such as *Rhinolophus* and *Chilonycteris* which emit constant frequency signals. This is largely due to an enormous development of the basal turn

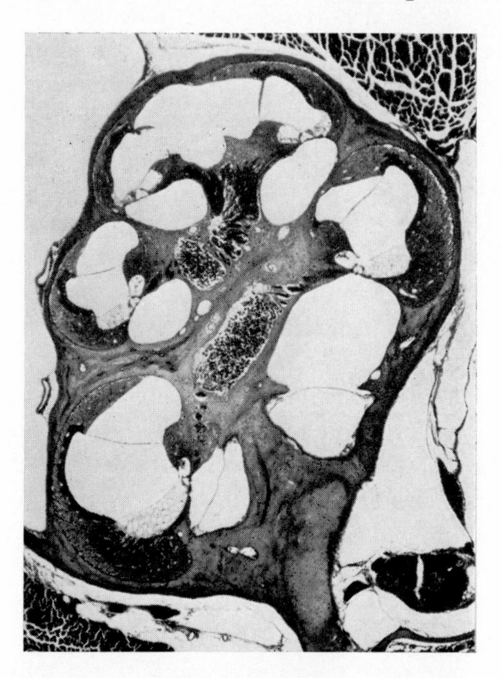

FIG. 6. Horizontal section through the cochlea of *Natalus tumidirostris* (Natalidae), showing two and a half functional turns. In the basal turn there is enlargement of the spiral ligament and supporting cells, and the basilar membrane is very narrow. ×34

in these forms. Henson (1967) comments that the oval window in the species he examined is placed further along the basal turn than in most other mammals. A. Pye finds this feature to be very variable. The stapes footplate is inserted at the vestibulo-cochlear junction in some species and there is often considerable variation between different specimens of the same species.

Within the cochlea the basilar membrane is very narrow at the basal end and is supported by an unusually large spiral ligament (Fig. 6). This ligament is strengthened in *Rhinolophus* and *Hipposideros* by a strip of bone, the secondary spiral lamina (Fig. 7). In most species the basilar membrane steadily increases in width from base to apex, but in *Rhinolophus* it decreases over the first two turns and then increases again normally. This is reflected by a similar decrease in the height, but not in the width, of the spiral ligament and in the size of the secondary spiral lamina.

The basilar membrane is also thickened in most bats, usually in two parts, from the spiral lamina to the outer rods of Corti and again beneath the

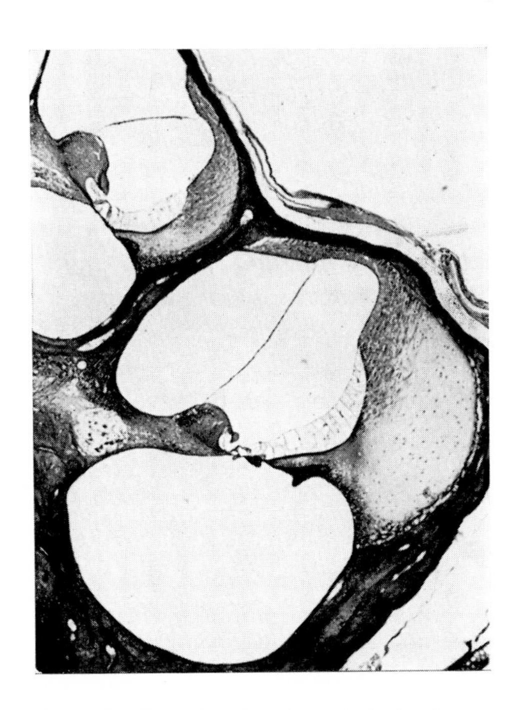

FIG. 7. Horizontal section through the basal turn of the cochlea of *Rhinolophus ferrumequinum* (Rhinolophidae), showing the secondary spiral lamina in the enormous spiral ligament, thickenings of the basilar membrane and greatly enlarged supporting cells. ×45

tunnel of Corti. These thickenings appear to consist of hyaline tissue and are most pronounced at the basal turn, decreasing towards the apex. They are largest in *Rhinolophus* (Fig. 7), in *Hipposideros*, where they reach 50 μm. in the basal turn, and in *Chilonycteris*. Similar thickenings are found in the apical turns of the cochleas of Heteromyid rodents (Webster, 1961; A. Pye, 1965) which show pronounced steep peaks in their cochlear microphonic responses at quite low frequencies (Webster, 1960). This tempts one to suggest that the thickenings are responsible for narrowly tuned responses and that their presence in the basal turn is therefore especially significant in those bats which emit constant high frequency signals. Their development, however, is only slightly less in forms, such as *Natalus*, in which no constant frequency component has been found. In *Pteropus* the basilar membrane is much wider and shows a slight thickening at the apical end.

Sections of the cochlea also show a pronounced development of the supporting cells, especially those of Claudius. In *Natalus* and *Hipposideros* these cells are 120 µm. high in the basal turn, reducing to about 35 µm. at the apex. The space of Nuel, between the outer hair cells and the arch of Corti, is well developed, and the "outer tunnel" forms a large space beneath the reticular membrane and between the outer hair cells and the cells of Hensen. The latter are therefore deflected outwards at their bases and inwards at their tips to connect with the reticular membrane.

No explanation can yet be put forward to account for these peculiarities of structure. A. Pye has attempted to correlate many of the features with various characteristics of the sounds used for echolocation, but has so far been unsuccessful beyond the somewhat equivocal associations with constant frequencies mentioned above. Recently (personal communication) she has applied the cochlear microdissection technique of Engström, Ades and Andersson (1966) to various species of bats. This shows the normal mammalian pattern of three rows of outer hair cells and one row of inner hair cells (Fig. 8), but the cells themselves are rather large and show certain peculiarities which have not yet been fully defined.

Cochlear microphonic potentials of bats were first investigated by Galambos (1941, 1942) who obtained responses up to 98 kHz, the limit of the stimulating equipment, in Vespertilionids. He found the greatest responses between 25 kHz and 45 kHz and showed that the intra-aural reflex produces attenuation in transmission. Wever and Vernon (1961*a*) made further observations of this kind in the range of 100 Hz—100 kHz. They compared the sensitivity of *Myotis lucifugus* rather unfavourably with that of several other species of mammals and the pigeon. However, the bat was found to have narrow peaks of sensitivity between 10 kHz and 60 kHz, generally around 40 kHz. These responses were used to measure the directionality of the external ear as described above.

Pye (unpublished observations) found rather smooth changes in sensitivity of cochlear responses of *Phyllostomus hastatus* from 10 kHz to over 100 kHz. It may be remarked here that bats generally form very convenient subjects for cochlear microphonic investigations. The round window is easily accessible through the back of the bulla. But since the bulla is often restricted to a ring of bone on the outer surface of the cochlea, much of the basal turn and even higher turns can be drilled for the insertion of miniature wire electrodes without opening the bulla at all. This greatly simplifies dissection and makes it unnecessary to seal the sound source to the external meatus.

Henson's observations of cochlear microphonic potentials in active bats have been mentioned above.

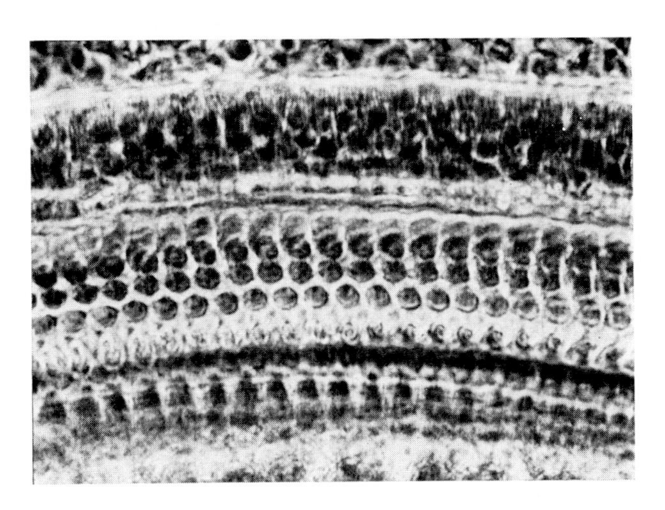

FIG. 8. Surface view of the organ of Corti in *Pipistrellus pipistrellus* (Vespertilionidae) obtained by a microdissection technique and photographed with phase-contrast microscopy by A. Pye. The arrangement of inner and outer hair cells and their supporting structures is clearly shown. × 540

CENTRAL NERVOUS SYSTEM

As might be expected, the auditory nerve and its central connexions are well developed in bats and the relative sizes of the cochlear nucleus, trapezoid, olivary complex and posterior colliculus are much greater than in other mammals, although their absolute sizes remain quite small. The forebrain centres of the medial geniculate and auditory cortex, however, are not well developed and retain a primitive degree of organization. The literature in this field is not very extensive and has been well reviewed by Henson (1967).

Electrophysiological investigations of the neural responses of Vespertilionids (mainly *Myotis lucifugus*) to various types of ultrasonic stimulation have been described in eleven important papers by Grinnell (1963*a*–*d*), Grinnell and McCue (1963), Suga (1964*a*, *b*, 1965*a*, *b*), Harrison (1965) and

Friend, Suga and Suthers (1966). Grinnell (1963*a*) first used gross electrodes to follow evoked potentials at various levels up to the auditory cortex. Using stimulating sounds and pulses of up to 150 kHz, he identified four successive responses ($N_1 - N_4$) with latencies from 0·8 msec. to 3 msec. N_1 was interpreted as the first-order response and N_4 as the collicular input. The evoked responses appeared to indicate the auditory sensitivity of the bat at various frequencies. They appeared only between 700 Hz and 150 kHz in *Myotis*, with lowest threshold at 40 kHz, while *Plecotus* responded from 700 Hz to 110 kHz with greatest sensitivity at 15–35 kHz and again at 55–65 kHz. These figures agree well with the acoustic spectrum of orientation sounds used by each species, including the presence of a prominent second harmonic component in *Plecotus* pulses.

Single unit responses, observed with microelectrodes, showed variable but generally longer latencies. A variety of responses was found in different isolated units. Some showed extremely sharp tuning curves, as steep as 35 db/kHz (Grinnell, 1963*d*) and within the colliculus the most ventral units responded to the highest frequencies (Grinnell, 1963*a*). Some responses were effectively amplified and reached maximum amplitude for only slightly suprathreshold intensities (0·2–0·5 db, Grinnell, 1963*d*), while a large proportion of units showed intensity-gating, with an upper as well as a lower threshold of intensity (Grinnell, 1963*a*). Grinnell (1963*c*) also investigated binaural interaction at the colliculi for sound sources in various positions around the bat's head. He found the N_4 evoked potential, derived from both ears, to be more "directional" than N_1 from a single ear, indicating inhibition for certain directions and suggesting that localization within 1–2° is possible in the horizontal plane. Single units often showed sharp directionality, with threshold changes as sharp as 9 db/degree. The use of collicular responses by Grinnell and Grinnell (1965) for measuring directionality in directions other than the horizontal plane has already been described.

All these responses suggest means by which bats can reduce the effects of interference and "clutter" (Grinnell, 1963*d*, 1967) to which they are remarkably resistant (Griffin, McCue and Grinnell, 1963; Webster, 1967). But in summarizing this research Grinnell (1967) stresses that the most interesting and unusual responses involved temporal effects (Grinnell, 1963*b*). When paired pulses, representing "call" and "echo", were presented, a second response could be obtained when the interval between the two was as little as 1 msec. The recovery of N_4 is faster than for earlier potentials and it appears that successive stages of neural processing progressively improve the detection of "echo" signals with short delays. Indeed the response at N_4 was often supranormal in amplitude for the second stimulus, apparently being potentiated by the first, unlike the situation in most mammals where echo responses are usually suppressed. Similar effects were obtained, with only 50–75 per cent reduction in N_4, when the "echo" followed 1–2 msec. after a "call" which was 40–50 db more intense. Many single units showed similar temporal facilitation and some were capable of responding twice within 0·5–1 msec.

Since the Vespertilionids used in the above experiments use frequency modulated (f.m.) signals for echolocation, Grinnell and McCue (1963) made further observations with a frequency-sweep oscillator driving the acoustic pulse stimulator. Some regions of the colliculus responded to particular frequencies when these occurred in sweeps and responded twice at full amplitude to repeated passes at intervals as short as 0·5 msec. Evoked responses to very loud f.m. pulses were weaker than to less intense pulses, suggesting the presence of an echo-selecting mechanism. Some units had lower thresholds to constant frequency (c.f.) pulses and others had lower thresholds at the same frequencies when these occurred during sweeps. One single unit responded only to f.m. sweeps. With paired pulses the potentials evoked by a c.f. second pulse were potentiated by an f.m. first pulse.

Suga (1964*a*, *b*) elaborated this work by investigating responses in the cochlear nucleus as well as the posterior colliculus. A variety of sharply tuned responses was found in the cochlear nucleus, including tonic responses which were excited or suppressed by particular frequencies or by stimuli from certain directions. No neurons were found with a lower threshold for f.m. sweeps that included their best frequency than for c.f. pulses of this best frequency alone. Some sharply tuned phasic units in the colliculus (but not in the cochlear nucleus) were found to differentiate between upward and downward sweeps over the same frequency range. Such responses were explained by postulating patterns of successive inhibition and excitation between units sharply tuned to different frequencies within the sweep. This was supported by an elegant series of further experiments in which paired pulses were used to explore the excitatory and inhibitory areas of response for single units (Suga, 1965*a*, 1967). The general picture is of a progressive sharpening of response and recovery of second-pulse responses as the signals pass towards the higher centres of the auditory nervous system.

Within the auditory cortex Suga (1965*b*) found a variety of responses, with high frequency neurons occurring more anteriorly than the low frequency neurons. There was a much lower proportion of neurons which responded equally well to upward f.m. sweeps than in the colliculus. This suggests that still more selection and elaboration of adaptively significant responses occurs in the forebrain centres, despite their apparently undeveloped condition.

Friend, Suga and Suthers (1966) investigated further the recovery of echo sensitivity within the colliculus. They used paired f.m. pulses of which the second was less intense, or alternatively tape-recordings of actual *Myotis* signals followed by real echoes. Recovery was found to be related to the relative intensity of the two pulses, a louder first pulse causing more protracted recovery. Rapid recovery, sometimes with actual potentiation, was found in only 7 per cent of the units investigated. The rest showed suppression or inhibition for various periods up to 26 msec.

Finally Harrison (1965) investigated the peculiar behaviour of collicular evoked potentials as a function of body temperature. As might be expected, the responses decreased in magnitude with cooling and disappeared at about 12°C. But the responses to high frequencies declined more sensitively than those to lower frequencies, so that the most sensitive frequency changed with temperature. Facilitation of second responses increased with cooling below 25°C but behavioural responses could not be elicited by external sounds at such temperatures. Cochlear microphonic potentials showed similar but less marked changes, and changes in the N_1 evoked potential appeared to be independent of this effect to some extent. The changes were found in both anaesthetized animals and those with chronically implanted electrodes.

BEHAVIOUR

Surprisingly little work has been done in investigating the hearing abilities of bats by behavioural responses, although they are clearly disturbed by high frequency sounds. Dijkgraaf (1957) used training methods to investigate all the major senses of various Vespertilionids. He concluded that there was an upper limit of hearing between 175 kHz and 400 kHz, and a lower limit of 2–3 kHz.

Dalland (1965) investigated hearing in *Eptesicus fuscus* and *Myotis lucifugus* by an operant conditioning technique. *Eptesicus* showed a lowest threshold of −68 db re 1·0 μbar (about 4×10^{-4} μbar) at 20 kHz, with a second peak of sensitivity at 60 kHz. The second peak, like that of *Plecotus* (Grinnell, 1963*a*) corresponds to the often prominent peak of the second harmonic of orientation sounds in this species. *Myotis* showed a lowest threshold of −64 db re 1·0 μbar (about 6×10^{-4} μbar) at 40 kHz. It "clearly heard" 120 kHz but gave no responses below 10 kHz and showed unstable responses below 15 kHz. The slope of the low frequency end of this audiogram was 266 db/octave over a limited range.

Finally, a clue to the differences which may be expected between the hearing responses of different species may be found in the signals they use for echolocation. Radar theory (Skolnik, 1962; Berkowitz, 1965) shows that f.m. and c.f. signals must produce unambiguous information of different kinds, the former giving range accurately and the latter giving relative velocity. Cahlander (1967) has actually computed the radar ambiguity diagrams of some Vespertilionid pulses and shown that the shorter pursuit pulses trade velocity information for better range information. The signals of a large number of Microchiroptera have been classified intuitively and discussed on this basis by Pye (1967*a*,*b* and in preparation).

Unfortunately most of the work on hearing responses has been performed on "pure f.m." bats such as *Myotis*. Henson (1967) found little response in the brain of *Chilonycteris* to the c.f. part of its pulses. He argued that only the f.m. part is significant in echolocation and the c.f. part may serve merely to mask later echoes. This does not seem likely in the longer pulse bats such as *Rhinolophus* or for the "pure c.f." pulses produced under certain conditions by several species, including *Nyctalus* of the Vespertilionidae. It seems much more likely that these signals are utilized by the animals for measuring velocity by Doppler shift.

How these bats could discriminate long echoes with only slight frequency changes remains problematical. It seems unlikely that progressive sharpening of neuronal tuning curves at higher levels of the nervous system could separate simultaneous sounds which had been combined in less selective lower levels. The suggestion that peripheral intermodulation could produce echo transformations without loss of information (Pye, 1960, 1961, 1963; Kay, 1962) still seems to offer a simple explanation here, although the finding of elaborate temporal processing by neural mechanisms makes it seem less likely for f.m. pulses. At present, however, there is no information on the masking of similar frequencies in c.f. bats or on the degree of intermodulation in the ears of any bats.

SUMMARY

The ability to hear sounds above the human frequency limit of 17–20 kHz is known in many mammals, but the hearing of bats is especially interesting because of their ultrasonic echolocation. Such a large and diverse Order shows considerable variation in aural anatomy, although few forms have been studied in detail. The external ears are usually large, and sometimes highly mobile. The tympanum and ossicles are light, but the middle-ear muscles are large, especially the stapedius. The cochlea shows considerable variations, especially in the basilar membrane, supporting cells and spiral ligament. Finally the auditory nervous system is large and complex up to the posterior colliculus but is relatively simple in the fore-brain.

Functional studies have concentrated mainly on the middle-ear and nervous system of a few species. The stapedius is under central control, and acts as a variable attenuator to compensate for variations in echo intensity with distance. Cochlear microphonic potentials have been recorded to more than 100 kHz. Neural responses have been much studied and often show great rapidity and specificity for certain signal parameters. But the biological solutions to the problems of echo discrimination and interpretation in active flight are still far from being solved.

Acknowledgements

The author's work on bats is supported by the Air Force Office of Scientific Research under Contract no. AF61(052)–876, through the European Office of Aerospace Research, OAR. Thanks are also due to his wife, Dr. A. Pye of the Institute of Laryngology and Otology, for the use of Figs. 5–8.

REFERENCES

Berkowitz, R. S. (ed.) (1965). *Modern Radar: Analysis, Evaluation and System Design.* New York: Wiley.

Cahlander, D. A. (1967). In *Animal Sonar Systems: Biology and Bionics*, pp. 1052–1081, ed. Busnel, R.-G. Jouy-en-Josas: INRA-CNRZ.

Dalland, J. I. (1965). *Science*, **150**, 1185–1186.

Dijkgraaf, S. (1957). *Acta physiol. pharmac. neerl.*, **6**, 675–684.

Engström, H., Ades, H. W., and Andersson, A. (1966). *Structural Pattern of the Organ of Corti.* Stockholm: Almqvist and Wiksell.

Friend, J. H., Suga, N., and Suthers, R. A. (1966). *J. cell. Physiol.*, **67**, 319–332.

Galambos, R. (1941). *Am. J. Physiol.*, **133**, 285.

Galambos, R. (1942). *J. acoust. Soc. Am.*, **13**, 41–49.

Gould, E., Negus, N. C., and Novick, A. (1964). *J. exp. Zool.*, **156**, 19–38.

Griffin, D. R. (1953). *Proc. natn. Acad. Sci. U.S.A.*, **39**, 884–893.

Griffin, D. R. (1958). *Listening in the Dark.* New Haven: Yale University Press.

Griffin, D. R., Dunning, D. C., Cahlander, D. A., and Webster, F. A. (1962). *Nature, Lond.*, **196**, 1185–1186.

Griffin, D. R., McCue, J. J. G., and Grinnell, A. D. (1963). *J. exp. Zool.*, **152**, 229–250.

Grinnell, A. D. (1963*a*). *J. Physiol., Lond.*, **167**, 38–66.

Grinnell, A. D. (1963*b*). *J. Physiol., Lond.*, **167**, 67–96.

Grinnell, A. D. (1963*c*). *J. Physiol., Lond.*, **167**, 97–113.

Grinnell, A. D. (1963*d*). *J. Physiol., Lond.*, **167**, 114–127.

GRINNELL, A. D. (1967). In *Animal Sonar Systems: Biology and Bionics*, pp. 451–481, ed. Busnel, R.-G. Jouy-en-Josas: INRA-CNRZ.
GRINNELL, A. D., and GRINNELL, V. S. (1965). *J. Physiol., Lond.*, **181,** 830–851.
GRINNELL, A. D., and MCCUE, J. J. G. (1963). *Nature, Lond.*, **198,** 453–455.
HARRISON, J. B. (1965). *Physiol. Zool.*, **38,** 34–48.
HARTRIDGE, H. (1945). *Nature, Lond.*, **156,** 690–694.
HENSON, O. W. (1961). *Kans. Univ. Sci. Bull.*, **42,** 151–255.
HENSON, O. W. (1965). *J. Physiol., Lond.*, **180,** 871–887.
HENSON, O. W. (1967). In *Animal Sonar Systems: Biology and Bionics*, pp. 949–1003, ed. Busnel, R.-G. Jouy-en-Josas: INRA-CNRZ.
KAY, L. (1962). *Anim. Behav.*, **10,** 34–41.
KELLOGG, W. N. (1959). *J. acoust. Soc. Am.*, **31,** 1–6.
KELLOGG, W. N. (1961). *Porpoises and Sonar*. Chicago: Chicago University Press.
KULZER, E. (1960). *Z. vergl. Physiol.*, **43,** 231–268.
MEDWAY, Lord (1959). *Nature, Lond.*, **184,** 1352–1353.
MÖHRES, F. P., and KULZER, E. (1956). *Z. vergl. Physiol.*, **38,** 1–29.
NOIROT, E. (1966). *Anim. Behav.*, **14,** 459–462.
NORRIS, K. S., PRESCOTT, J. H., ASA-DORIAN, P. V., and PERKINS, P. (1961). *Biol. Bull. mar. biol. Lab., Wood's Hole*, **120,** 163–176.
NOVICK, A. (1958). *J. exp. Zool.*, **137,** 443–462.
NOVICK, A. (1959). *Biol. Bull. mar. biol. Lab., Wood's Hole*, **117,** 497–503.
PIERCE, G. W., and GRIFFIN, D. R. (1938). *J. Mammal.*, **19,** 454–455.
PUMPHREY, R. J. (1948). *Ibis*, **90,** 171–199.
PUMPHREY, R. J. (1950). *Nature, Lond.*, **166,** 571.
PYE, A. (1965). *J. Anat.*, **99,** 161–174.
PYE, A. (1966*a*). *J. Morph.*, **118,** 495–510.
PYE, A. (1966*b*). *J. Morph.*, **119,** 101–120.
PYE, A. (1967). *J. Morph.*, **121,** 241–254.
PYE, A., and HINCHCLIFFE, R. (1968). *Med. biol. Illust.*, **18,** 122–127.
PYE, J. D. (1960). *J. Lar. Otol.*, **74,** 718–729.
PYE, J. D. (1961). *Endeavour*, **20,** 101–111.
PYE, J. D. (1963). *Ergebn. Biol.*, **26** (Symp. Animal Orientation, Garmisch-Partenkirchen), 12–20.
PYE, J. D. (1967*a*). In *Animal Sonar Systems: Biology and Bionics*, pp. 43–65, ed. Busnel, R.-G. Jouy-en-Josas: INRA-CNRZ.
PYE, J. D. (1967*b*). In *Animal Sonar Systems: Biology and Bionics*, pp. 1121–1136, ed. Busnel, R.-G. Jouy-en-Josas: INRA-CNRZ.
PYE, J. D., FLINN, M., and PYE, A. (1962). *Nature, Lond.*, **196,** 1186–1188.
RALLS, K. (1967). *Anim. Behav.*, **15,** 123–8.
SCHNEIDER, H. (1961). *Zool. Jb. Abt. Anat. Ontog.*, **79,** 93–122.
SCHNEIDER, H., and MÖHRES, F. P. (1960). *Z. vergl. Physiol.*, **44,** 1–40.
SEWELL, G. D. (1967). *Nature, Lond.*, **215,** 512.
SKOLNIK, M. I. (1962). *Introduction to Radar Systems*. New York: McGraw-Hill.
SUGA, N. (1964*a*). *J. Physiol., Lond.*, **172,** 449–474.
SUGA, N. (1964*b*). *J. Physiol., Lond.*, **175,** 50–80.
SUGA, N. (1965*a*). *J. Physiol., Lond.*, **179,** 26–53.
SUGA, N. (1965*b*). *J. Physiol., Lond.*, **181,** 671–700.
SUGA, N. (1967). In *Animal Sonar Systems: Biology and Bionics*, pp. 1004–1020, ed. Busnel, R.-G. Jouy-en-Josas: INRA-CNRZ.
WASSIF, K. (1946). *Nature, Lond.*, **157,** 877.
WEBSTER, D. B. (1960). *Anat. Rec.*, **136,** 299.

Webster, D. B. (1961). *Am. J. Anat.*, **108**, 123–148.
Webster, F. A. (1967). In *Animal Sonar Systems: Biology and Bionics*, pp. 673–713, ed. Busnel, R.-G. Jouy-en-Josas: INRA-CNRZ.
Wever, E. G., and Vernon, J. A. (1961*a*). *J. aud. Res.*, **2**, 158–175.
Wever, E. G., and Vernon, J. A. (1961*b*). *Ann. Otol. Rhinol. Lar.*, **70**, 5–17.

DISCUSSION

Tumarkin: I was interested to hear Dr. Pye's comments on the function of the large middle-ear muscles of bats. It is generally held that the middle-ear muscles act as a defence mechanism against loud noise. As loud noise is to a large extent a product of civilization, it is difficult to believe that the tympanic muscles were developed a hundred million years ago for that purpose! I have always felt that in some way they improve the mechanism of hearing; I suggested some years ago that if they do interfere with the transmission of sound it is in order to render the ear more sensitive subsequently. The deafening effect of the contraction of the muscles is not to "protect" the ear against damage by loud noise but to prepare it for the subsequent reception of the physiological sound.

J. D. Pye: I think this is so. The intensity measured just in front of a bat is very close to its assumed pain threshold, but of course we do not know quite what the tissue conduction at the cochlea actually is at that time. But in any case there is a lot of sound, and this device probably saves too many units firing, so that they can fire again a very short period later.

Lowenstein: In the bat I thought this device was to exclude hearing its own vocalization in order not to confuse it with incoming echoes. Is that not so?

J. D. Pye: Yes, in the sense that it protects the threshold.

Engström: It has long been the practice to promote—or try to promote—a contraction of the middle-ear muscles by giving a sound stimulation before exposure to impact noise. It has been supposed that such stimulation could reduce or prevent damage. Today, however, it is widely assumed that the middle-ear muscles have an important role in improving hearing in a noisy environment.

Salomon: We showed that the middle-ear muscles in man contract about 100 msec. before a sound is delivered by the throat (Salomon, G., and Starr, A. [1963]. *Acta neurol. scand.*, **39**, 161–168). Sound intensity in the back of the mouth at normal speech level is around 95 db above normal threshold. The timing of the contraction indicates that the middle-ear muscles will prevent us from hearing our own voices too loudly during vocalization.

Tumarkin: In our otological investigations we make a loud noise and then watch the tensor tympani or stapedius jerk. I have always maintained that this is quite unphysiological. It tells us nothing about the real function of these muscles, just as the knee jerk tells us nothing about the real function of the thigh muscles. What Dr. Pye describes seems far more like what I have always envisaged. The muscles contract not after, but before the onset of the noise, and their purpose is not to prevent damage to the ear, but rather to improve its performance.

Schwartzkopff: Recent studies on middle-ear muscle contraction show that if you do not use extreme situations with very high noises, but sounds of only medium level, you may by activation of both muscles (tensor tympani and stapedius) actually improve the perception of sound. It looks as if this system only in rare cases works defensively, and that normally it works to adapt the ear to the level of intensity at which communication takes place and thus finds the best way to transfer information. In the bat there is a very extreme situation; practically only its own voice has to be excluded.

Lowenstein: Are you suggesting that there is a two-way system? And that, apart from attenuating sound, it can also facilitate it?

Schwartzkopff: Attenuation is good only for one's own voice and not for external sounds; it comes too late for that.

Salomon: It is quite normal to find facilitation of transmission through the middle ear at moderate sound levels (Salomon, G. [1966]. *Proc. R. Soc. Med.*, **59**, 966–971). We normally find that there is spontaneous activity in the stapedius muscle. This activity can be inhibited by moderate sound levels.

Davis: Professor Schwartzkopff told us that time-resolution in birds is more rapid than in man. In bats we are dealing not only with higher frequencies but also presumably with very small time-differences between signals from the two ears, in order to make localization possible. Dr. Pye mentioned Grinnell's findings that some of the physiological responses at the higher levels of the nervous system showed rapid recovery periods, which would favour resolution in time. It would be interesting to know whether there are any electrophysiological studies of primary sensory units of bats with respect to their recovery process or refractory period—that is, the minimum interval between successive discharges—and also with respect to the variability of their latency. In the cat and many other mammals there is a statistical distribution of latencies of successive responses or across units over a rather wide range. The cat depends on statistical averaging in order to extract this bit of time information, which is actually very accurate. Is it possible that bats are more precise than cats in regard to their recovery process and have a narrower range of neural latencies?

J. D. Pye: Grinnell has made some observations on this (Grinnell, A. D. [1963]. *J. Physiol., Lond.*, **167**, 38–66). Assuming that the N_1 evoked potential was a primary response, he found that its latency was not very variable. It is only when you get single units within the colliculus that the latency becomes variable, mainly according to sound intensity; the N_1 is not so variable, and it does recover more quickly than in other mammals (0·8 msec. ±0·2 msec.). He also obtained quite marked evoked potentials to the second of paired sound-pulses (Grinnell, A. D. [1963]. *J. Physiol., Lond.*, **167**, 67–95), again depending on the relative intensity of the first and second sounds. If the two sounds were of equal intensity, then the N_4 recovered within a millisecond or so, which is quite a lot faster than in cats or guinea pigs.

Davis: The minimal interval in the cat of about 1 msec. represents a rather extreme performance for a single peripheral sensory unit. Of course the stability of latency of the group in the larger mammals is well known. I was really wondering about the performance of the single unit as opposed to the average of a group of units, which would make possible still finer discrimination.

J. D. Pye: The fastest recoveries of evoked potentials are found at the higher levels, but the fastest single unit recoveries are at the cochlear nucleus. There N. Suga ([1964]. *J. Physiol., Lond.*, **175**, 50–80) reported single units that will respond even when a second sound is 13 db softer than the first sound and follows it within half a millisecond, although in the colliculus J. H. Friend, N. Suga and R. A.

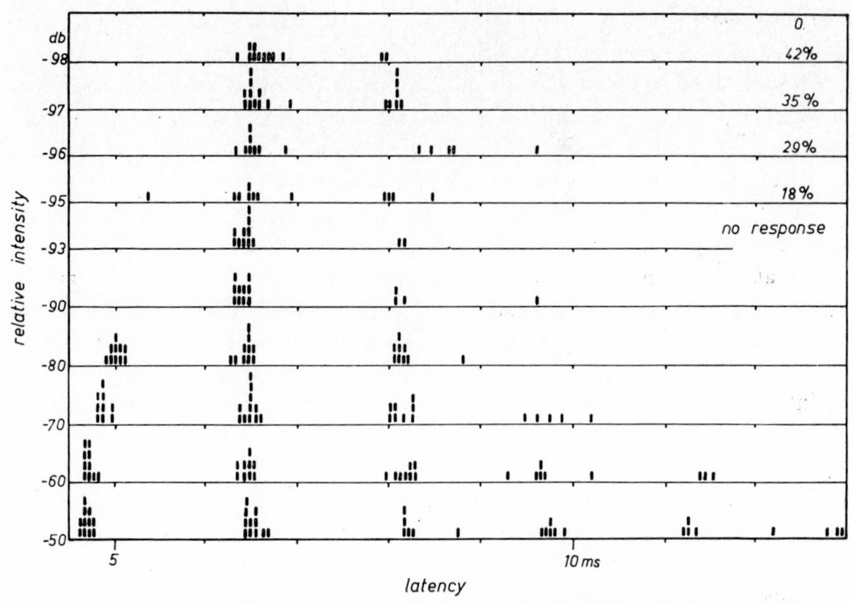

FIG. 1. (Schwartzkopff). Spike-interval histogram from an auditory neuron of the cat's medulla, stimulated by clicks.

Suthers ([1966]. *J. cell. Physiol.*, **67**, 319–332) found that such fast recoveries are not typical, even in unanaesthetized bats.

Schwartzkopff: May I comment on the variability of the latencies of single units, which are shown in Fig. 1 for the cat, recorded from the medial segment of the superior olive. It can be seen that at a certain time different spikes are added. At this time (1 millisecond) the unit will respond with a variability of $\pm 0 \cdot 2$ msec., which makes it a very good instrument. I do not want to comment on the grouping of the signals, which is another interesting question, but on the fact that the variability is so small.

Johnstone: What is the upper limit of the cochlear microphonic that has been recorded in bats?

J. D. Pye: About 150 kilocycles has been detected, though I seldom went up to this level myself. This is perhaps not the ultimate limit, but there was a difficulty here in getting suitable transducers, and when I made my measurements, ionized-air transducers were not yet available.

Johnstone: Did you pick up any summating potentials?

J. D. Pye: I have not seen any, since I used high-pass amplifiers (>10 kHz) only. This ought to be done, but as far as I know nobody has reported it.

Johnstone: I am trying to imagine how a current of 150 kilocycles could release the transmitter substance. This is the sort of frequency which would be carried solely by capacity current through a membrane. Perhaps a summating potential should be looked for, as far as an actual receptor potential is concerned.

Davis: Do you think that the summating potential, or possibly the rectifier, is intracellular?

Johnstone: Yes; at least the negative summating potential must be.

Lowenstein: You would suggest electrical transmission?

Johnstone: No, because a current of 150 kilocycles through any normal membrane will go through via the capacity channel; that is, it will be a displacement current and there will be no real movement of ions.

Lowenstein: So the rectifier mechanism might be the answer?

Johnstone: Or the summating potential, which may be the same thing (Johnstone, J. R., and Johnstone, B. M. [1966]. *J. acoust. Soc. Am.*, **40**, 1405–1413).

Davis: The summating potential is essentially a mechanical rectification, depending on unsymmetrical nonlinear distortion; that is the fundamental basis for the generation of the summating potential, and it could be a very effective physiological rectifier and thus enable the cochlear microphonic to do work and release the chemical transmitter substance.

Tumarkin: Dr. Pye, does the auditory bulla ever contribute to resonance in bats? You said that it varied in structure and was sometimes restricted to a ring of bone on the cochlea.

J. D. Pye: As far as we know the bulla does not contribute to resonance. There are resonant responses which appear to be neurally produced. The cochlea is very loosely suspended in the skull, and O. W. Hensen ([1961]. *Kans. Univ. Sci. Bull.*, **42**, 151–255) suggested that fatty layers and connective tissue between it and the skull may reduce tissue conduction of sound from the larynx. But as far as we know, the actual air-space of the middle ear does not contribute any useful resonances to the system.

Bosher: We have been considering single bats but in fact they are surely there in hundreds in caves, all emitting sounds at different times. Granted that the ability of the bat to pick up echoes is quite remarkable, picking up the faint echoes of its own signal against a background of quite high intensity sound and multiple echoes must be a considerable achievement. Is it possible to jam a bat's echolocatory system by a signal of high intensity?

J. D. Pye: It is not possible by a single sine wave; one must use "white noise" of rather large intensities. D. R. Griffin, J. J. G. McCue and A. D. Grinnell ([1963]. *J. exp. Zool.*, **152**, 229–250) have measured this: Grinnell feels that many of his neural responses suggest that the bat is locking the nervous system each time on to the noise just made, and that if the pattern of another bat's pulse differs very slightly, the information can be rejected. I am not quite sure about this myself. Under many of the conditions in which millions of bats live in caves, while they are flying through the cave they are in familiar territory, and bats are known to have a very good topographical memory which enables them to fly through narrow gaps even when the edges that form the gap have been removed. This has been shown recently in work in Tübingen (Neuweiler, G., and Möhres, F. P. [1967]. In *Animal Sonar Systems: Biology and Bionics*, pp. 129–140, ed. Busnel, R.-G. Jouy-en-Josas: INRA-CNRZ), in which bats flying through a doorway strung across with wires so close together that they had to fold their wings to get through, persisted in folding their wings even when the wires were taken away and replaced by light beams; they still folded their wings and flew between where the wires used to be. They cannot be doing this by echo-location or any other sense, because the wires are not there any more, so that when they are crowded together in caves perhaps they are producing sounds but the reception of echoes does not matter so much. Outside the cave, they disperse.

But I still feel there is a problem here and it is one to which we have no real answer.

Fex: N. Suga sent me several years ago a drawing of a vestibular–cochlear anastomosis in a bat, in all probability carrying efferent cochlear fibres. Has any electrophysiological work been done? And has Dr. Ade Pye looked for efferents in bats?

J. D. Pye: Their use has been speculated upon by both Grinnell and Suga but no one has identified efferent activity. We have not looked for efferents ourselves.

Engström: The person to ask would be R. S. Kimura of Boston, because he is doing an electron microscope study of the bat ear, and as far as I know, he finds an efferent system at the periphery.

Lowenstein: Dr. David Pye mentioned that there are not many morphological features of the bat's inner ear which can be specifically connected with ultrasonic reception. Has not Dr. Ade Pye found that the height of certain cells is characteristic for each different species?

Ade Pye: In a number of bats the cells of Hensen and Claudius are very high. We have done some computer correlations and there seems to be a factor that the narrower the basilar membrane at the base of the cochlea, the higher the supporting cells. But this correlation happens in only a few species of bats and those few seem to be the ones that produce the long constant frequency signals rather than the frequency modulated ones. We are still working on this.

ULTRASTRUCTURE AND PERIPHERAL INNERVATION PATTERN OF THE RECEPTOR IN RELATION TO THE FIRST CODING OF THE ACOUSTIC MESSAGE

H. SPOENDLIN

Otorhinolaryngologische Klinik und Poliklinik der Universität, Kantonsspital Zurich

THE cochlear receptor has three principal components: (1) The supporting structures, mainly concerned with the transformation of acoustic energy into the proper stimulus for the receptor cells; (2) The receptor (sensory) cells as mechano-electrical transducers; (3) The nerve endings and peripheral nerve fibres responsible for the first coding of the acoustic message.

The supporting structures, represented by the basilar membrane, pillar and Deiters' cells and the reticular membrane, build up a very elaborate framework. With the evolution of the cochlear receptor in vertebrates these supporting elements became increasingly differentiated. Pillar cells, Deiters' cells and the reticular membrane hold the sensory cells in a stable connexion to the basilar membrane. The sensory cells therefore follow the movements of the basilar membrane exactly without much loss by mechanical distortion. Because of this stable connexion the movements of the basilar membrane result in an effective shearing motion between the receptor pole of the sensory cells and the tectorial membrane, which appears to be the proper stimulation mechanism.

Deiters' and pillar cells are basically the same type of cell. Their principal supporting elements are intracellular fibrils of tubular appearance with no visible periodicity and a diameter of 200 Å. Arranged in bundles of several hundreds, they form intracellular struts, which originate in a large basal foot on the basilar membrane and end in the framework of the reticular membrane, where the cuticular plates of the sensory cells are tightly fixed (Figs. 1, 3). The pillars are inclined in a radial, and the Deiters' cell extensions in a longitudinal apical-basal direction, whereas the sensory cells are slightly inclined apical-wards. Such an arrangement of the supporting struts gives the whole system high resistance against compression and shearing motion. There is an obvious effort of Nature to

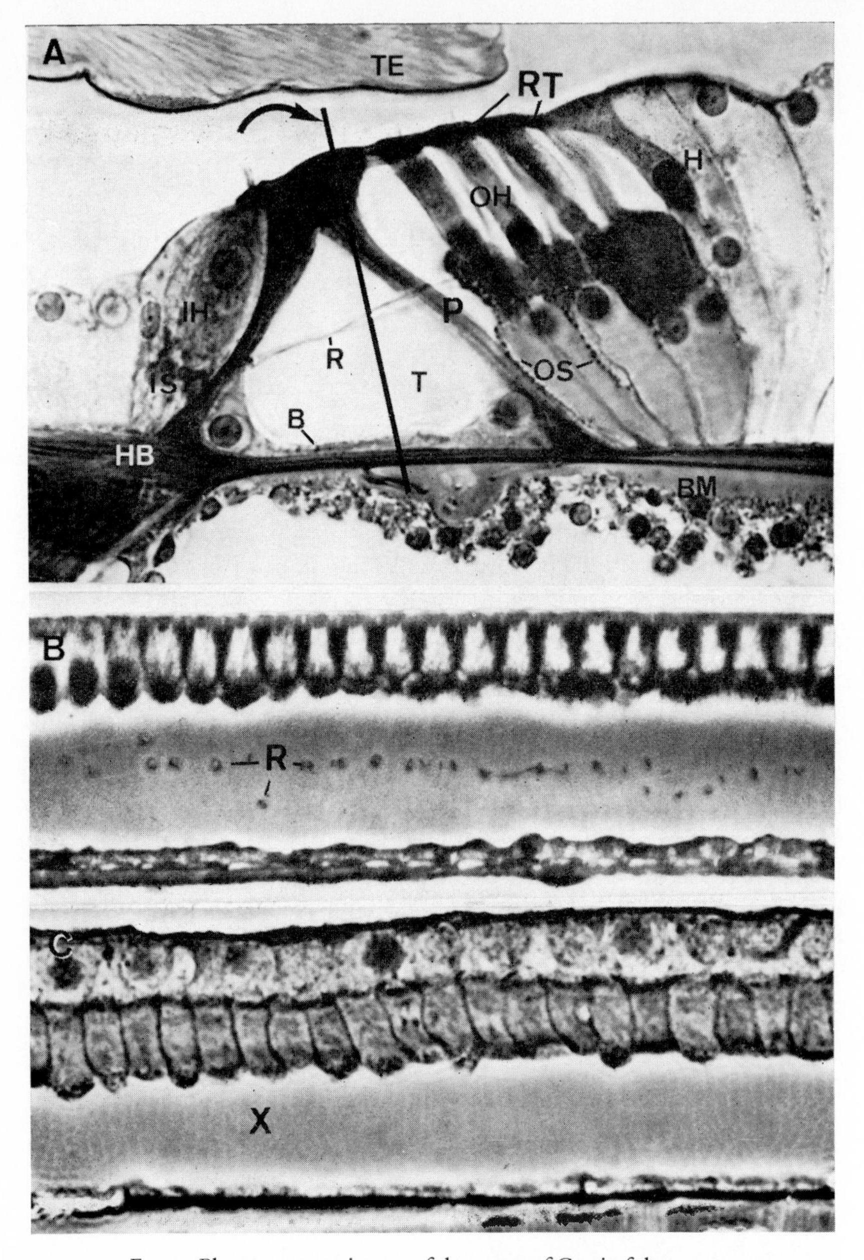

Fig. 1. Phase-contrast pictures of the organ of Corti of the cat.

A. Radial section. The different elements of the organ of Corti, such as the basilar membrane (BM), the pillars (P), the reticular membrane (RT) as well as the outer (OH) and the inner (IH) hair cells, are seen. The upper tunnel radial fibres (R) and the outer spiral fibres (OS) are clearly outlined. The inner spiral fibres (IS) and the basilar fibres at the bottom of the tunnel (B) are more difficult to distinguish. The nerve fibres penetrate the organ of Corti at the habenula perforata (HB). Tunnel (T), and tectorial membrane (TE). The vertical black line through the tunnel indicates the tangential plane of section represented in B and C.

B. Phase-contrast picture of a tangential section through the tunnel where the tunnel radial fibres are transversely cut and appear as a row of black points (R).

C. Phase-contrast picture of a tangential section through the tunnel of a cat in which the olivo-cochlear bundle has been sectioned two weeks previously. The tunnel space is completely empty and no tunnel radial fibres are visible (X).

reduce inner distortion in the organ of Corti, to make shearing motion at the receptor pole as effective as possible.

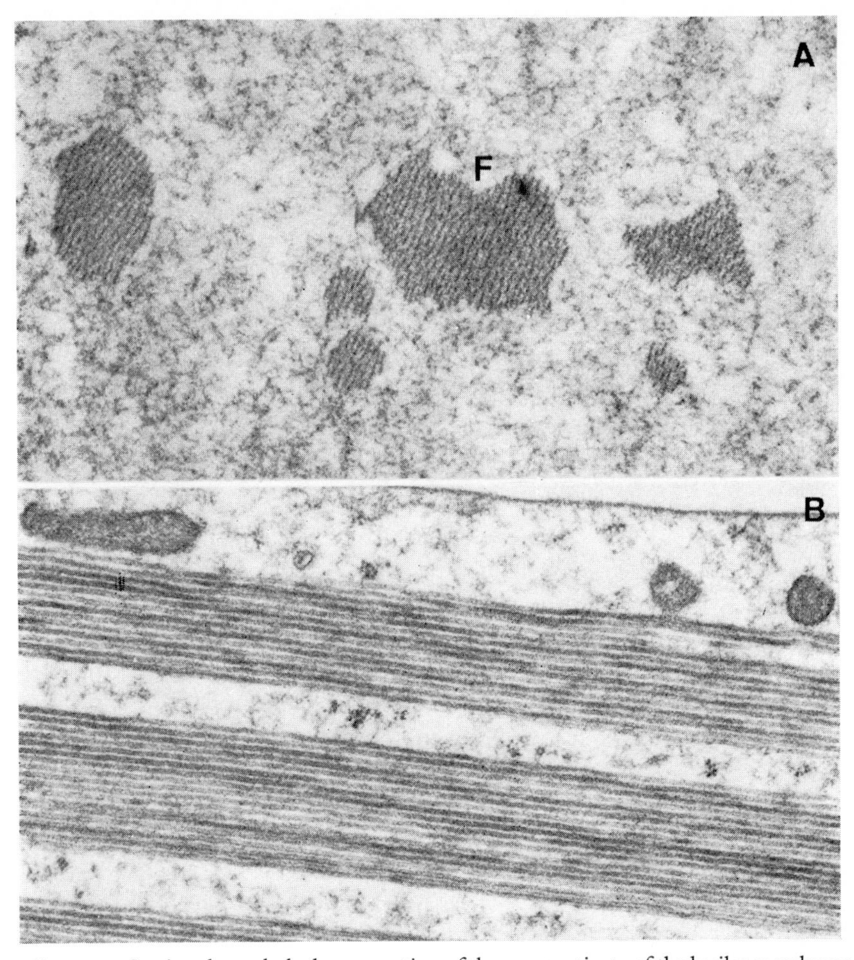

FIG. 2. A. Section through the lower portion of the pars pectinata of the basilar membrane with individual irregular fibres (F) embedded in a rather loose ground substance. The fibres are composed of a dense amorphous material and a great number of small fibrils with a diameter of approximately 50 Å.

B. Longitudinal section through two bundles of intracellular fibrils in an outer pillar. The fibrils are tubule-like and show no apparent periodicity. Their diameter is approximately 200 Å.

The basilar membrane seen in phase-contrast microscopy seems to be formed by fibres arranged in parallel. Electron micrographs, however, show a single continuous layer in the pars arcuata, and in the pars pectinata an upper continuous layer and a lower portion with irregular radial fibres. It consists in all parts of a fine granular substance and small fibrils (Fig. 2).

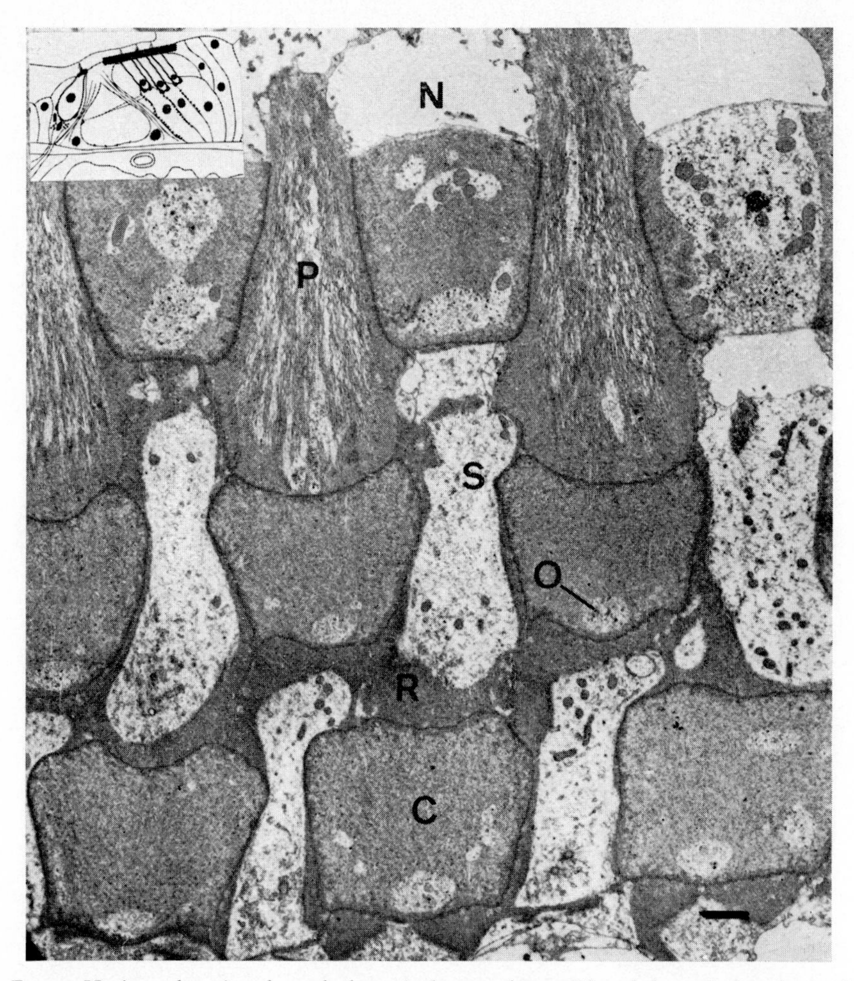

FIG. 3. Horizontal section through the reticular membrane (R) and the cuticular plates of the sensory cells (C) which are tightly fixed all round by the reticular membrane. The pillars (P) radiate into the reticular membrane. Cytoplasmic portions of the reticular membrane (S). Nuel's spaces (N). Each cuticular plate has a round opening (O) at one pole of the cell surface. In young animals kinocilial basal bodies can be found in this area. In the adult cat, however, such basal bodies are no longer found. Scale bar 1 μm.

The very loose meshwork of the tympanic lamina cells constitutes no continuous cellular barrier against the scala tympani, in contrast to the upper surface of the organ of Corti, where there are no gaps between the cells, which are connected by tight junctions (Iurato, 1967).

The tectorial membrane has always been a puzzling structure. It is made entirely of a spongy extracellular material which appears after fixation as an irregular texture of fine mucopolysaccharide-containing fibrils (Fig. 4). It is quite elastic and remarkably solid. The whole membrane can be torn

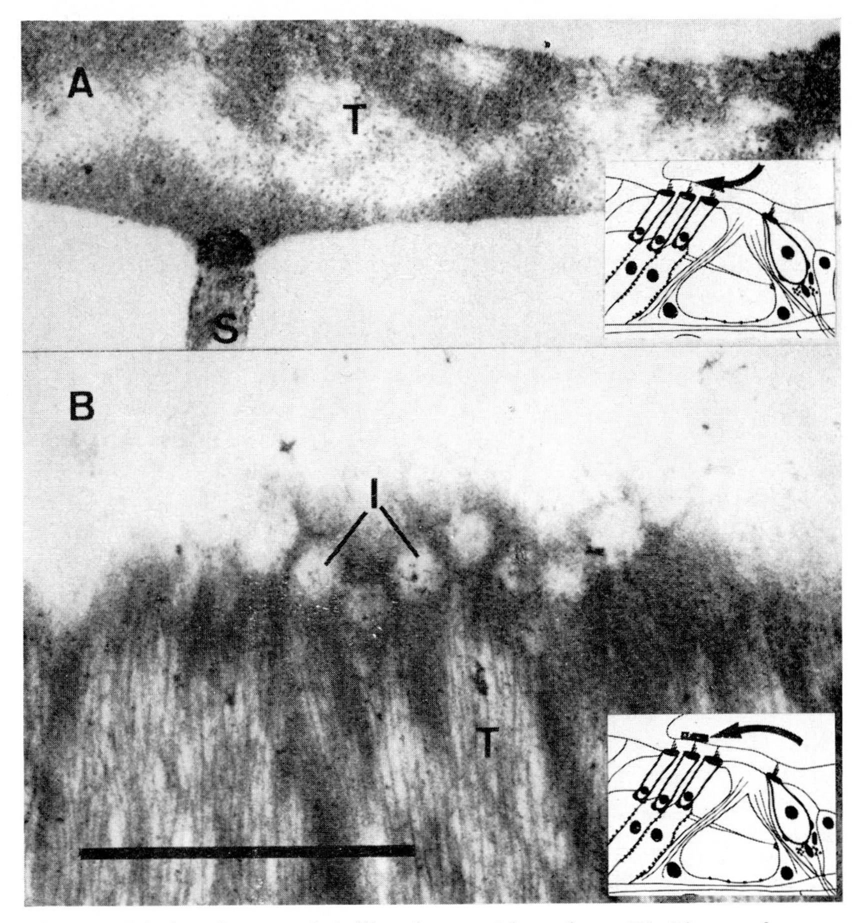

FIG. 4. A. Relation of a sensory hair (S) to the tectorial membrane (T). The top of a sensory hair is slightly embedded in the lower surface of the tectorial membrane.
B. A series of impressions (I) of sensory hairs in the lower surface of the tectorial membrane (T). The arrangement of the impressions shows also the W-pattern of the sensory hairs. Scale bar: 1 μm.

off in one piece during preparation of the cochlea. Its relation to the sensory cells has been the source of many arguments. In fresh vitally stained and fixed tissue the tectorial membrane is always slightly adherent to the surface of the organ of Corti but it can easily be lifted off without much damage to the underlying cells.

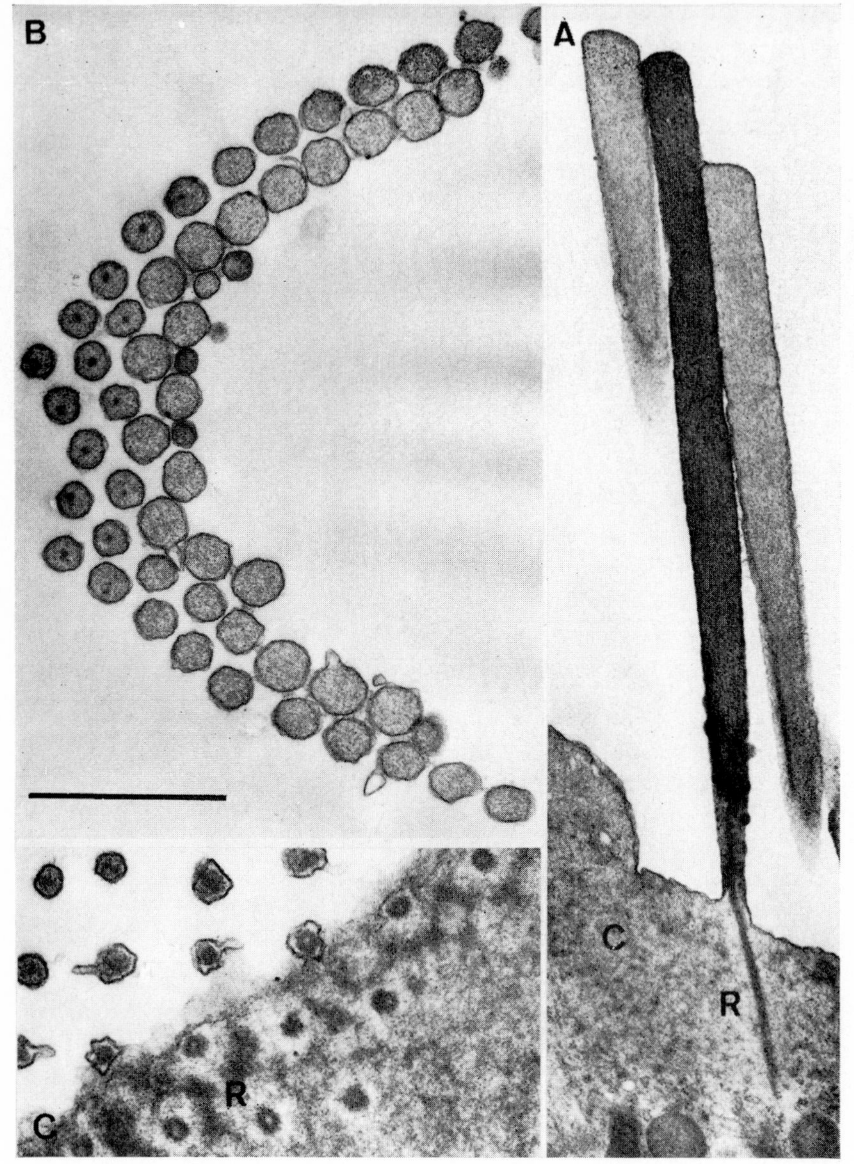

FIG. 5. A. Longitudinal section through a stereocilium of an inner hair cell with the rootlet (R) in the cuticular plate (C). The tops of the hairs are usually flattened. The ground substance shows a very fine longitudinal fibrillar texture.

B. Horizontal section through the stereocilia of one outer hair cell of a guinea pig. Typical W-arrangement of the stereocilia, in three rows. Close to the cell surface a central dense core is visible within the hairs (left side). Scale bar: 1 μm.

C. Horizontal section through the cuticular plate and the neck portion of some sensory hairs. The rootlets (R) are small tubules surrounded by a light zone in the cuticular plate. The sensory hairs are limited by a typical unit membrane.

The sensory hairs of the hair cells do not penetrate deeply into the tectorial membrane; their tips are slightly embedded in its lower surface (Fig. 4). This may explain why the tectorial membrane is easily lifted off in the course of preparing the tissue, without apparent damage to the sensory hairs. It is obvious that such a superficial connexion between sensory hairs and tectorial membrane is mainly effective against horizontal forces and therefore allows the bending of the sensory hairs by shearing movements between the tectorial membrane and the surface of the organ of Corti.

The second group of structural elements in the cochlear receptor is represented by the sensory cells. In the outer hair cells different zones can be distinguished: the hair-bearing upper end as receptor pole, the cell body, and finally the recepto–neural junction at the base of the cell. Especially in the upper cochlear turns they are very elongated slender cells with the receptor pole at one end and the connexion to the nerve endings at the other. This polar organization certainly supports the view that the sensory cells are specialized neurons (Fig. 7).

There has always been much discussion about the place where the first step in the mechano-electrical transduction in the cochlear receptor occurs. Electrophysiological evidence has been provided that the cochlear microphonics originate at the upper end of the sensory cells (Tasaki, Davis and Eldredge, 1954). From the structural point of view the receptor pole is certainly the most specialized and highly differentiated part of the cell. Each cell bears around 100 stereocilia which are anchored with tubule-like rootlets in the cuticular plate. The sensory hairs are in origin simple protrusions of the cell surface. The extremely regular, geometrical arrangement of the stereocilia in three rows in a W-shape facing the modiolus is striking (Fig. 5). The cuticular plate regularly has an opening at the base of this W, where, especially in young animals, a kinociliar basal body, which connects to a rudimentary kinocilium, can be found. In the adult cat, however, I have never been able to find such basal bodies (Fig. 3). Thus it is very unlikely that the basal body as such is involved in the receptor mechanism. Its definite and constant position, however, seems to determine the differentiation of the receptor pole and the polarized arrangement of the stereocilia during the development of the cell. That the basal body or centriole has such a determining influence on the differentiation of the receptor pole is illustrated by the occasional finding of a centriole at the wrong, basal end of a vestibular sensory cell, inducing there the development of a kinocilium and a cuticular plate (Spoendlin, 1964).

The stereocilia are stiff rods surrounded by a triple-layered unit membrane. The ground substance of the hair shows a distinct longitudinal

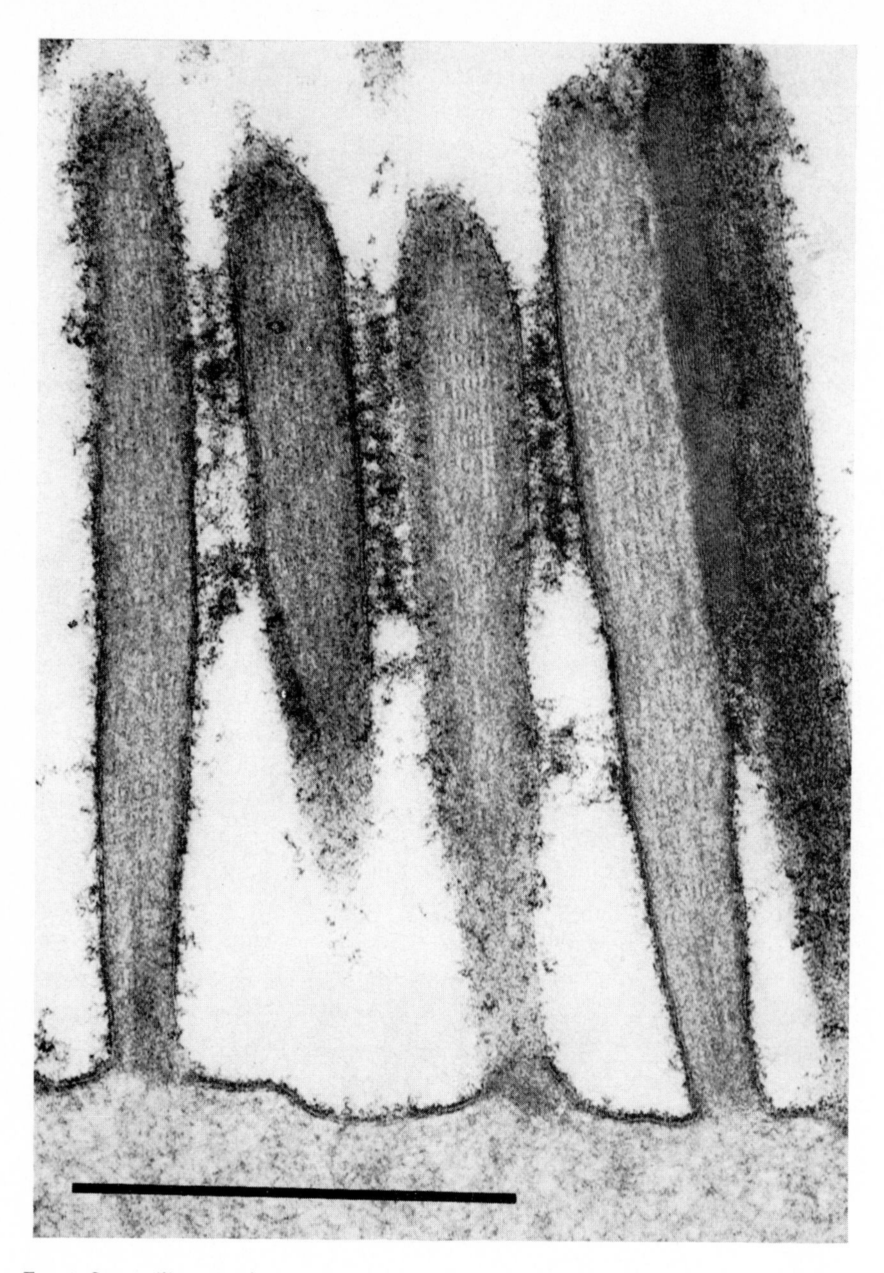

FIG. 6. Stereocilia, stained with ruthenium red. The space between the sensory hairs appears to be filled with acid mucopolysaccharides which stain with ruthenium red. The unit membrane of the stereocilia and the fibrillar ground substance are clearly visible. Scale bar: 1 μm.

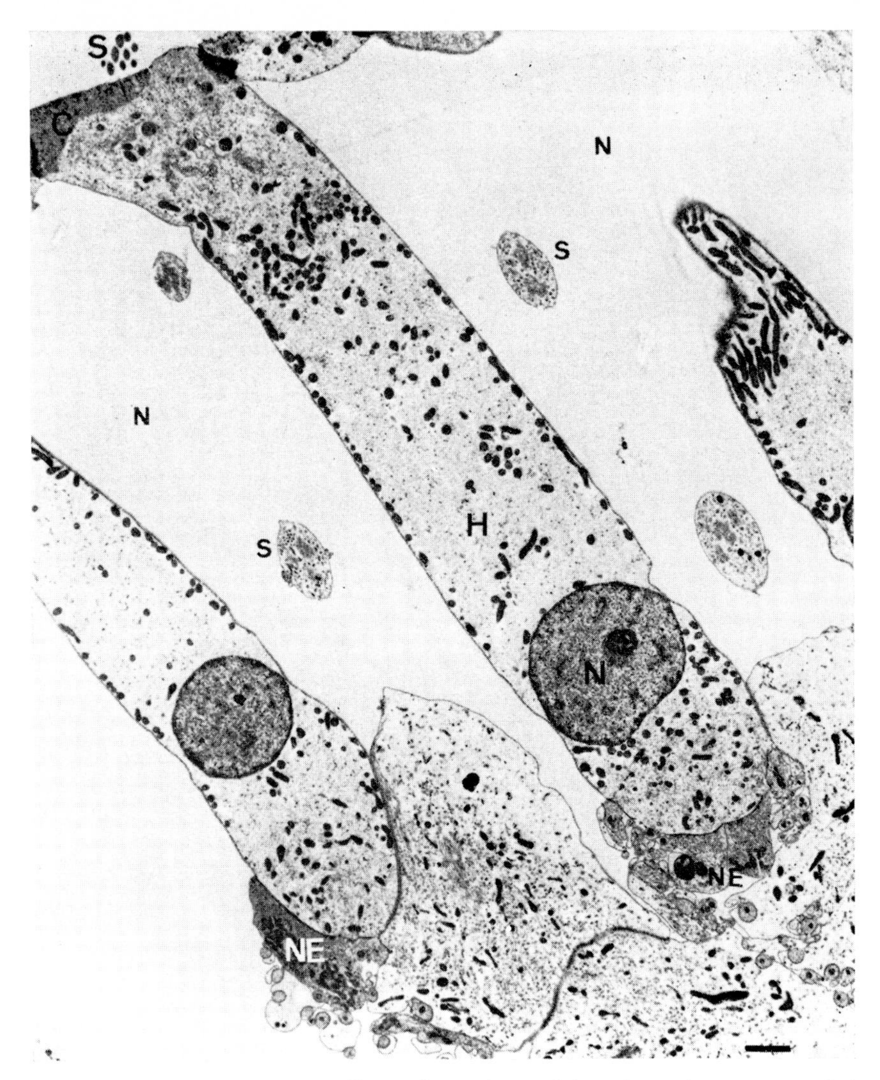

FIG. 7. Survey of an outer hair cell (H) of a cat in a radial section. The receptor pole is characterized by the cuticular plate (C), the sensory hairs (S) and an accumulation of mitochondria and endoplasmic membranes. The nucleus (N) sits in the basal portion of the high cylindrical cell. The nerve endings (NE) are adjacent to the base of the hair cell. The upper extensions of the supporting cells (S), which run obliquely in respect of the sensory cells, are seen here in cross-section in the Nuel's spaces (N). Scale bar: 1 μm.

fibrillar structure, condensed at the basal end to a central core which continues as the rootlet in the cuticular plate. These rootlets only occasionally penetrate the entire thickness of the cuticular plate (Fig. 5).

The stereocilia could possibly serve simply as passive levers to transmit the shearing forces to the cuticular plate, the tilting of which would lead to distortion of the sensory cell and produce a change in electrical resistance

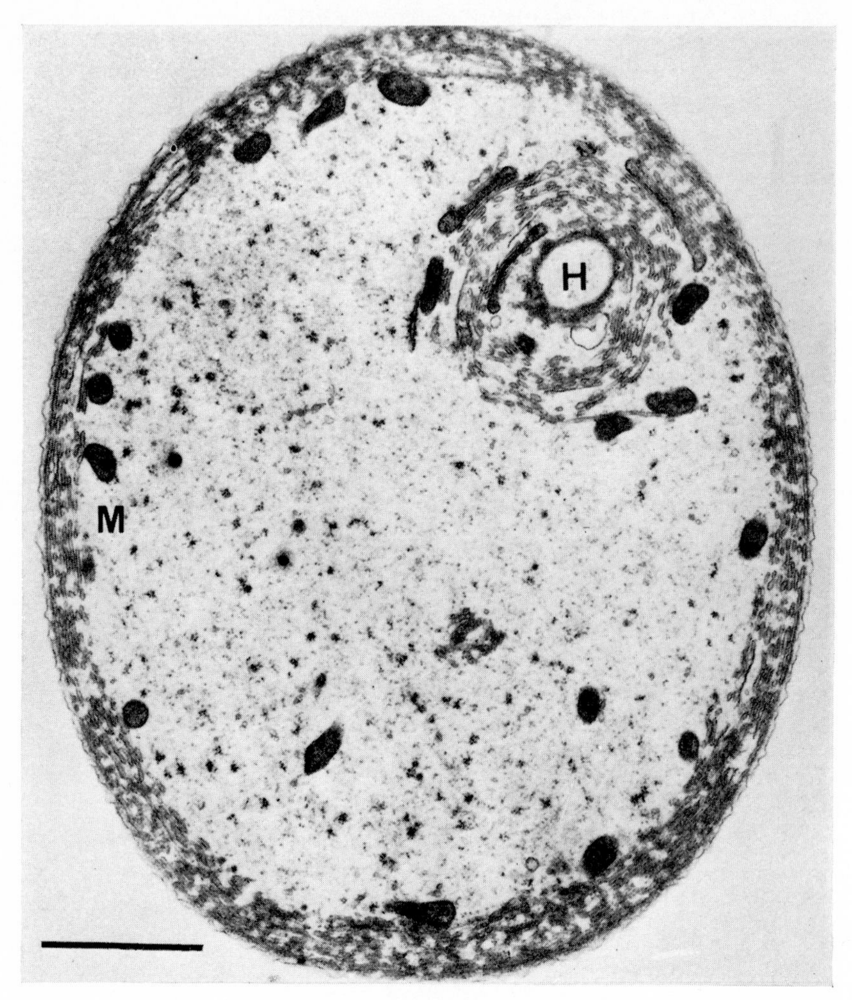

FIG. 8. Transverse section through the cell body of an outer hair cell of a guinea pig. There are six to eight layers of cisternal membranes (M) along the cell wall. The same type of membranes form the so-called Hensen's body (H). The cytoplasm is relatively empty and the mitochondria are concentrated near the membranes. Scale bar: 1 μm.

and in current. The cuticular plate appears, however, to be tightly fixed all round within the solid framework of the reticular membrane, which would seem to prevent any appreciable tilting. On the other hand, it is conceivable that the stereocilia, with their longitudinally orientated fibrillar ultrastructure, have piezo-electric properties allowing them to serve as initial mechano-electrical transducers. The hypothesis that acid mucopolysaccharides could primarily be responsible for the first steps of such a mechanism was mainly suggested by Vilstrup and Jensen (1961) and further developed by Christiansen (1967). The histochemical demonstration of mucopolysaccharides with the PAS reaction in the cochlea provides only overall and inconsistent results. A new possibility is opened up, however, by the use of ruthenium red in electron microscope preparations (Luft, 1966). Ruthenium red in combination with osmium tetroxide appears to stain extracellular, normally invisible material, most probably acid mucopolysaccharides. Used in the cochlea it reveals that the spaces between the sensory hairs are indeed filled with mucopolysaccharides which even have a certain periodicity, indicating a regular molecular arrangement (Fig. 6). Such findings fit exactly with Christiansen's postulated concept that mucopolysaccharides sitting on the surface of the hairs induce the first potential changes under mechanical distortion.

The sensory cell body presents the least specific differentiation. The outstanding features in the outer hair cells are intracellular membrane systems along the cell wall, associated with an accumulation of mitochondria, which indicate the high metabolic activities of these membranes. (Fig. 8). They differ among different species. In guinea pigs, for instance, they are found in 6–8 layers, whereas in cats and man there is only one layer. Their functional significance is unknown. They might serve to maintain the potential difference between sensory cell and surrounding fluid spaces. Or if we take the sensory cell as a specialized neuron, they might perhaps be compared with the isolating myelin sheaths of nerve fibres.

The basal portion of the cell with the recepto–neural junction is again a zone of high metabolic activity, as expressed in an agglomeration of mitochondria and a great number of microtubules and vesicles within the cytoplasm. The plasma membrane of the cell is in close contact with several nerve endings (Figs. 10, 11).

The receptor poles of the inner and outer hair cells are similarly organized, except that the somewhat larger stereocilia of the inner hair cells are arranged in a flatter W. The cell body differs very much, however. The inner hair cell resembles more an unspecific cell, with fewer structural differentiations, a more regular distribution of mitochondria, rough endoplasmic reticulum and a less localized arrangement of the recepto–neural junctions over the greater part of the cell body (Fig. 9).

At the recepto–neural junction the third component of the cochlear

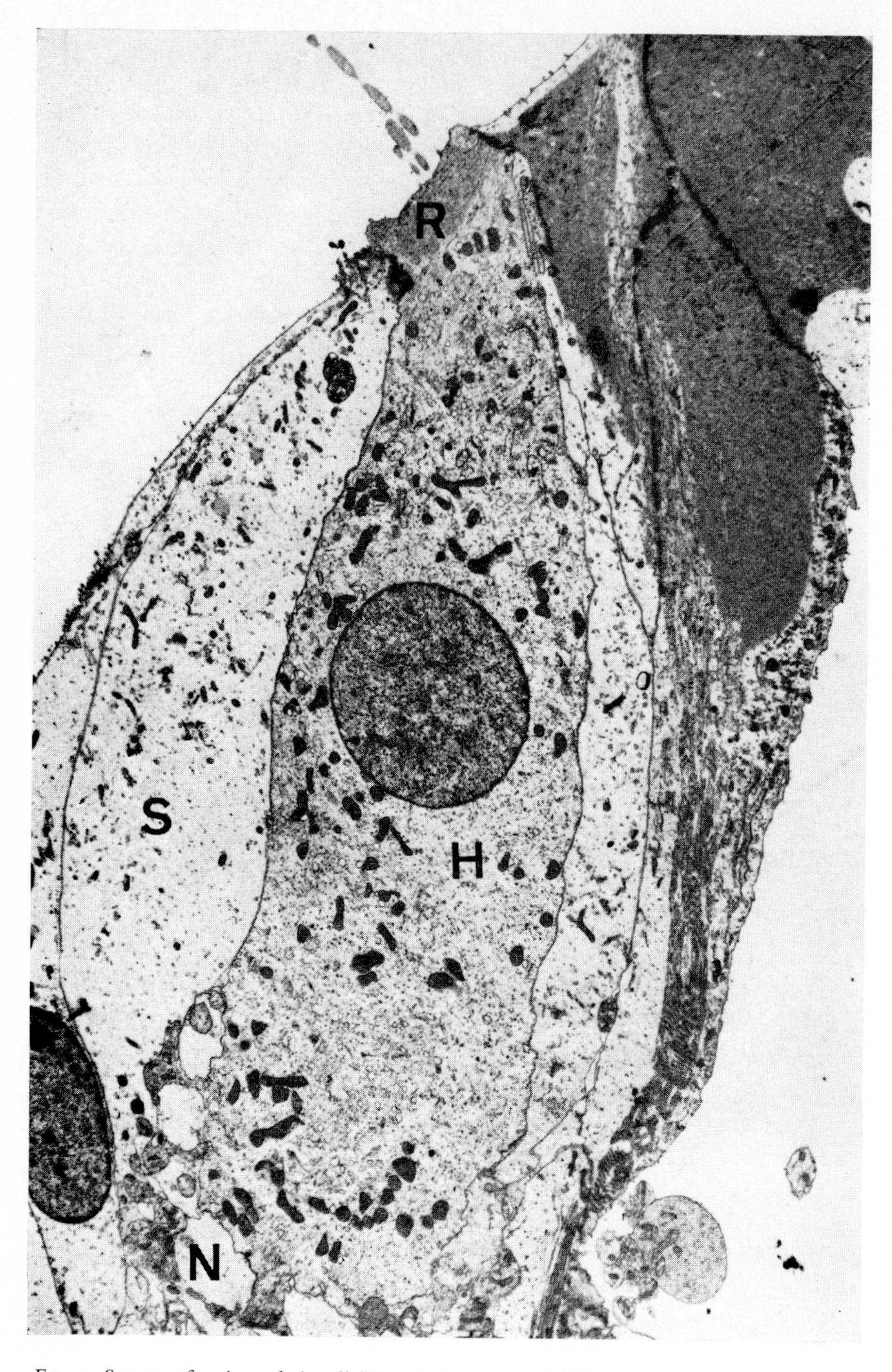

FIG. 9. Survey of an inner hair cell (H) entirely surrounded by supporting cells (S). With the exception of the receptor pole (R), the inner hair cell shows little special structural differentiation. The nerve endings of afferent dendrites (N) are grouped around the lower half of the cell.

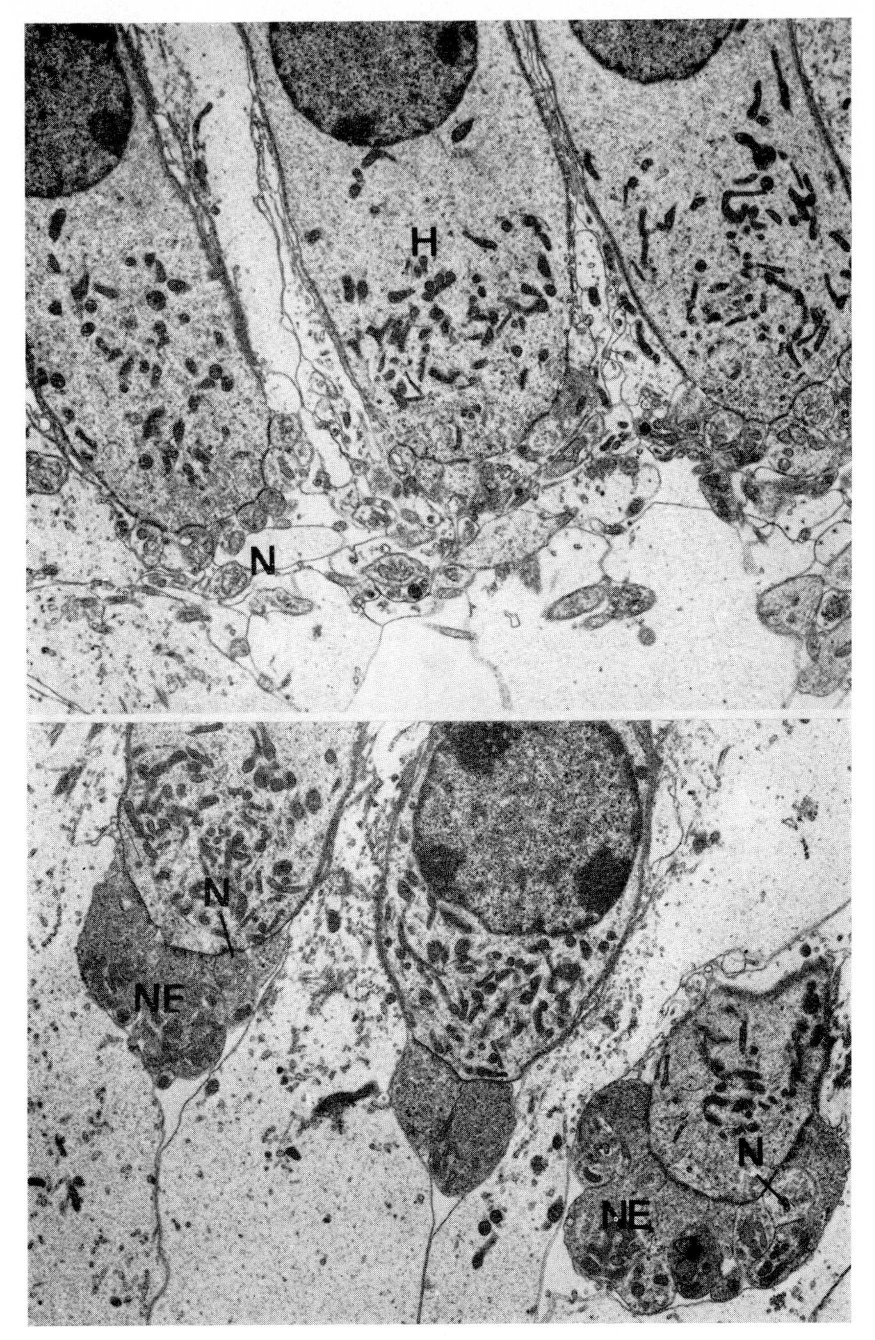

FIG. 10. A. Section in a tangential plane through the outer hair cells (H) of the third row from the third cochlear turn of a guinea pig. There are almost exclusively afferent nerve endings of type I (N).

B. Section through three outer hair cells of the first row in the same cochlear area as in A. The efferent nerve endings (NE) are clearly predominant and only a few afferent nerve endings (N) are visible.

receptor begins, with nerve endings and fibres where the initial electrical response of the receptor cell is transformed into nervous activity.

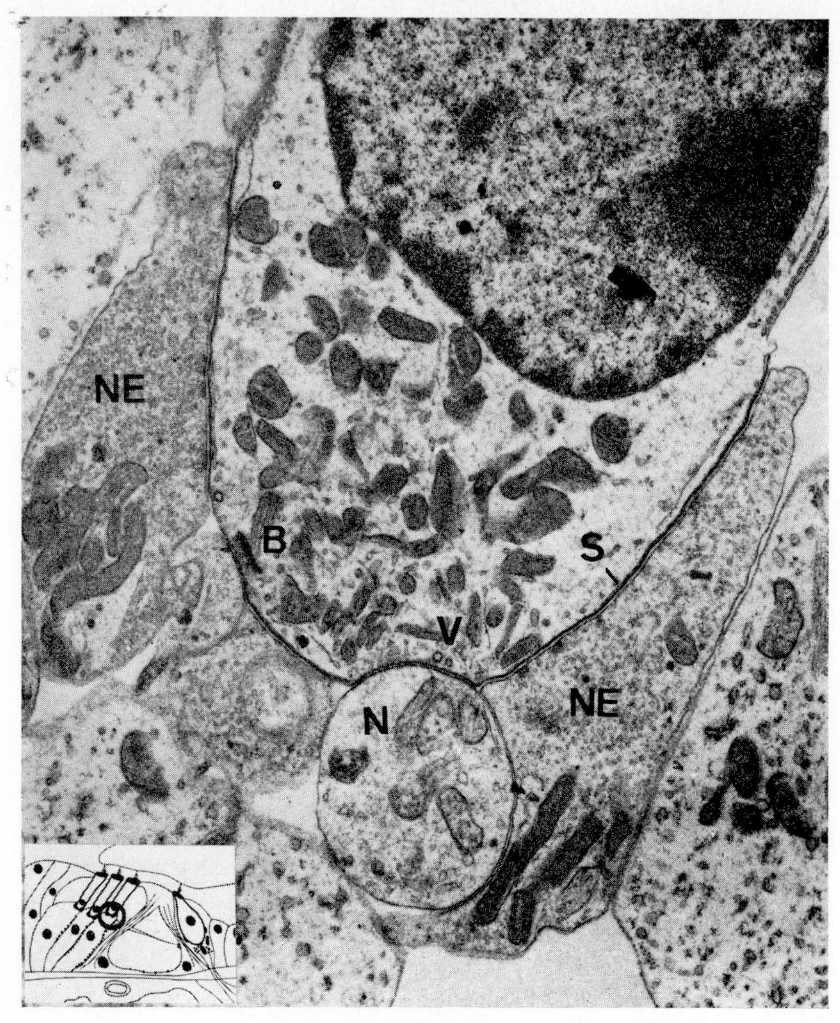

Fig. 11. Recepto–neural junction of an outer hair cell of the second cochlear turn from a guinea pig. Two types of nerve endings are distinguished. Type I, the afferent endings (N), which are relatively small, contain only a few irregular vesicles and are frequently associated with synaptic bars (B) and coated vesicles (V) in the adjacent portion of the sensory cell. Type II or efferent nerve endings (NE) are large and filled with synaptic vesicles condensed in certain areas of the presynaptic membrane. The postsynaptic membrane is associated with a subsynaptic cisterna (S).

Between the nerve ending at the base of the hair cells and the habenula perforata the nerve fibres frequently take a long course and show a definite

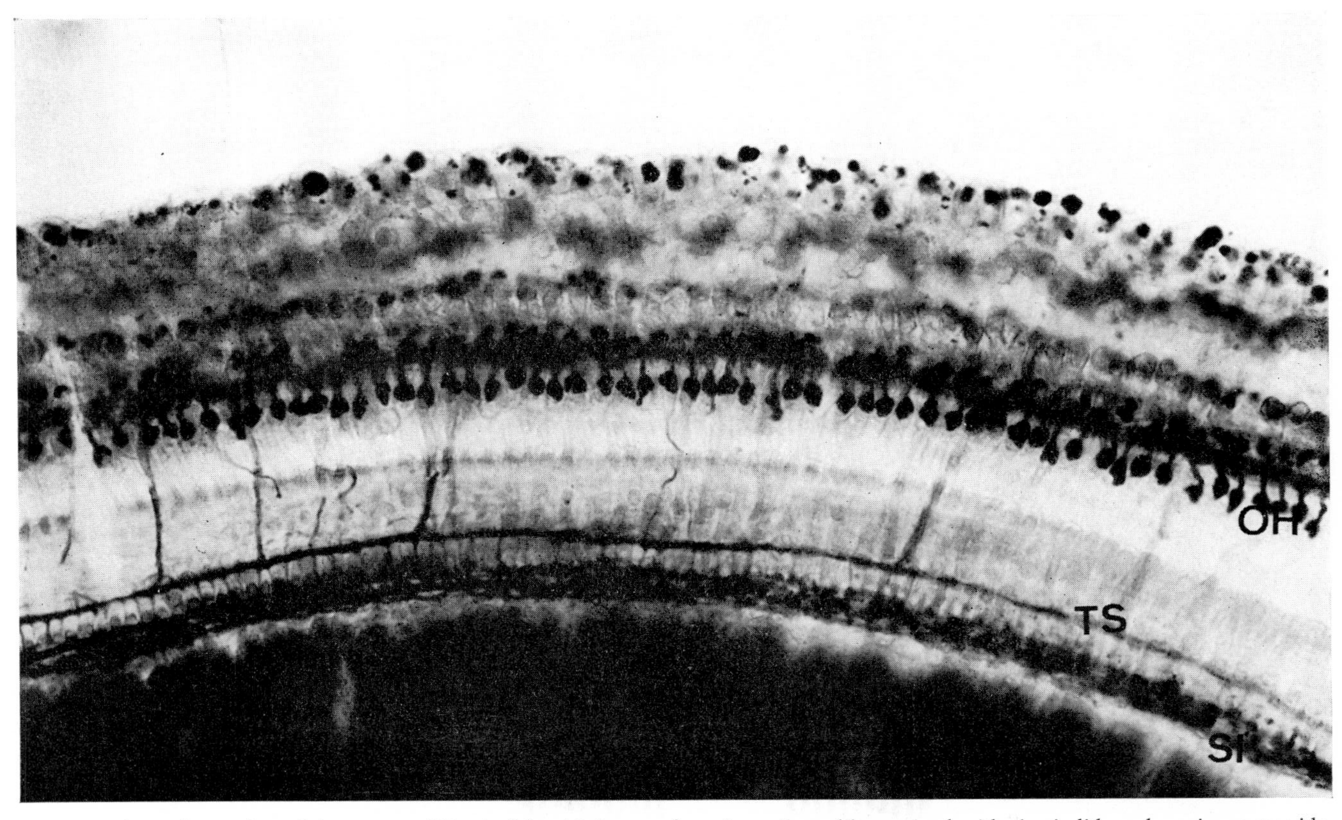

FIG. 12. View of a portion of the organ of Corti of the third turn of a guinea pig cochlea, stained with zinc iodide and osmium tetroxide. The efferent nerve fibres and endings are stained black. The predominant efferent innervation of the first row of outer hair cells (OH) is clearly visible. The third row has only a very small efferent nerve supply. The inner spiral plexus (SI) and the tunnel spiral fibres (TS) also belong to the efferent system.

pattern of distribution. Within the organ of Corti all nerve fibres are unmyelinated. They acquire their myelin sheaths only below the habenula

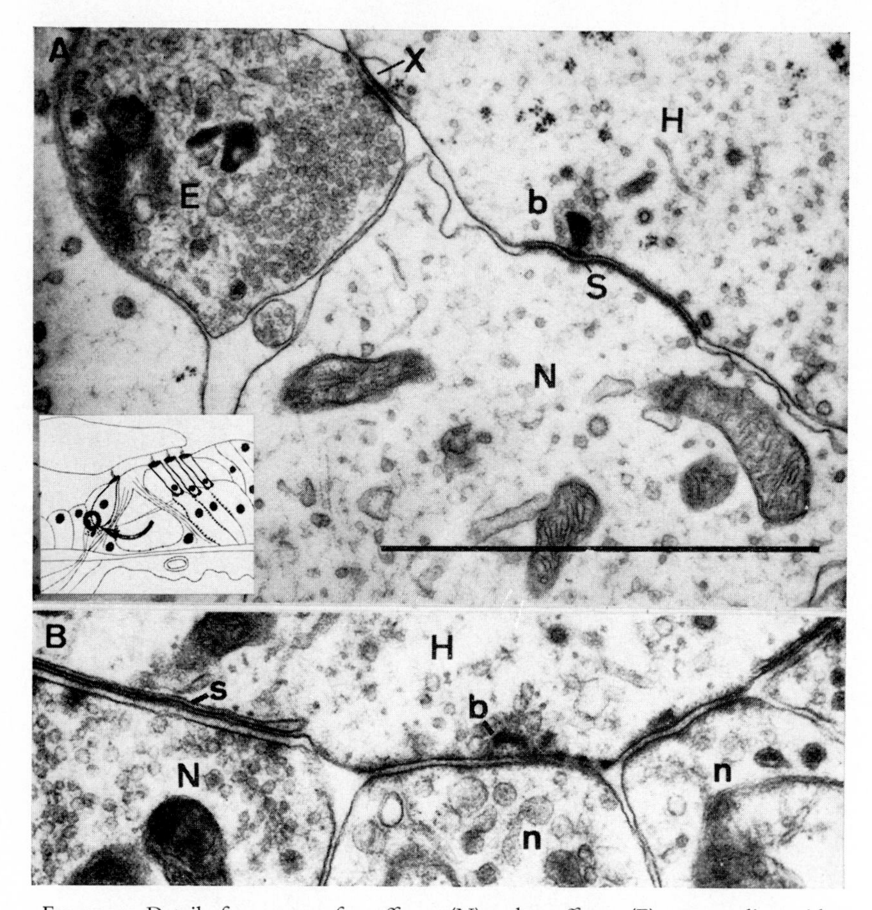

FIG. 13. A. Detail of a synapse of an afferent (N) and an efferent (E) nerve ending with an inner hair cell (H). The afferent synapse (S) is characterized by a very marked thickening of the postsynaptic membrane (S) and a pronounced synaptic bar (b) surrounded by synaptic vesicles. Synaptic contacts of efferent nerve endings with the inner hair cell (X) are exceptional. This synapse is characterized by relatively little thickening of pre- and postsynaptic membranes and by a subsynaptic cisterna (X). Scale bar: 1 μm.

B. Detail of synaptic zones between afferent (n) and efferent (N) nerve endings and an outer hair cell (H). The afferent ending contains only a few irregular vesicles. The pre- and postsynaptic membranes are clearly thickened and the presynaptic membrane is frequently associated with a synaptic bar (b). The efferent nerve ending (N) is filled with regular synaptic vesicles which are condensed at the presynaptic membrane in certain places. The subsynaptic cisterna (S) is always associated with the postsynaptic membrane.

perforata. The unmyelinated fibres in the organ of Corti are known to be arranged in different groups, such as the inner and the outer spiral fibres, the inner and the tunnel radial fibres, and the basilar fibres, usually not visible in the light microscope (Fig. 1).

Afferent and efferent fibres contribute to the innervation of the sensory cells. At the time Rasmussen described an efferent nerve supply to the cochlea through the olivo-cochlear bundle, these fibres could be followed only as far as the habenula perforata. Some years later two different types of nerve endings were found at the base of the outer hair cells (Engström, 1958; Smith and Sjöstrand, 1961; Spoendlin, 1959, 1966). The first type consists of small bud-like fibre endings containing very few cytoplasmic organelles, and is frequently associated with so-called synaptic bars at the membrane of the adjacent sensory cell. The second type is much larger, filled with small synaptic vesicles, and the adjacent portion of the sensory cell membrane is associated with a subsynaptic cisterna (Figs. 11, 13). According to the morphological similarity with endings of other known centrifugal fibres, these type II endings were assumed to be efferent in nature. The proof that they do belong to the olivo-cochlear efferent system was obtained in different animals by a number of authors by experimental transection of the olivo-cochlear bundle and subsequent electron microscopic examination of the cochlea. A few days after nerve section the large, vesiculated nerve endings were found obviously degenerating. In cats in which the entire olivo-cochlear bundle was selectively transected by cutting the vestibular nerve, all type II endings degenerated and disappeared. This brings up the surprising fact that we are dealing with an enormous efferent innervation of the cochlear receptor.

As shown by Engström, Ades and Andersson (1966), the efferent fibres and endings can be stained selectively with the mixture of zinc iodide and osmium tetroxide of Maillet. The selectivity is only relative. Afferent fibres can sometimes also be stained slightly; however, in a much less pronounced way than the efferent elements. The histochemical background of this reaction is not known. The reaction product consists of large crystalline deposits in axons and nerve endings. Such preparations give a good impression of the importance of the efferent innervation (Fig. 12). Unfortunately, this stain proved to be less reliable in cats than in guinea pigs. Nevertheless it was possible to demonstrate the association of these stained elements to the olivo-cochlear system, since they disappeared entirely after transection of the olivo-cochlear fibres in the vestibular root. As estimated from serial sections, the number of efferent nerve endings seems to be only slightly smaller than that of afferent endings. The contact area of the efferent endings with the hair cell is even greater. The estimated total number of efferent nerve endings at the level of the outer hair cells is about 40,000. In the basal turn all three rows of outer hair cells are provided with efferent endings. In upper turns they disappear gradually from the third and second turn (Fig. 10).

The same experiments showed that in the cat all upper tunnel radial fibres and the major part of the inner spiral plexus including the tunnel spiral fibres belong to the olivo-cochlear efferent system. They degenerate promptly after transection of the olivo-cochlear bundle. The total disappearance of the upper tunnel radial fibres can best be demonstrated in tangential phase-contrast sections through the tunnel, where the small fascicles of these fibres are transversely cut and normally appear as a row of black spots, which are gone after degeneration of the olivo-cochlear fibres (Fig. 1). Selective transection of the homolateral and contralateral olivo-cochlear fibres was done in eight cats. In none of these animals were we able to find remaining upper tunnel radial fibres. Iurato has provided some evidence that in the rat the inner spiral fibres originate in the homolateral and the efferent fibres to the outer hair cells in the contralateral olivo-cochlear fibres (personal communication).

The enormous number of efferent fibres and endings in the organ of Corti obviously necessitates a considerable branching of the relatively small number of about 500 olivo-cochlear fibres. This branching occurs at different levels—in the osseous spiral lamina (Nomura and Schuknecht, 1965), in the inner spiral plexus and at the level of the outer hair cells.

After degeneration of the olivo-cochlear efferent fibres and nerve endings, the small type I endings, the inner radial fibres and the outer spiral fibres remain unchanged. They must therefore be considered to belong to the afferent sensory neurons. In the cat they cross the tunnel exclusively at the bottom as basilar fibres, in most instances entirely embedded within the cell bodies of the pillar cells. They reach the outer hair cells by penetrating between the outer pillars, with an average number of one to two fibres between two pillars. This means that only relatively few fibres (probably not more than 5,000) are destined for the outer hair cells. At the level of the Deiters' cells they run for a long distance unbranched, spirally and basalwards, climbing gradually up to the base of the outer hair cells. Each fibre finally gives off collaterals to many sensory cells and each cell receives collaterals from different dendrites, according to the principle of multiple innervation.

The average spiral extension of the outer spiral fibres can be roughly estimated on the basis of the relative numbers of outer spiral fibres and basilar fibres penetrating between the outer pillar feet. There is an average of 100 outer spiral fibres at any given place. Each fibre is the continuation of one basilar fibre. If only 1 or 2 basilar fibres penetrate between two outer pillars, we must assume that each fibre runs in a spiral direction over a distance of at least 50–100 pillars (about 0·6 mm.) if the number of 100 outer spiral fibres at any one place is to be reached (Fig. 21).

The total number of cochlear sensory neurons is found to be about 50,000 in the cat. Only approximately 5,000 fibres lead to the outer hair cells. The great majority therefore seem to be destined for the inner hair cells. In transverse sections through the area below the internal hair cells in animals in which the efferent nerve supply has been eliminated we find an average of 10–20 fibres in each habenular opening, the number of which corresponds roughly to the number of inner hair cells (Fig. 15). The majority of the fibres in one habenular opening are found to lead directly to the inner hair cells, where they end unbranched, each fibre having one ending (Figs. 13, 18).

Efferent elements appear to have only an occasional direct contact to the internal hair cells, unlike the extensive efferent connexions to the outer hair cells. In the inner spiral plexus, however, there is an intimate interrelationship between afferent and efferent elements.

Contact alone between two nervous structures does not imply synaptic activity. The presence of synaptic contacts is usually indicated by membrane differentiations and agglomerations or condensations of synaptic vesicles at the presynaptic membrane. Such morphological evidence of synaptic activity is found at the level of the outer hair cells almost exclusively between efferent nerve endings and sensory cells and only rarely between nerve endings and afferent dendrites. The opposite is the case at the level of the inner spiral plexus, where the efferents only rarely have synaptic contacts with the sensory cells but have a great number of contacts with afferent dendrites. The efferent fibres of the inner spiral bundle present no actual nerve endings but merely fibre enlargements which are filled with vesicles and surround the afferent dendrites (Fig. 17).

These very slender fibres, with diameters down to 0·1 μm., are usually grouped in two main bundles (Fig. 14). The fibres pass from the lower to the upper bundle where they form vesicle-filled enlargements with synaptic contacts to afferent dendrites (Fig. 16). The spiral extension of these fibres could not yet be determined. The intimate interrelations between afferent and efferent fibres with many synaptic contacts suggests that this area is a kind of co-ordination and integration centre with a steady interaction between afferent and efferent fibres.

The pattern of distribution of afferent and efferent nerve fibres in the organ of Corti of the cat can be summarized as follows. The afferent

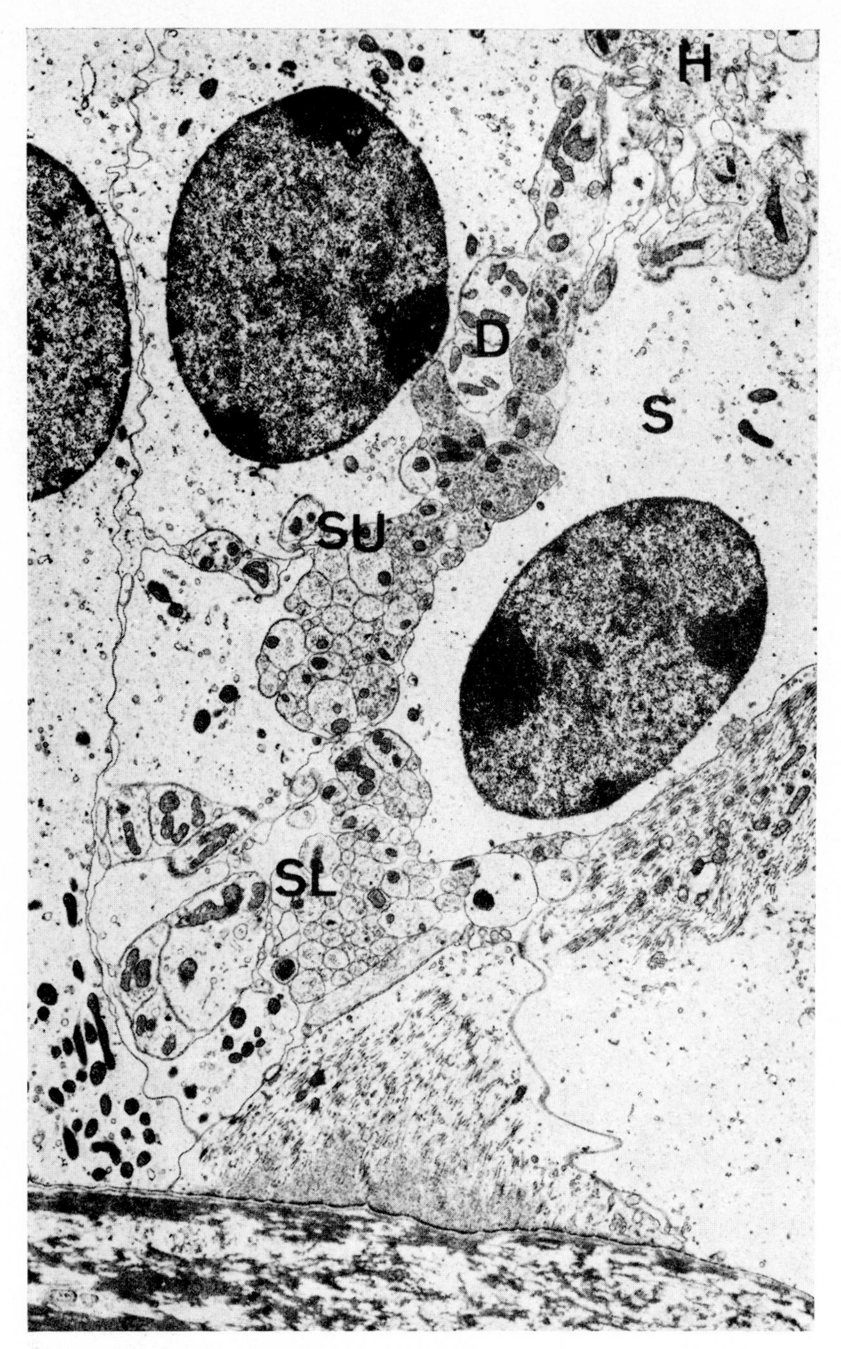

FIG. 14. Survey of the area below the inner hair cell (H) with the inner spiral plexus between the supporting cells (S). The inner spiral fibres are usually arranged in a lower bundle (SL) and an upper bundle (SU) where the fibres present numerous enlargements filled with synaptic vesicles and making synaptic contacts with the adjacent afferent dendrites (D).

dendrites show an almost exclusively spiral distribution at the level of the outer and a radial distribution at the level of the inner hair cells. The

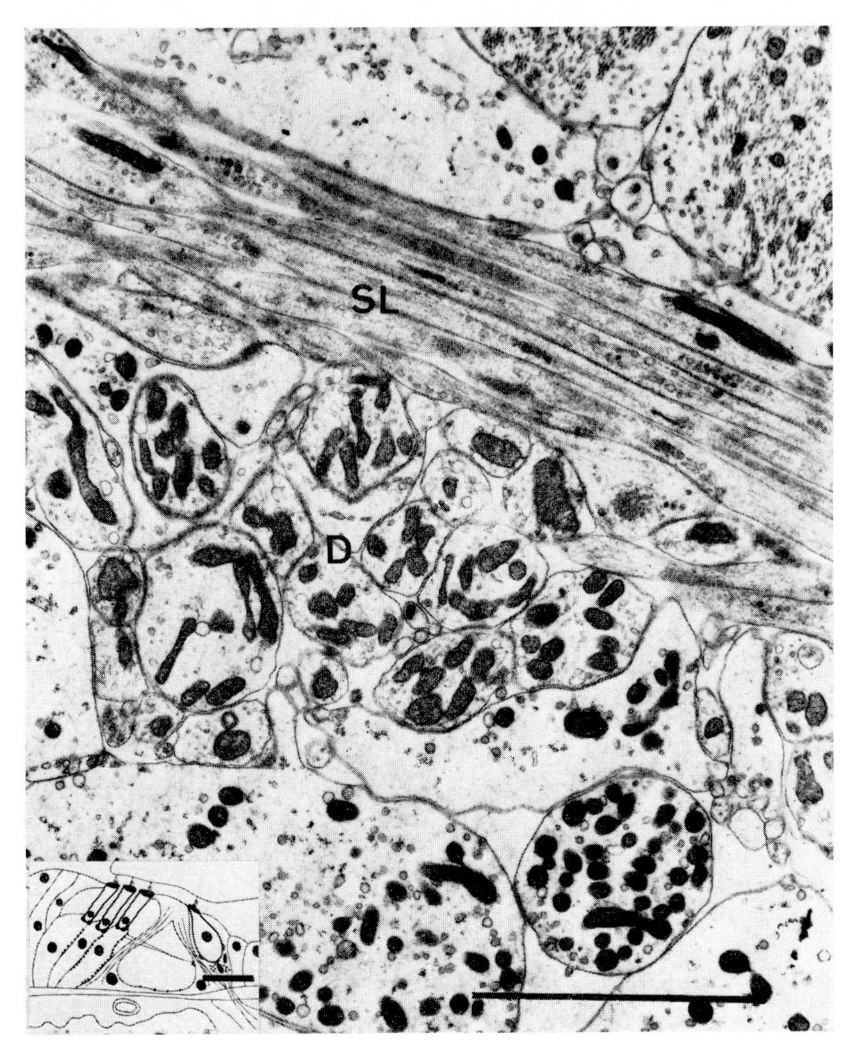

FIG. 15. Horizontal section through the inner spiral plexus immediately above the habenula perforata with the lower bundle of inner spiral fibres (SL) and the afferent dendrites (D) which have emerged from a habenular opening. At this level the efferent inner spiral fibres and the afferent radial dendrites have very little contact. Scale bar: 1 μm.

efferent fibres on the other hand present a predominantly radial distribution at the level of the outer hair cells and a wide spiral extension in the inner spiral plexus (Fig. 20). The efferent fibres to the outer hair cells make almost exclusively synaptic contacts with the sensory cells, whereas the

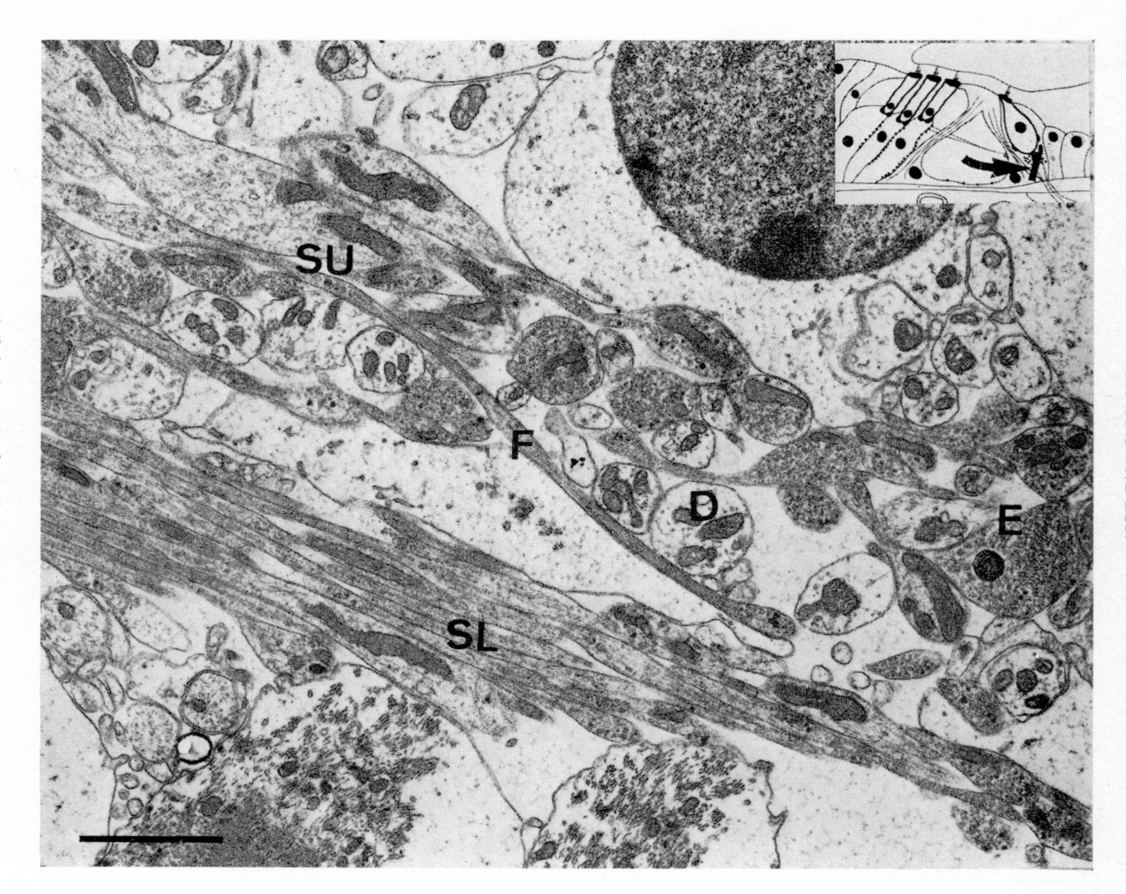

FIG. 16. Tangential section through the inner spiral plexus with the lower bundle of inner spiral fibres (SL) and the upper bundle (SU), where the small inner spiral fibres show a great number of sack-like enlargements filled with synaptic vesicles (E) surrounding the afferent dendrites (D). One fibre is seen passing from the lower to the upper bundle (F).

Scale bar: 1 μm.

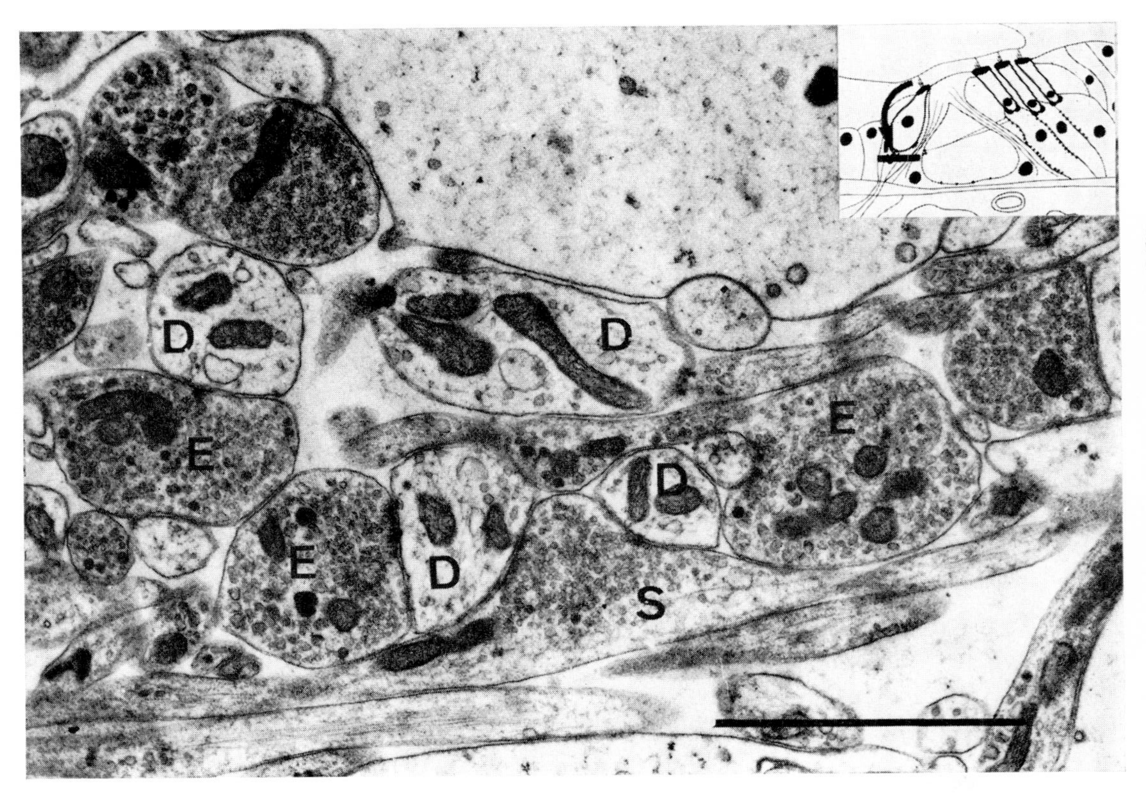

FIG. 17. Detail of the upper portion of the inner spiral plexus with many different fibre enlargements (E) filled with synaptic vesicles making frequent synaptic contacts with afferent dendrites (D). It is clear that the efferent fibre enlargements are not actual endings, since the fibre continues in both directions (S).
Scale bar: 1 μm.

efferent inner spiral fibres have their synaptic contacts almost exclusively with afferent dendrites.

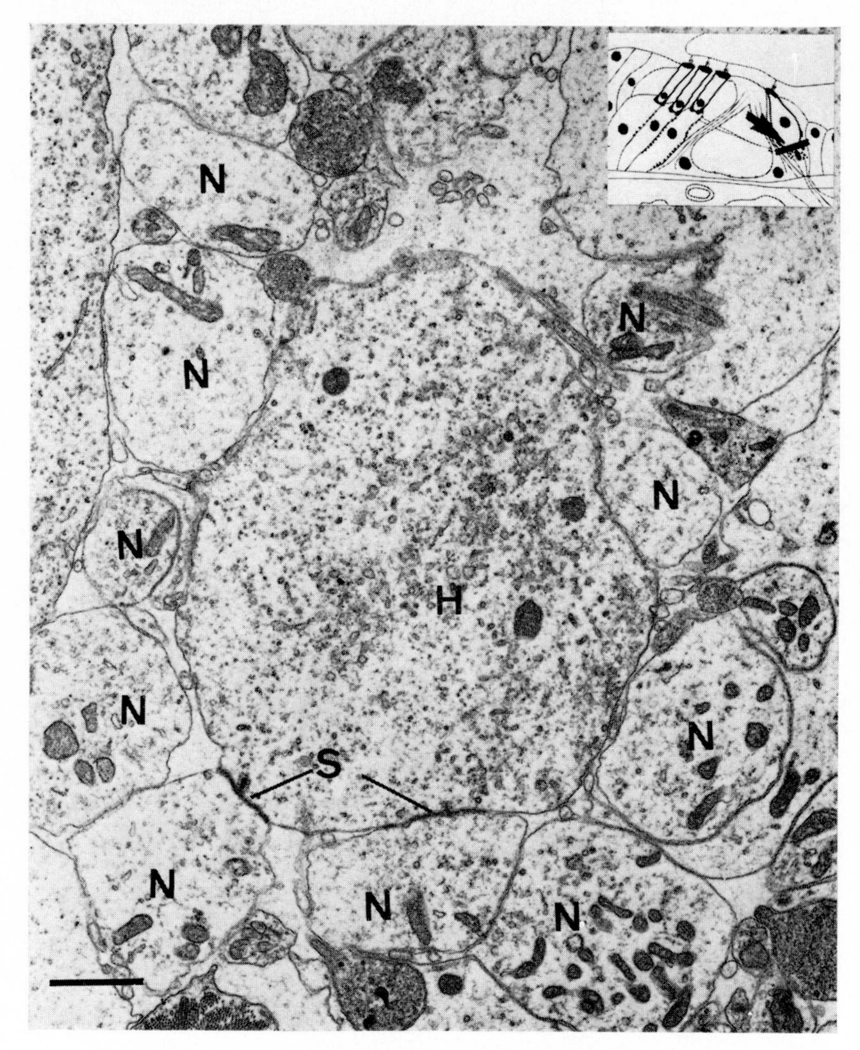

Fig. 18. Horizontal section through the basal portion of an inner hair cell (H) surrounded by a great number of relatively large afferent nerve endings (N) with synaptic complexes (S) at several places. Each inner hair cell is associated with 10–20 nerve endings, a number corresponding roughly to the number of afferent dendrites coming through each habenular opening. Scale bar: 1 μm.

The efferent and afferent fibres in the organ of Corti of the cat cannot only be distinguished topographically; they appear also to have a different ultrastructure. Whereas the efferent axons of the tunnel radial fibres

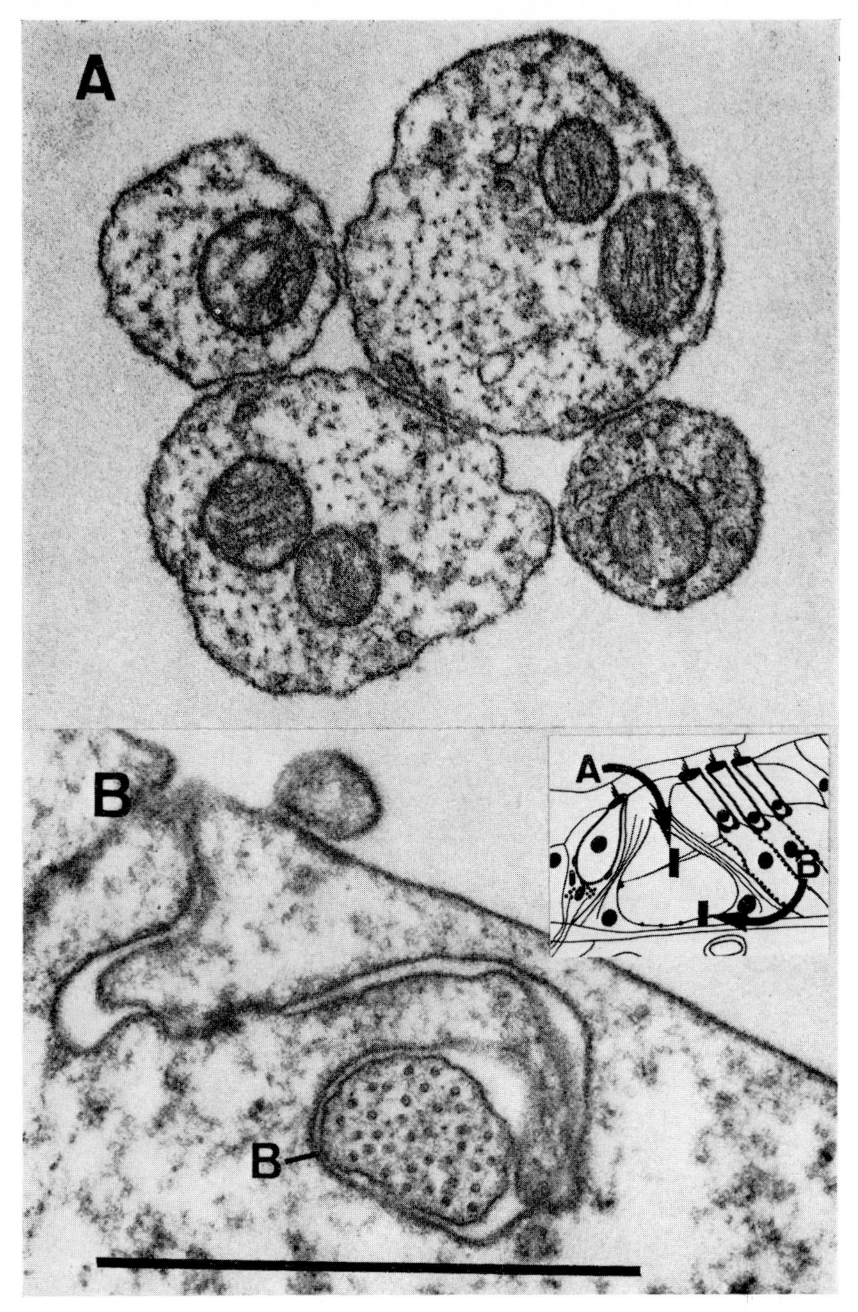

FIG. 19. Comparison of tunnel radial and basilar fibres of the cat after glutaraldehyde fixation.

A. Transverse section through a fascicle of tunnel radial fibres. They contain exclusively neurofilaments in their axoplasm.

B. Transverse section through one basilar fibre (B) which is entirely embedded in the cytoplasm of the pillar cells and contains exclusively neurocanaliculi. Scale bar: 1 μm.

contain neurofilaments almost exclusively, the afferent dendrites, such as the outer spiral or basilar fibres, are characterized by large neurocanaliculi in their axoplasm (Fig. 19). This structural differentiation is very distinct in basilar and outer spiral fibres but less clear in the inner spiral plexus, and it is entirely absent in the myelinated fibres. One wonders whether

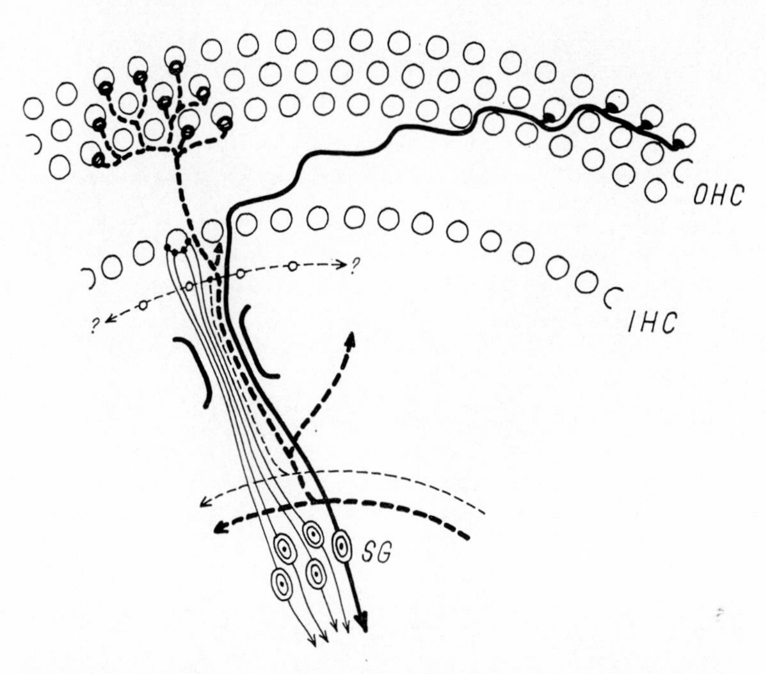

FIG. 20. Schematic representation of the pattern of innervation of the organ of Corti with afferent and efferent nerve fibres. The efferent fibres are shown by interrupted lines, the afferent fibres by full lines. The fibres destined for the outer hair cells are represented by thick and the fibres for the inner hair cells by thin lines. In order to simplify this schema, only the fibres of one habenular opening are drawn. The reciprocal innervation modus of efferent and afferent endings is evident. The number of fibres depicted does not correspond to their actual numbers. The relative numbers only of the different fibres are indicated.

such structural differences between efferent axons and afferent dendrites are related to different functional behaviour. The same ultrastructural characteristics are found in neurons of the central nervous system.

A third type of innervation—adrenergic innervation—is probably also involved in the process of coding the acoustic message. With a method described by Falck and co-workers (1962) it is possible to demonstrate adrenergic nerve fibres histochemically and very selectively. In this way a rich adrenergic terminal nerve plexus could be demonstrated, independent of blood vessels, at the level of the habenula where the unmyelinated fibres from the organ of Corti get their myelin sheaths, a place which probably corresponds to the initial segment of the neurons (Spoendlin and Lichtensteiger, 1966).

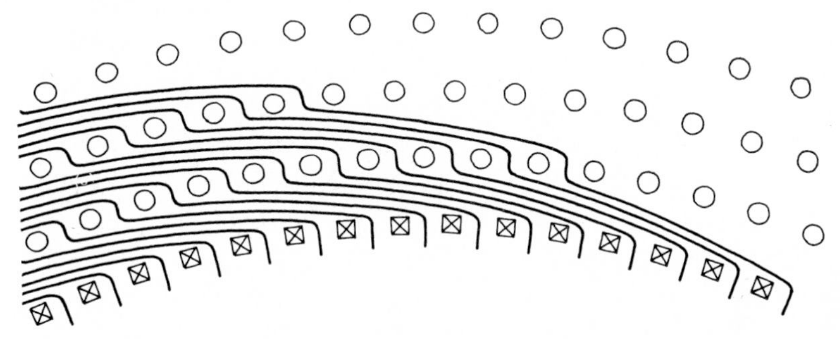

FIG. 21. Schema showing the evaluation of the average spiral extension of outer spiral fibres. The outer pillars are represented by crossed squares and the outer hair cells by circles. The fibres forming the outer spiral fibres penetrate between the outer pillars, one fibre between every two pillars. The number of outer spiral fibres found at any one place therefore corresponds to the number of pillars over which the nerve fibres run in a spiral direction. If there were 15 outer spiral fibres, they would extend over a distance of 15 pillars, or if we find 100 outer spiral fibres, as in the cat, they extend over a distance of 100 pillars, which corresponds approximately to 0·6 mm.

DISCUSSION

The innervation pattern of the organ of Corti—that is, the distribution and interrelations of the nerve fibres—has doubtless great functional significance for the first coding of the acoustic message. Taking into account some known electrophysiological phenomena, a number of possible mechanisms for the coding of the acoustic message can be deduced from the structural organization of the cochlear receptor.

The electrical activity of a typical dendrite is, according to the present concept (Davis, 1961), a graded potential without a threshold, whereas the activity of axons consists in spikes which originate with a threshold as an all-or-none response. In contrast to spikes in axons, the graded potential in dendrites can be summated and modified by inhibitory impulses.

The unmyelinated peripheral ramifications of the cochlear neurons correspond in an anatomical and most probably also functional sense to typical dendrites. We assume that spatial and temporal summation take place within these dendrite ramifications. The initial segment of the cochlear neurons, where the spikes originate, is most probably below the habenula at the beginning of the myelin sheath of the nerve fibres. Electrical activity in the dendrite ramification would therefore be collected in the main dendrite according to the principles of spatial and temporal summation, until the threshold for firing an action potential at the initial segment is reached. Judging from the structural appearance of the synapses, with synaptic bars and vesicles at the presynaptic membrane (Fig. 13), the activation of the nerve ending at the sensory cell is most likely to be a chemical process with a still unknown transmitter substance, which leads to depolarization of the postsynaptic membrane. It is not excluded that acetylcholine plays a certain role in this transmission, because it seems to be affected by tubocurarine (Katsuki, Tanaka and Miyoshi, 1965).

The efferent inhibitory innervation system might however not be cholinergic in a pharmacological sense, in spite of the positive histochemical acetylcholine reaction (Schuknecht, Churchill and Doran, 1959). It is probably similar to the inhibitory system in the central nervous system, since it is also sensitive to strychnine and brucine (Desmedt and Monaco, 1962). The synaptic activity of the efferent fibres on sensory cells and afferent dendrites is likely to be hyperpolarizing—inhibitory. According to Gray (1967), excitatory and inhibitory synapses might have different structural characteristics. The excitatory synapse would be characterized by a much more pronounced thickening of the synaptic membrane than the inhibitory synapse, whereas the inhibitory nerve ending would contain flattened vesicles instead of round ones. If this concept proves to be correct the synapses of the efferent endings in the organ of Corti should be considered as inhibitory synapses, since their synaptic membranes show very little thickening compared with the afferent synapses (Fig. 13). The summated activity within the afferent dendrites is presumably constantly corrected by inhibitory action of the efferent system, so that the final depolarization at the initial segment is the product of afferent summation effects modified by efferent inhibitory influences. According to the predominant axodendritic synapses, the efferent fibres will exert postsynaptic inhibition at the level of the inner spiral plexus. At the level of the outer hair cell, however, the majority of synaptic contacts are found between nerve endings and sensory cells, suggesting presynaptic inhibition.

We would assume that spatial summation is an important factor within the extensively ramified dendrites associated with the outer hair cells, whereas the unbranched fibres to the inner hair cells allow only temporal, and no spatial summation. Because of the possibility of spatial summation, the sensory units of the outer hair cells with their associated dendrites should have a much higher sensitivity than the inner hair cells. A weak stimulus might not be sufficient to activate the single nerve endings of the unbranched neurons associated with the inner hair cell sufficiently for the threshold of the action potential at the initial segment to be reached, whereas the same weak stimulus could be summated in the dendrite ramifications of the outer spiral fibres so that the resulting depolarization reaches the threshold to trigger off an action potential at the initial segment. Stronger stimuli, however, might not be adequately transmitted in these fibres because their capacity is limited by the refractory period of the axon. But they might activate sufficiently the great number of unbranched single neurons at the inner hair cells, so that action potentials occur in each of them. The possibility of temporal summation is presumably equal in the dendrites for inner and outer hair cells. The lack of spatial summation, however, increases the time required to build up the necessary degree of depolarization in the neurons associated with the inner hair cells.

It has been directly demonstrated that the olivo-cochlear fibres have a relatively mild inhibitory influence on the building up of action potentials in the cochlear neurons (Desmedt, 1962; Fex, 1962). Their enormous anatomical representation within the cochlear receptor makes it unlikely, however, that their function is restricted to such a limited inhibitory action. I would think that the elaborate efferent innervation at this level is directly involved in many important steps of the first coding of the acoustic message, such as intensity or frequency discrimination, adaptation and other aspects.

If the activation of this feedback system went entirely over a reflex arc through the central nervous system, its functional efficiency would be limited by the relatively large latency and by the small number of individual neurons. One wonders whether efferent elements could be activated directly in the periphery by electrical activity in the afferent dendrites, without the detour over the central nervous system. Such an activation would presumably rely on electrical rather than chemical phenomena.

There is finally the adrenergic innervation of the cochlea, which probably has a direct influence on the formation of the acoustic message. The localization of its terminal plexus in the area of the initial segments of the cochlear neurons suggest the possibility that it influences the threshold of the action potentials directly at the initial segment (Spoendlin and Lichtensteiger, 1966).

In conclusion, it appears that the cochlear receptor is far more than a simple energy transducer. By means of the innervation pattern of the sensory cells, the interaction between afferent and efferent neurons and possibly the influences of the adrenergic innervation, an important integration and modification of the acoustic message is most probably accomplished within the cochlear receptor.

SUMMARY

Three components are essential in the cochlear receptor:

(1) The supporting structures and the tectorial membrane, mainly concerned with the stimulating mechanism.

(2) The sensory cells as mechano-electrical transducers. The polarization of the receptor pole is probably determined by the original polar position of the centriole. The stiff, polysaccharide-containing stereocilia connected to the tectorial membrane and anchored in the immobile cuticula, might serve as the initial transducer. Specific membrane systems characterize the cytoplasm of the outer hair cells.

(3) The nerve endings and fibres within the organ of Corti where the first coding of the acoustic message occurs. Besides the afferent and efferent nerve supply there is an adrenergic innervation. Distal to the habenula all fibres are small and unmyelinated. The extensive efferent nerve supply decreases from base to apex. The afferent and efferent fibres show a reciprocal radial or spiral distribution at the level of the outer and inner hair cells. Only a minority of afferent fibres reach the outer hair cells and each fibre sends collaterals to a great number of sensory cells. The majority of the fibres lead to the inner hair cells. Numerous synaptic contacts between afferent dendrites and efferent fibres are found within the inner spiral plexus. Such fibre arrangements and interrelations suggest the possibility of spatial and temporal summation, and might explain a difference in sensitivity between inner and outer hair cells. The enormous anatomical representation of the efferent innervation indicates its important role for the primary coding of the acoustic message.

REFERENCES

Christiansen, J. A. (1967). *Acta oto-lar.*, **57**, 33.
Davis, H. (1961). *Physiol. Rev.*, **41**, 391–416.
Desmedt, J. E. (1962). *J. acoust. Soc. Am.*, **34**, 1478.
Desmedt, J. E., and Monaco, P. (1962). *Proc. 1st Int. pharmac. Meet.*, **8**, 183.
Engström, H. (1958). *Acta oto-lar.*, **49**, 109.
Engström, H., Ades, H. W., and Andersson, A. (1966). *Structural Pattern of the Organ of Corti*, pp. 1–172. Stockholm: Almqvist and Wiksell.
Falck, B., Hillarp, N.-A., Thieme, G., and Torp, A. (1962). *J. Histochem. Cytochem.*, **10**, 348.
Fex, J. (1962). *Acta physiol scand.*, Suppl. 189.
Gray, E. G. (1967). *Science Jl*, **3:5**, 66–72.
Iurato, S. (1967). *Arch. exp. Ohr.-Nas.-u. KehlkHeilk*, **189**, 113–126.
Katsuki, Y., Tanaka, Y., and Miyoshi, T. (1965). *Nature, Lond.*, **207**, 32–34.
Luft, J. (1966). *Fedn Proc. Fedn Am. Socs exp. Biol.*, **25**, 1773–1783.
Nomura, Y., and Schuknecht, H. F. (1965). *Ann. Otol. Rhinol. Lar.*, **74**, 289–302.

Schuknecht, H. F., Churchill, J., and Doran, R. (1959). *Archs Otolar.*, **69**, 549.
Smith, C. A., and Sjöstrand, P. S. (1961). *J. Ultrastruct. Res.*, **5**, 184.
Spoendlin, H. (1959). *Practica oto-rhino-lar.*, **21**, 34–48.
Spoendlin, H. (1964). *Z. Zellforsch. mikrosk. Anat.*, **62**, 701–716.
Spoendlin, H. (1966). *The Organization of the Cochlear Receptor. Advances in Oto-Rhino-Laryngology*, vol. 13. Basel and New York: Karger.
Spoendlin, H., and Lichtensteiger, W. (1966). *Acta oto-lar.*, **61**, 423–434.
Tasaki, J., Davis, H., and Eldredge, O. (1954). *J. acoust. Soc. Am.*, **26**, 765.
Vilstrup, Th., and Jensen, C. E. (1961). *Acta oto-lar.*, **63**, 42.

DISCUSSION

Davis: These extremely interesting observations on the ratio of the number of afferent fibres to the number of hair cells provide a possible answer to a question that has long been bothering me. It is implicit in your finding that each internal hair cell has multiple innervation and that a greater dynamic range is thereby provided for the cell. We must make the additional hypothesis that the different fibres which innervate a particular inner hair cell have a graded set of thresholds. If we may assume that the afferent endings of some fibres are very sensitive and others are less sensitive, then the single hair cell, which occupies a small distance along the basilar membrane (which is important for frequency analysis) may have a much greater effective range of response by stimulating first one, then two, then three and finally four or more different afferent neurons all at once. We know from the study of single neurons that the range of intensity between the first increase in rate of discharge up to the maximal rate of discharge is only about 30 db, whereas the range between the thresholds of the most sensitive fibre unit and the least sensitive unit at a given characteristic frequency may be 50 or 60 db. The model that you have given, with this additional postulate of graded thresholds among the fibres that innervate a single internal hair cell, would account for this proposition. It is a very useful theoretical suggestion.

Batteau: I am very pleased by what Dr. Spoendlin has said, because I have looked at the need on mathematical grounds to construct certain kinds of time-dependent networks, and it seems that what he has described is consistent with this. The question of localization of sound I have looked at as one of a computer-steered array—the outer ear has the array which you compute in order to steer. This is not new in sonar or radar. The efferent innervation would be consistent with my hypothesis, and you could steer the array by patterns that influence reception.

Fex: I like Dr. Spoendlin's story very much too, but some warnings should be given. First, whether or not dendrites can respond to a stimulus with a full spike is not settled yet. Also, since fibres going to inner and outer hair cells are close to each other in the nest of fibres underneath the inner hair cells, how does one know that afferent fibres do not branch there, providing inner and outer hair cells with shared spiral ganglion cells for their afferent innervation? Furthermore, no work seems to have been done on the branching of auditory nerve fibres central to the habenula perforata. Such branching also could provide inner and outer hair cells with afferent innervation from shared, or common, spiral ganglion cells.

There is an incompatibility between Dr. Spoendlin's conclusions, those made by S. Iurato ([1964]. *Atti Soc. ital. Anat.*, **72**, 60) and findings by myself, which will be discussed in my paper (p. 169).

I would add that there is no strict physiological evidence that transduction takes place at the sensory pole of the hair cells. My findings have a bearing on this. By stimulating the crossed cochlear efferents in the brainstem, an evoked potential can be recorded in the cochlea. This potential shows a phase reversal when the recording electrode is passed through the organ of Corti into the scala media. And actually the main argument (Davis, H. [1965]. *Cold Spring Harb. Symp. quant. Biol.*, **30**, 181–190) that the cochlear microphonics are generated at the sensory pole of the hair cells is that the cochlear microphonics show a phase reversal when the recording electrode passes from the organ of Corti to the scala media. My point is that wherever there is a radical change of the state of the hair cell membrane, at the sensory pole or at the synaptic pole of the cell, this would be recorded as a potential that would show a phase reversal when you go from the organ of Corti into the scala media with your electrode.

Spoendlin: I studied the afferent dendrites in animals where the efferents have been eliminated by cutting the vestibular nerve (in cats one can do this without damage to the cochlea). I found branching fibres only at the level of the outer hair cells. The afferent fibres to the inner hair cells always appeared unbranched. The only branching fibres at the level of the inner spiral plexus were found in animals with intact efferent innervation. There is a lower and an upper bundle of inner spiral fibres. In the upper bundle we find vesicle-filled enlargements which are predominantly in contact with dendrites going to the inner hair cells. Dendrites to the outer hair cells tend to go off at a lower level, in the cat. (In the guinea pig things seem to be more complicated and less clearly organized.)

Fex: Have you seen branching central to the habenula?

Spoendlin: I have never seen this, nor have histologists found it. Branching in the myelinated portion of afferent cochlear neurons has not been observed, but the efferent fibres are well known to have ramifications in their myelinated portions.

Johnstone: You say, Dr. Spoendlin, that efferents in the upper spiral plexus make synaptic connexions with dendrites, but I was not sure from your pictures which way the signals will go. How do you know which is the forward direction?

Spoendlin: I assume that it goes from the efferent to the afferent dendrite. In the efferent fibre enlargements, which closely surround afferent dendrites, we frequently find agglomerations and condensations of synaptic vesicles adjacent to the dendrites, indicating synaptic activity going from the efferent fibre to the dendrite. According to E. G. Gray's recent findings ([1967]. *Science Jl*, **3**:**5**, 66–72), one should be able to differentiate inhibitory synapses from excitatory synapses on a morphological basis. He says that if the synaptic membrane thickenings are not very marked the synapse is inhibitory, hyperpolarizing, and if the membranes are much thickened, the synapse is excitatory. In the organ of Corti there are only very slight membrane thickenings at places of synaptic contact between efferent fibres or endings and afferent dendrites or hair cells, which would indicate inhibitory hyperpolarizing activity of these synapses.

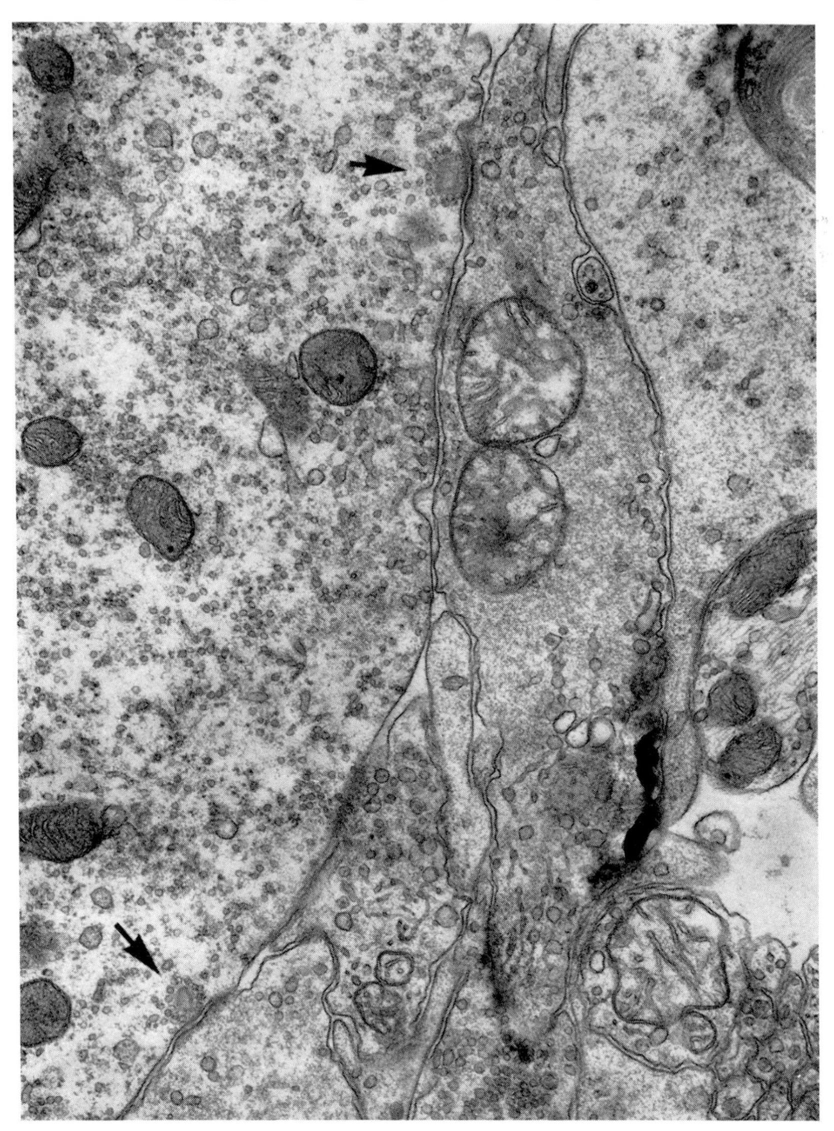

FIG. 1 (Engström). Inner hair cells (left) with two afferent endings and two synaptic bodies (arrows). Squirrel monkey.

Fex: A year or so ago we read (Uchizono, K. [1965]. *Nature, Lond.*, **207**, 642–643) that inhibitory synapses had vesicles which were ellipsodial, and the excitatory synapses had spherical ones.

Spoendlin: In the organ of Corti the synaptic vesicles usually appear to be spherical.

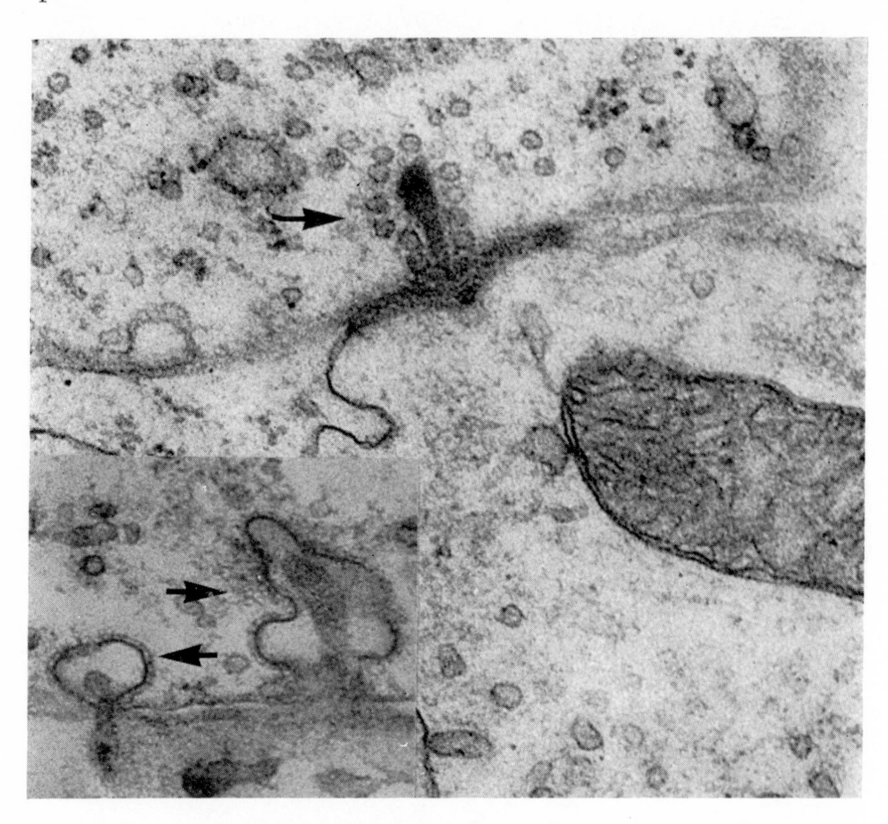

Fig. 2 (Engström). Inner hair cell (upwards) with a synaptic bar (arrow). In the lower left corner (*insert*), are synaptic invaginations (arrows) into an inner hair cell. Guinea pig.

Engström: My first question is: are there different kinds of afferent nerve fibres in the organ of Corti, and how can they be distinguished? There is clear evidence (cf. Kellerhals, B., Engström, H., and Ades, H. W. [1967]. *Acta oto-lar.*, Suppl. 226) that the spiral ganglion contains two distinct populations of ganglion cells. Ninety per cent of the ganglion cells are of one type, while 10 per cent are of another kind, in the guinea pig. They differ in volume and in their myelin sheaths. These two populations could be related to inner and outer hair cells respectively, but this seems not to be the case; there is more evidence that they are both related to afferent fibres. As there is evidence of different kinds of ganglion cells I want to repeat my question and rephrase it a little: are we correct in our terminology when we talk of afferent and efferent fibres as the only two kinds of fibres? Could there, as Lorente de Nó supposed, be several kinds of afferents and several kinds of efferents? In my material of guinea pigs and squirrel monkeys I have clearly observed that the course of the afferent and efferent fibres is different from what Dr. Spoendlin has found in the cat.

A second question is: are there at the bases of cochlear outer and inner hair cells different kinds of synaptic areas; in particular, are the synaptic invaginations and synaptic bars or bodies (Fig. 1) of one and the same kind? The invaginations (Fig. 2) often found specially at inner hair cells have no synaptic vesicles on the hair cell side. On the other hand they may contain an extension from an afferent ending, very much resembling the synaptic contacts seen in the eye. My question therefore is: are synaptic invaginations, synaptic bars and synaptic bodies all a kind of invagination into the sensory cell and are you familiar with different forms of synaptic bars?

Spoendlin: I think there are different forms of afferents. In the cat the synaptic zones between afferents and inner hair cells are quite different from the synaptic zones between afferents and outer hair cells. In the inner hair cell the afferent synapses are characterized by a very important thickening of the postsynaptic membrane, and a synaptic bar surrounded by vesicles. Of course, the section is frequently outside the synaptic bar or ribbon of a synapse and then you do not see them.

Engström: In the inner hair cells, the invaginations seem to increase with stimulation.

Spoendlin: The outer hair cell I find is quite different, exactly as Dr. Engström has shown. The afferent synapses at the outer hair cell show much less thickening of the postsynaptic membrane and very frequently so-called coated vesicles. They are found as invaginations of the presynaptic membrane or already detached from the surface. Such vesicles have also been seen in motor end-plates, where they are thought to bring the inactivated transmitter back to the cell, where it is going to be re-formed. Such a mechanism might also take place at the afferent synapses of the outer hair cells. It might then be that there are two types of synapses for outer and inner hair cells, associated with different types of dendrites. The invaginated coated vesicles are perhaps homologous to the synaptic bars in the inner hair cells. I have, however, never found transitional stages between coated vesicles and synaptic bars, which are not surrounded by a unit membrane, in contrast to the coated vesicles.

Engström: We found synaptic bars and round objects which we call synaptic bodies rather than synaptic bars, along the inner hair cells, and the invaginations also; and the question is whether they are of one kind. But a more important question is whether there are different afferents to outer hair cells and whether the spironeurons are the only afferents to the outer hair cells, or can there be more than one kind of afferent fibre to them?

Spoendlin: After section of the olivo-cochlear fibres at the level of the vestibular root or the vestibular nerve in eight cats we never found any remaining upper tunnel radial fibres. All the afferents therefore appeared to cross the tunnel at the bottom as basilar fibres which continue as outer spiral fibres. Hitherto I was not able to find direct afferent radial fibres to the outer hair cells in the animals in which the efferents have been eliminated.

Engström: In the squirrel monkey and in the guinea pig also the radiating fibres contain both afferent and efferent fibres. So there is a species difference which is rather considerable.

Spoendlin: I fully agree that in the guinea pig afferent and efferent fibres are not as clearly separated as in the cat. Both types of fibres cross the tunnel, together and free.

Bosher: Dr. Spoendlin suggested that the initial segment, the site at which the action potential is initiated, probably corresponds to the region of the habenula perforata where myelination commences. From a morphological viewpoint alone this must certainly be one possibility but there are some more general aspects which merit further consideration. The concept of electrotonus has been well established for other sensory end-organs and it now seems clear that the electrical changes are due essentially to alterations in the permeability of the nerve fibre membrane to sodium. In the Pacinian corpuscle, which has been extensively investigated, the initial segment is the first node of Ranvier, not the commencement of myelination (Diamond, J., Gray, J. A. B., and Inman, D. R. [1958]. *J. Physiol., Lond.*, **142**, 382–394). On the other hand there seems to me to be no reason why it should not be in the unmyelinated portion of the nerve. Presumably the action potential will be initiated at the site where the necessary rapid changes in axon membrane permeability to sodium and potassium can first take place, and we all know that action potentials do occur in unmyelinated nerves.

The crux of the matter is that electrotonus shows a relatively large decrement with distance and Dr. Spoendlin has shown us that there is a considerable length of nerve fibre between the habenula perforata and the hair cells, the equivalent of 90 pillar cells—about 0·6 mm. The functional significance is that if the initial segment is at the habenula, the outer hair cells will, in all probability, make much the same contribution to the action potential, because the decrement which will occur between one hair cell and another will be small in relation to the distance between the hair cells and the habenula perforata. But if the initial segment is nearer the hair cells, in the unmyelinated portion, the contribution made by the individual hair cells will be significantly different. The nearest hair cell will be much more important in producing the action potential in its nerve fibre than the furthest, as the electrotonus produced by its action will suffer substantially less decrement. In this respect it would be interesting to know firstly, if the spacing of the hair cells along the nerve fibre is regular and secondly, if any evidence of topographical organization is present; for example, whether the first outer hair cell supplied by the afferent fibre is situated in the first row? In the efferent system Dr. Spoendlin has told us that differences do exist between the rows, the efferent nerve endings being distributed more numerously to the first row than to the other two rows. It seems to me that the site of the initial segment is of fundamental importance in the hair cell–nerve fibre mechanism outlined by Dr. Spoendlin.

Spoendlin: I would not say that the initial segment could not be somewhere in the unmyelinated portion. However, we are inclined to associate some kind of structural change of the nerve fibre with the initial segment where such an important functional change is taking place. As far as we now know the afferent dendrites change two of their structural characteristics in the area of the habenula perforata: the fibre starts to be myelinated and most neuro-tubules in the axoplasm of the peripheral dendrite ramifications are replaced by neuro-filaments. No other major structural transitions can be found along the dendrites within the organ of Corti. Furthermore, it is probably convenient if the initiation of the all-or-none response of the action potential coincides with the beginning of saltatory conduction, which depends on the myelin sheaths.

Bosher: Saltatory conduction will certainly occur at the first node of Ranvier, which is central to the habenula perforata. It is difficult, I believe, to make accurate recordings from myelinated nerves but the results show that basically the same mechanism is involved in the production of the action potential as that which occurs in unmyelinated nerves, the differences between the two being merely speed of conduction and economy of energy expenditure (see Hodgkin, A. L. [1958]. *Proc. R. Soc. B*, **148**, 1–37; Katz, B. [1966]. *Nerve, Muscle and Synapse*. New York: McGraw-Hill). The principal difference between the action potential and electrotonus is simply that in the electrotonic portion of the nerve fibre the rapid changes in membrane permeability cannot take place.

COCHLEAR STRUCTURE AND HEARING IN MAN

GÖRAN BREDBERG

Öronkliniken, Göteborgs Universitet, Göteborg, Sweden

THERE have been several studies in human material in which cochlear structure and function have been compared. Pioneers in this field were Crowe and Guild and their co-workers at the Johns Hopkins Hospital (Guild *et al.*, 1931; Guild, 1932; Crowe, Guild and Polvogt, 1934; Guild, 1935), and since then many investigators have worked along similar lines. During the last decades, Schuknecht and his group in particular have made important contributions in this field (Schuknecht, 1955; 1964; Schuknecht, Benitez and Beckhuis, 1962; Schuknecht *et al.*, 1962; Schuknecht and Igarashi, 1964; Benitez, Schuknecht and Brandenburg, 1964; Igarashi, Schuknecht and Myers, 1964). From the investigations carried out we have gained a lot of information on the structural background of different forms of hearing loss. However, much more information is necessary if we are to draw more definite conclusions regarding cochlear function from studies on the histopathology of the human inner ear.

There is a lack of definite information regarding both the innervation pattern and the sensory cell population of the organ of Corti in man. Earlier investigations in this field give rather vague descriptions, and since Retzius (1884) made his counts of the sensory and supporting elements in the organ of Corti, no thorough, quantitative evaluations have been made of the number of sensory cells in the normal or the functionally impaired human cochlea.

There are several reasons for this lack of information, but the most important ones are presumably the technical difficulties inherent in the preparation. During this century, the method almost exclusively used until now has been the method of serial sectioning of decalcified specimens. There is no doubt that this one-sided use of a single technique, with the disadvantages inevitably associated with any technique, has impeded the progress of our knowledge of the relation of structure to function in the human cochlea.

To try to overcome some of these problems, the surface specimen technique developed by Engström, Ades and Hawkins (1962, 1964) and

Engström, Ades and Andersson (1966) was applied and modified for the study of the human cochlea (Bredberg, Engström and Ades, 1965; Bredberg, 1967). The method is described in detail in a more extensive publication now in preparation by the author (1968). A short description will, however, be given here.

As soon as possible after death (within 10 hours) the middle ear is explored via an endaural approach and the stapes is extracted and the round window membrane opened. Through the round window, the cochlea is perfused

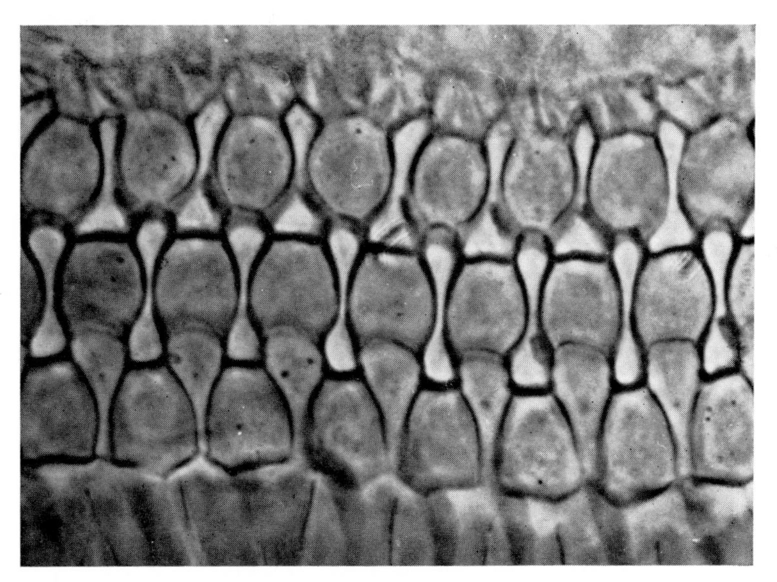

Fig. 1. Surface preparation of the organ of Corti of the rabbit, showing the very regular pattern of the outer hair cells.

by a veronal-buffered osmic acid solution, and the fixative is left until an autopsy is performed. The bone surrounding the membranous cochlea is removed with the aid of dental diamond burrs. The organ of Corti with its associated spiral osseous lamina is extracted in its entirety and studied by phase or light microscopy without any sectioning. In this way the surface of, or structures within, the organ of Corti are easily studied and the pattern of sensory and supporting cells can be recorded throughout the entire cochlea.

My total material consists of around 150 cochleas from foetuses, children and adults of varying ages. In 15 cases (25 cochleas) it has been possible to assess the hearing before death.

When this study began we found a very striking irregularity of the pattern of sensory cells in the organ of Corti of man as compared with other mammals (guinea pig, cat, rabbit, bat). The rabbit for instance, has a very regular, almost geometric pattern of the outer hair cells (Fig. 1). Only occasional and single sensory cells are out of this regular mosaic. Thus the place of every sensory cell can be predicted, so that, if there is degeneration of cells, this can be recorded with almost complete confidence.

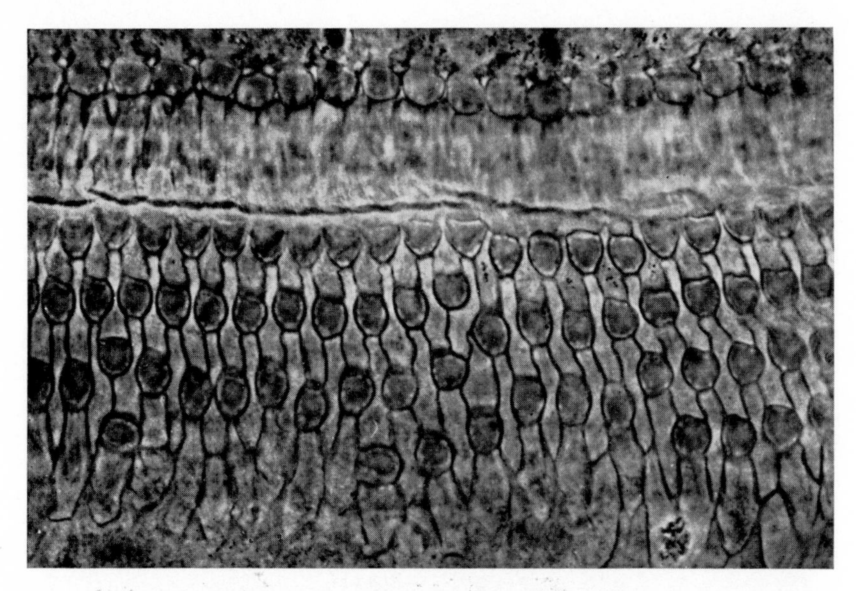

FIG. 2. Surface preparation of the organ of Corti from the apical coil of a squirrel monkey cochlea, showing one row of inner hair cells and three to four rows of outer hair cells. Note the irregularity of the pattern of the outer hair cells.

To try to explain the background for the irregularity of the pattern in man, the organ of Corti was studied during foetal development. Short descriptions of this development have been given previously (Bredberg, Engström and Ades, 1965; Bredberg, 1967).

At a foetal age of four months the pattern of sensory cells in the organ of Corti is regular, in agreement with the pattern in other mammals (guinea pig, cat, rabbit, bat). At a foetal age of five to six months the pattern shows a deviation from this regularity and becomes irregular, especially in the middle and apical coils. An interesting feature is that we have found a similar, but less pronounced, irregularity in the squirrel monkey (Fig. 2), and this is also described in the rhesus monkey (Johansson and Hawkins,

1967). What the significance of this irregular pattern in man and monkey may be, we do not know. In man a fourth and sometimes a fifth row of outer hair cells occurs. These cells are distributed irregularly and are found most often in the middle and apical coils. Although the "disturbed" cellular pattern may be found in the first, second or third row, the cells can almost always be referred to a particular row, so that the number of cells in each row is in agreement with that of the other rows. This makes it possible to give an accurate account of the ratio of intact and degenerated cells in the three first rows of outer hair cells. In the fourth and fifth row, however, the extent of degeneration must be calculated from an average obtained from normal cochleas.

What then is a "normal" cochlea from a histopathological point of view? In the literature there are frequent reports of a "normal" sensory cell population, or that no loss of sensory cells was noticeable in persons over 50 years of age. In the present material it was found, however, that in not one cochlea out of 120 cochleas studied was there a completely intact sensory cell population. Thus even in cochleas from young persons there were a certain number of degenerated cells. This finding makes it difficult to define the "normal" sensory cell population, unless one were to have a large amount of material from different age-groups so that all hearing losses but presbycusis could be excluded, and thus the "normal" loss of sensory cells for a specific age-group could be calculated.

However, if we regard as a "normal" cochlea one in which no degeneration of sensory cells has occurred, which means a cochlea with a full complement of sensory cells, then we have a basic value which can be used as a reference for the other material studied. Again, this study was directed toward the foetal cochlea, where, in the present investigation, we have not found any signs of sensory cell degeneration. This made the foetus extremely suitable as material for estimating the full complement of sensory cells, including a calculation of the fourth and fifth row of hair cells. Naturally, this is possible only if the foetus is studied at such a mature age that its cochlear sensory cells are all developed.

In the present study, 10 cochleas from human foetuses, at ages between 17 and 24 weeks, were used for this purpose. There was no sign of an increase in the number of sensory cells within this age range. Kolmer (1927), in man, and Ruben (1967), in the mouse, found that the mitoses of the sensory cells occurred at a very early age and that later there were no signs of further cell divisions.

In two cochleas every sensory cell was counted. This, however, is a very time-consuming procedure, so, in the other specimens, a sample of sensory cells has been counted at several levels of the cochlea and a graph representing the density of cells in those areas has been constructed (Fig. 3). The number of sensory cells has been calculated from this graph. The average number of outer hair cells was estimated as 13,400, with a range of variation between 11,200 and 16,000. The inner hair cells showed an average

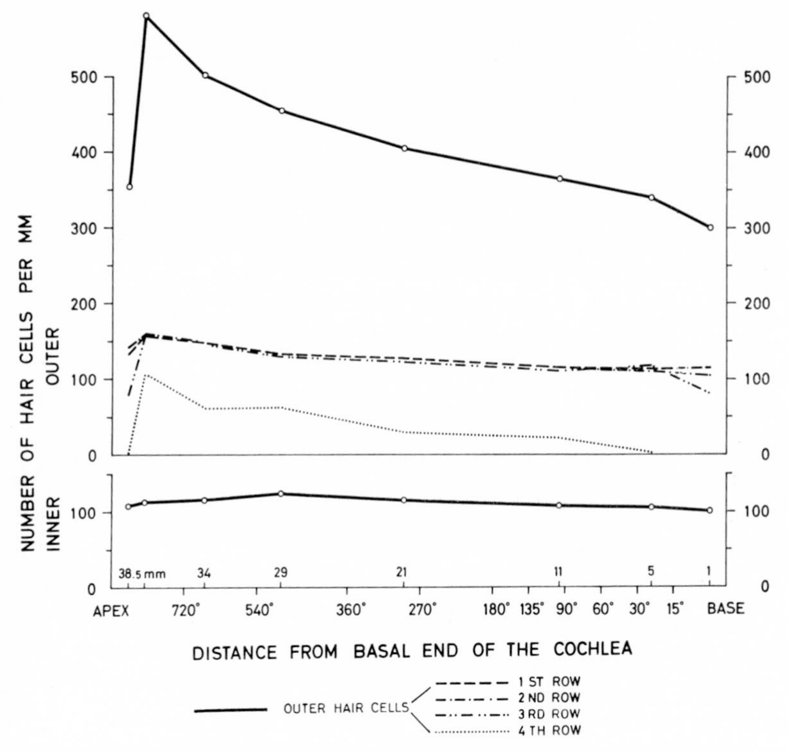

FIG. 3. Graph representing the density of hair cells at different levels of the cochlea of a human foetus (24 weeks). There is an increased density of the outer hair cells toward the apex, while the inner hair cells show a rather uniform distribution.

number of 3,400, with a range of variation between 2,800 and 4,400 cells. This variation in numbers may seem rather large, but when the numbers are plotted against the length of the organ of Corti, a certain correlation is found, so that the number of sensory cells is proportional to the length of the organ of Corti. The values almost always referred to in the literature are those of Retzius (1884). He found 12,000 outer and 3,500 inner hair cells. He had calculated the numbers with a technique similar to that used in the present investigation, but he had measured only one sample of cells from every cochlea. In addition he described a much lesser variation in length of the organ of Corti than is found in the present material, where the variation lies between 29·9 and 39·5 mm. with an average of 34·0 mm. Hardy (1938) used a method of reconstruction of sectioned material and calculated somewhat smaller values with an average length of 31·5 mm.

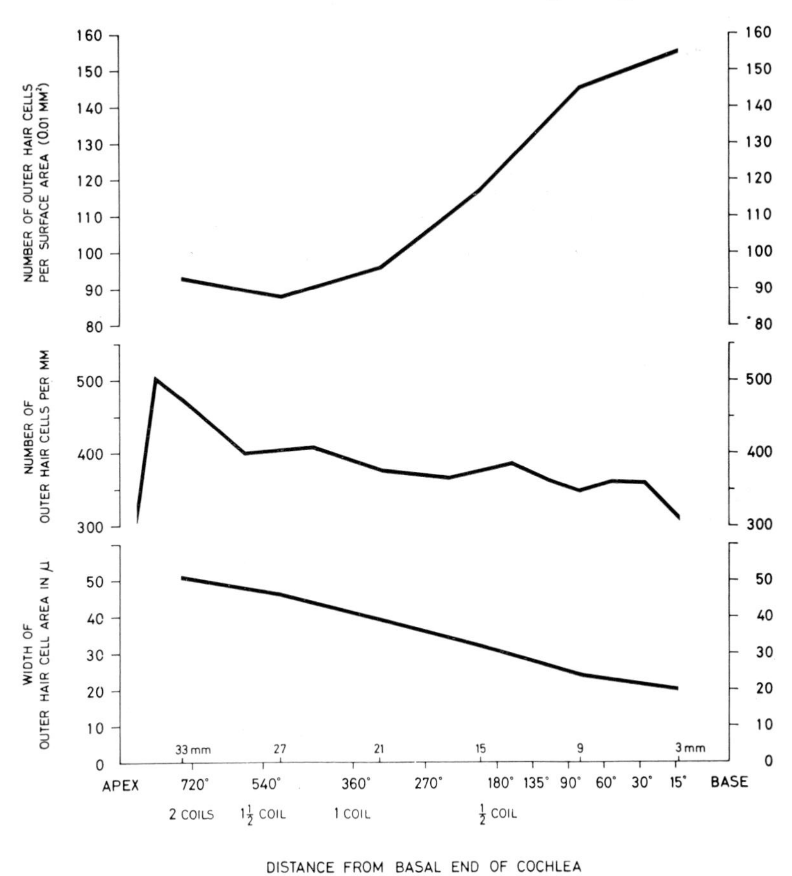

FIG. 4. Graph representing, from above, the density of outer hair cells per surface area, the density per unit length (cells per mm.), and the width of the outer hair cell area. See text for further explanation.

(25·3–35·5 mm.). When we see these variations in the number of sensory cells it is tempting to speculate that an individual with a larger number of cells might have better hearing than one with a smaller number.

If the number of outer hair cells is calculated per unit length of the organ of Corti a higher density of cells is found in the apical region than at the base. The value close to the apex is around 60 per cent higher than that at the base (Fig. 4). The inner hair cells, however, do not show the same increase in density toward the apex (Fig. 3).

The width of the outer hair cell area—that is, the distance between the first and the third or the fourth row of cells—increases toward the apex, so that when the density of hair cells per unit of surface area is calculated, it becomes apparent that there is a considerably higher density toward the base (Fig. 4). The functional significance of these two density parameters of the distribution of the outer hair cells has not been discussed previously. The width of the basilar membrane and the mass of the organ of Corti both increase toward the apex, factors of importance for the pattern of vibrations following tonal stimulation. The spreading out of the outer hair cells over a larger area is probably another factor of significance in the stimulation of the organ of Corti. The high linear density toward the apex is another point to be taken into consideration. In contrast to the variation in density of the outer hair cells from the base toward the apex is the rather uniform density of the inner hair cells.

As mentioned earlier, it has not been possible to find a full complement of cochlear sensory cells in any one of the cochleas studied. The cellular population of a five-year-old boy showed a slight loss in the third row of outer hair cells. An 18-year-old boy with normal pure-tone audiogram had a loss of about 10 per cent outer and 5 per cent inner hair cells, and a 34-year-old diabetic, with almost normal hearing tests, also showed a slight overall degeneration of sensory cells (Fig. 5, A and B).

The question then arises of at what age a degeneration of sensory cells occurs depending purely on increasing age. This is a very difficult question, because there are many factors that must be ruled out before an answer can be given. However, data in the present study indicate that a degeneration of sensory cells in the cochlea, dependent on ageing, might begin in the third decade of life (or even earlier). This agrees well with the facts that the threshold sensitivity for pure tones has its maximum during the second decade of life, and that subsequently there occurs a loss of nerve cells in the central nervous system (Brody, 1955).

To estimate the extent of degeneration of sensory cells attributed purely to age, we should need a great number of cases of different ages, which were all audiologically tested, so that other forms of hearing losses than presbycusis could be excluded. The present material does not permit such a calculation. However, in a group of 40 cochleas which have been studied in detail, certain types of degenerations can be discerned. This analysis is made purely according to the findings in the cochlea and not from an aetiological point of view.

(1) A diffusely spread loss of hair cells, just discussed, which begins early in life and becomes more pronounced with age. In unselected material it

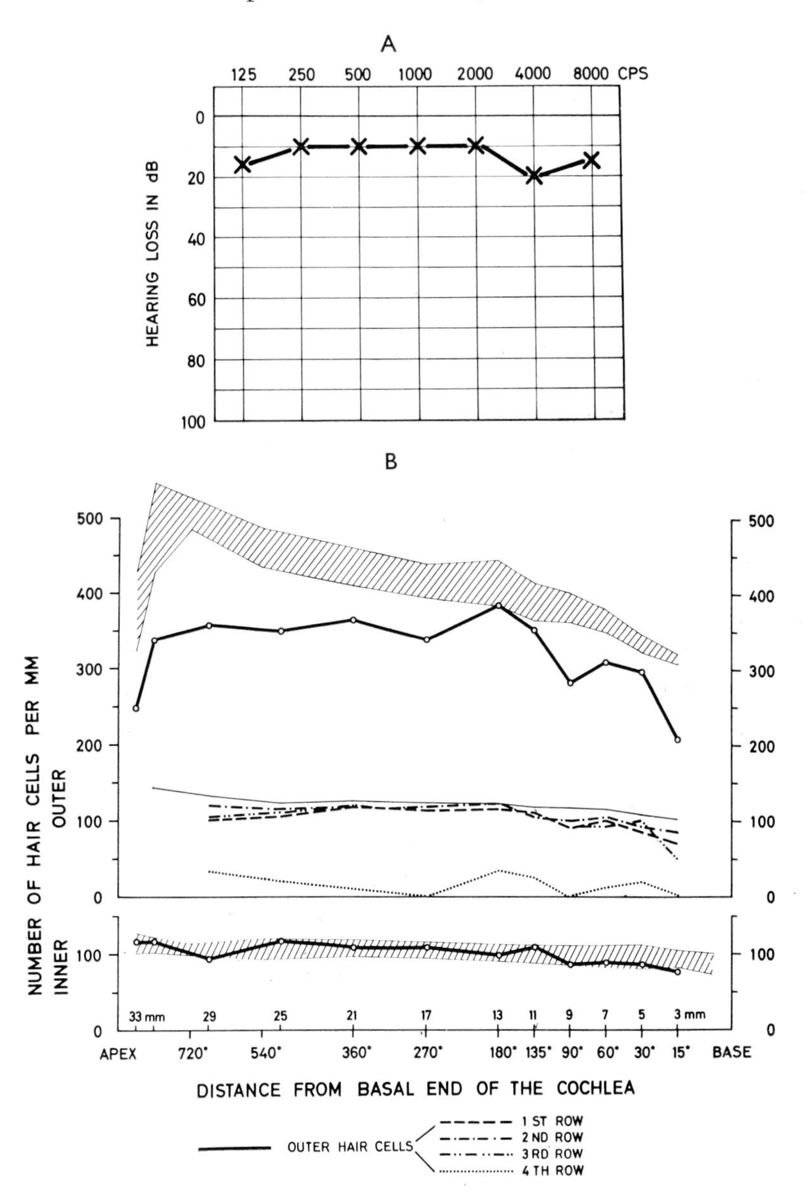

FIG. 5. Audiogram (A), and a graph (B) representing the cellular population in the organ of Corti of a 34-year-old diabetic. The hatched areas represent the full complement of sensory cells as obtained from foetal material. Note the diffusely spread loss of outer hair cells. The audiogram is normal for the age.

shows considerable variations, and when the degeneration is rather extensive, there is an associated loss of nerves in the spiral osseous lamina.

(2) A basal degeneration, which is almost always associated with the diffusely spread loss of sensory cells. It begins at the basal end, usually extending a few millimetres toward the apex, and consists of a loss of sensory cells and nerves in the spiral osseous lamina. It thus coincides with sensory presbycusis, as described by Schuknecht (1955, 1964). Although the degeneration is usually rather small, in a few cases a basal degeneration has been observed extending throughout the lower basal turn.

(3) An apical loss of sensory cells, which is not infrequently seen (Bredberg, 1967). Whether this type is related purely to the age factor or whether other factors may be present is not quite clear. Although degeneration of nerves and ganglion cells in the apical coil of the cochlea has been described earlier (Guild, 1935; Lindsay and von Schultess, 1958; Bernstein and Schuknecht, 1967), no mention of a specific degeneration of sensory cells in this region has been noted. This type of degeneration is illustrated in Fig. 6, A and B, which shows the cellular population and the pure-tone audiogram from a farmer aged 93 years. The audiogram reveals an overall loss of threshold sensitivity, most pronounced for the high tones. There is an extensive reduction of the cellular population of especially the outer hair cells. This degeneration is most pronounced in the apical coil, so that the main bulk of intact sensory cells is located in the basal half of the cochlea. The hearing for pure tones is better in the lower range of frequencies.

(4) Circumscribed degenerations, not associated with the basal or apical type, are frequently seen and may vary in extent and location. One kind of these is characterized by a loss of sensory cells, while the nerve fibres are intact. It is illustrated by the case of a 62-year-old mechanic who had been exposed to the noise in an engine-room of a ship for many years. The pure-tone audiogram showed an overall loss of threshold sensitivity of about 20 db. In addition to a diffusely spread and a slight apical degeneration, there was a 70 per cent loss of outer hair cells in the 21 mm. region; that is, one coil from the basal end of the cochlea. A similar loss, which in the graph of the cellular population appears as a dip, has been noted in several cases, who have all been males and some of whom are known to have been exposed to noise. Maximum damage has been found at levels varying from the 7 mm. to the 21 mm. regions. Thus this dip is often localized far above the region of the frequency localization for 4,000 cyc./sec., which according to von Békésy's (1949) experiments on the pattern of

vibrations in the cochlea, is located just above the 10 mm. region. A localized loss of sensory cells has been reported in the human organ of Corti following exposure to noise (Guild *et al.*, 1931; Igarashi, Schuknecht and Myers, 1964). The damage was found in the 7–13 mm. region.

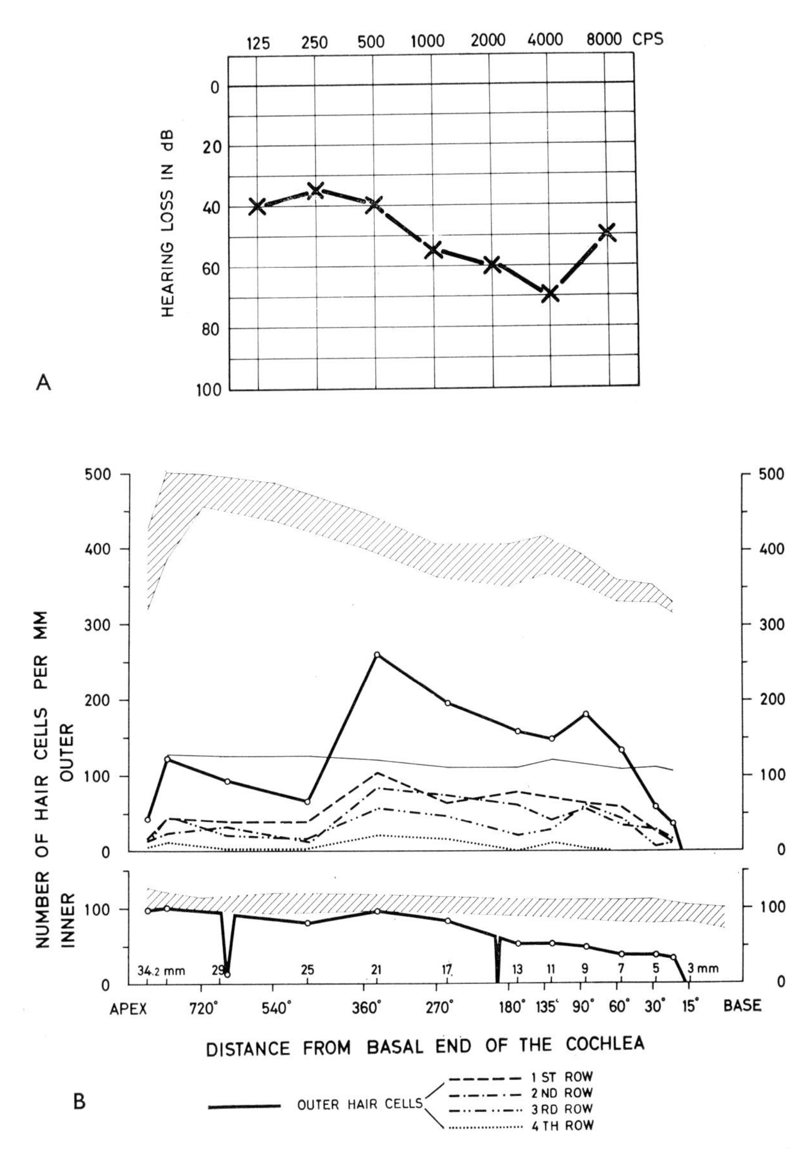

FIG. 6. Audiogram (A), and graph (B) of cellular population showing the apical type of degeneration, from a 93-year-old farmer. In addition to this loss of sensory cells there is a diffusely spread and a basal degeneration.

The cases reported in the present material were mostly exposed to an industrial type of noise and they did not all show the characteristic noise-induced (4,000 cyc./sec.) dip in the audiogram. It is still possible, however, that the loss of sensory cells located above the 10–12 mm. region was induced by noise.

When the noise damage is extensive there may be degeneration of the organ of Corti as well as of the nerves in the spiral osseous lamina in a circumscribed region. This kind of degeneration was found in the cochlea of a man aged 71 years, who had been exposed to high-intensity noise from

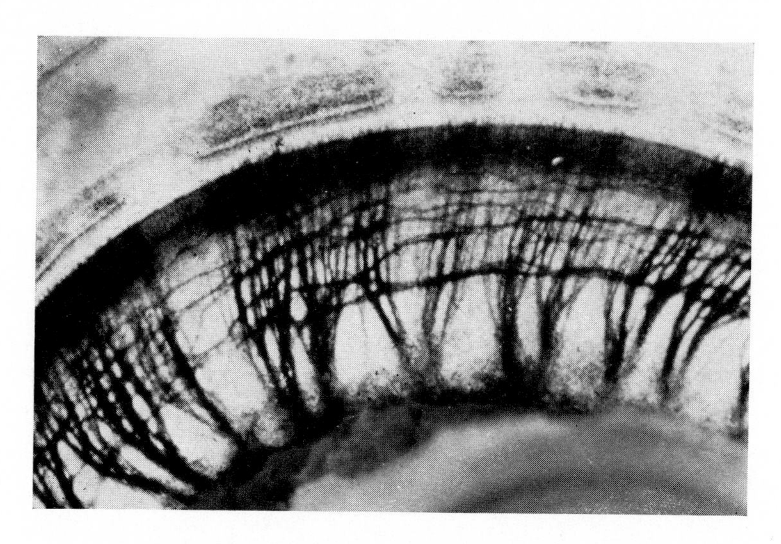

FIG. 7. Spiral osseous lamina and associated organ of Corti, showing four areas of circumscribed degenerations of both the organ of Corti and the nerves in the spiral osseous lamina.

a hammer saw for many years. The audiogram showed a pronounced dip at the frequencies 2,000 cyc./sec. to 4,000 cyc./sec. and the histopathological finding in one ear was a complete degeneration of both the nerves and the organ of Corti, in a region 4 mm. in length, located 10–14 mm. from the basal end of the cochlea. Thus this cochlea contains two groups of sensory cells completely separated by a distance of 4 mm. From a functional point of view it would have been extremely interesting to have made a thorough audiological test before death.

The degeneration of the organ of Corti and of associated nerves in the spiral osseous lamina may show a patchy appearance (Guild, 1932; Crowe, Guild and Polvogt, 1934; Bredberg, 1967), so that there are multiple areas of degeneration separated by fairly intact regions (Fig. 7). This appearance is mainly seen in the basal coil. In general a considerable area of completely degenerated nerves is associated with a corresponding complete atrophy of the organ of Corti. In one case in the present material, however, it was found that in spite of a considerable degeneration of nerve fibres in several separate areas, no corresponding damage was evident among the cells of the organ of Corti. This indicates that different forms of damage may differently affect the organ of Corti and its nerve supply.

From a functional point of view there are many problems relating to the histopathological findings. Some of these have already been mentioned, but many more remain. For instance, how do the circumscribed degenerations, multiple or single, influence the different parameters of hearing? That a single area of pronounced degeneration in the organ of Corti in the 10–14 mm. region may correspond to a 4,000 cyc./sec. dip in the pure-tone audiogram is quite clear, but what is the functional significance of degenerations located higher up in the cochlea, or of multiple degenerations? Many similar questions could be asked, and we have dealt only with the peripheral part of the auditory analysing system. The higher auditory pathways and centres have not been taken into consideration.

Many of these problems might be solved by experiments in animals, but as some of the more complex hearing tests are impossible, or very difficult, to carry out in animals, it seems quite clear that studies in man can contribute valuable information on the relation of structure to function.

SUMMARY

The full complement of sensory cells in the organ of Corti has been estimated in the human foetal cochlea. The outer hair cells average 13,400 in number (11,200–16,000), and the inner hair cells 3,400 (2,800–4,400). The number of sensory cells is proportional to the length of the organ of Corti, which shows a range of variation between 29·9 and 39·5 mm., with an average of 34·0 mm.

The linear distribution of the outer hair cells shows an increase in density per millimetre from base to apex of about 60 per cent. The inner hair cells, however, have a rather uniform distribution. The density of the outer hair cells per unit surface area has its maximum at the base, with a decrease to almost half the density at the apex.

From a histopathological point of view, certain types of degenerations occur in the adult organ of Corti:

(1) A diffusely spread loss of hair cells, which has been found in every cochlea studied. This degeneration begins in the third decade (or even earlier) and becomes more pronounced with age.

(2) A basal degeneration, which is generally associated with the diffusely spread loss of sensory cells. It usually consists of a loss of cells and nerves in the most basal few millimetres, but sometimes the degeneration may extend further up in the basal coil.

(3) An apical loss of sensory cells, which is seen rather frequently and becomes more pronounced with increasing age.

(4) Circumscribed degenerations, which are frequently found and may vary in extent and location. At least one type of these degenerations may be associated with noise damage, and is manifested as a loss of sensory cells only, or in addition to a loss of nerves in the spiral osseous lamina.

The functional significance of some of these findings is discussed.

REFERENCES

Békésy, G. von (1949). *J. acoust. Soc. Am.*, **21,** 245–254.

Benitez, J. T., Schuknecht, H. F., and Brandenburg, J. H. (1964). *Archs Otolar.*, **75,** 192–197.

Bernstein, J. M., and Schuknecht, H. F. (1967). *J. Lar. Otol.*, **81,** 1–26.

Bredberg, G. (1967). *J. Lar. Otol.*, **81,** 739–758.

Bredberg, G. (1968). *Acta oto-lar.*, Suppl. In preparation.

Bredberg, G., Engström, H., and Ades, H. W. (1965). *Archs Otolar.*, **82,** 462–469.

Brody, H. (1955). *J. comp. Neurol.*, **102,** 511-556.

Crowe, S. J., Guild, S. R., and Polvogt, L. M. (1934). *Bull. Johns Hopkins Hosp.*, **54,** 315–379.

Engström, H., Ades, H. W., and Andersson, A. (1966). *Structural Pattern of the Organ of Corti.* Stockholm: Almqvist and Wiksell.

Engström, H., Ades, H W., Hawkins, J. E., Jr. (1962). *J. acoust. Soc. Am.*, **34,** 1356–1363.

Engström, H., Ades, H. W., and Hawkins, J. E., Jr. (1964). *Acta oto-lar.*, Suppl. **188,** 92–99.

Guild, S. R. (1932). *Acta oto-lar.*, **17,** 207–249.

Guild, S. R. (1935). *Ann. Otol. Rhinol. Lar.*, **44,** 738–753.

Guild, S. R., Crowe, S. J., Bunch, C. C., and Polvogt, L. M. (1931). *Acta oto-lar.*, **15,** 269–308.

Hardy, M. (1938). *Am. J. Anat.*, **62,** 291–311.

Igarashi, M., Schuknecht, H. F., and Myers, E. N. (1964). *J. Lar. Otol.*, **78,** 115–123.

Johansson, L. G., and Hawkins, J. E., Jr. (1967). *Archs Otolar.*, **85,** 599–613.

Kolmer, W. (1927). In *Handbuch der Mikroskopischen Anatomie des Menschen*, III/1, ed. Möllendorff, W. V. Berlin: Springer.

Lindsay, J. R., and Schultess, G. von (1958). *Acta oto-lar.*, **49,** 315–324.

Retzius, G. (1884). *Das Gehörorgan der Wirbelthiere*, vol. II: *Das Gehörorgan der Reptilien, der Vögel und der Säugethiere.* Stockholm: Samson and Wallin.

RUBEN, J. (1967). *Acta oto-lar.*, Suppl. 220, 1–44.
SCHUKNECHT, H. F. (1955). *Laryngoscope, St Louis*, **65**, 402–419.
SCHUKNECHT, H. F. (1964). *Archs Otolar.*, **80**, 369–382.
SCHUKNECHT, H. F., BENITEZ, J. T., and BECKHUIS, J. (1962). *Ann. Otol. Rhinol. Lar.*, **71**, 1039–1053.
SCHUKNECHT, H. F., BENITEZ, J. T., BECKHUIS, J., IGARASHI, M., SINGLETON, G., and RÜEDI, L. (1962). *Laryngoscope, St Louis*, **72**, 1142–1157.
SCHUKNECHT, H. F., and IGARASHI, M. (1964). *Trans. Am. Acad. Ophthal. Oto-lar.*, **68**, 222–242.

DISCUSSION

Ade Pye: Dr. Bredberg, how long can you leave the cochlea in the fixative? And do you store it in alcohol? Secondly, have you measured the length of the hairs on the outer hair cells: I wonder if there might be some variation in their length? In some animals (for example, the guinea pig) the outer hair cells seen in surface view change in shape from the base to the apex of the cochlea, and I wonder if there might also be a change in the length of the hairs.

Bredberg: We prefer to fix the specimens a rather short time, about 1–3 hours, but because of reasons of organization this cannot often be done and normally we have to fix our specimens for 12–24 hours. A few specimens have, however, been left for some days in fixative, and we have still been able to evaluate the cellular populations. The specimens are stored in 70 per cent alcohol; the best results are, however, obtained if dissection and mounting are done directly after fixation, without storage in alcohol.

The length of the hairs on the outer hair cells varies from base to apex in such a way that the hairs are longer toward apex. You can get a good impression of this in phase-contrast microscopy, but measurements can be more accurately made in electron microscopy, as done by R. S. Kimura ([1966]. *Acta oto-lar.*, **66**, 55–72).

Hallpike: Dr. Bredberg, you said that the counts of every cell in the cochlea took a considerable time and that you therefore went over to a sampling procedure. About how long does this take?

Bredberg: The time I need for the sampling procedure is about 4–6 hours. The complete study of one cochlea, in the way and to the extent described, takes 15–20 hours. This time includes the fixation procedure (but not time of fixation), microdissection, photography, recording on a cochleogram, counting and the construction of a graph of the sensory cell population.

Hallpike: I ask because this procedure may be an answer to one of the difficulties of conventional serial sectioning of the cochlea in celloidin, which has its advantages but is very lengthy. By your method, you examine a surface view of the organ of Corti, and get an idea of the sensory cells; but can you in the same sample see much of structures other than the organ of Corti—the stria vascularis or the limbus cells or displacements of the Reissner's membrane?

Bredberg: Provided you decide what to study in advance, so that you can do your preparation accordingly, almost any structure in the cochlea can be studied, either by direct observation in the stereo microscope, or as surface specimens, or as sections after embedding and sectioning portions of the cochlea. This is true of, for instance, the stria vascularis and the spiral ligament, the Reissner's membrane, the limbus and the spiral ganglion. The position of the Reissner's membrane is easily studied by direct observation in the stereo microscope.

Fex: If this turns out to be an important way of studying the cochlea, it might be possible to computerize some kind of scanning system, which would shorten the time taken considerably.

Bredberg: I am convinced that this is possible although some factors might present problems, as for instance the focusing of the microscope.

Fex: Dr. Bredberg, you spoke of the difficulty of deciding what "normal" material would be. I recall a paper some years ago on material in the Sudan (Rosen, S., Bergman, M., Plester, D., El-Mofty, A., and Satti, M. H. [1962]. *Ann. Otol. Rhinol. Lar.*, **71**, 727–743). It was found that men of 60 years had very nearly the same good hearing as young people of twenty—in short, perfect hearing. The investigators also found that these 60-year-old men did not show the change of blood pressure that usually occurs with age. So it is indeed difficult to really say what "normal" material is.

Bredberg: I have by-passed the problem by going to the foetal cochlea and looking at the full complement of sensory cells. But there are so many questions, like the ones raised by the findings in the Sudanese population, that must be answered before we can tell what presbycusis really is. Presbycusis in one area of the world might be different from that in another area.

Tumarkin: I had the same problem of what is meant by "normal" when I was associated with a ministerial enquiry into the effects of noise on industrial workers. I had to fight a losing battle with the physicists who were determined that an ear was normal if I said that the tympanic membrane looked normal, despite the fact that the audiogram might show anything up to 15 db of loss at any frequency and an average of 7·5 db loss overall. I have always believed that middle-ear pathology can involve the inner ear as well; I showed years ago that a vast proportion of our children, and certainly far more among the poorer classes, had some degree of middle-ear inflammation at some time quite early in their life, which caused a reduction of the pneumatization of the mastoid bone (Tumarkin, A. [1957]. *J. Lar. Otol.*, **71**, 65, 137, 211). I am not therefore surprised that Dr. Bredberg has found that degeneration of the inner ear is widespread. It would be interesting to ascertain to what extent his inner-ear changes are related to middle-ear pathology.

The other point I want to raise is that R. Plomp and D. W. Gravendeel ([1960]. *Acta oto-lar.*, **51**, 548–560) have described a low-tone perceptive deafness, and it is well known that in some cases of Ménière's syndrome there is a low-tone perceptive deafness which for a time is reversible. It improves when the patient's symptoms settle down, but eventually tends to become permanent. Dr. Bredberg, have you had a chance to study cochleas from patients with Ménière's disease and if so, have you found any association with this rather unusual degeneration which you found in some cases at the apex of the cochlea?

Bredberg: I have not yet had the opportunity to study cases of Ménière's disease. May I add that this apical degeneration is not unusual; it is in fact rather common, although the extent of the damage varies.

Rose: Did you make any counts of the cochlear fibres?

Bredberg: I have not made fibre counts.

Whitfield: You described considerable variations between different subjects. Have you done comparisons between the two ears, and would the kind of variation you showed be much less between the two ears of a given subject?

Bredberg: Yes, I have done this and have found the differences between the two ears of one subject to be smaller than between the ears of different individuals. M. Hardy has studied the length of the organ of Corti and she described corresponding findings ([1938]. *Am. J. Anat.*, **62**, 291–311).

Hood: We have always assumed that the outer hair cells are responsible for our perception of low-intensity sound, and the inner hair cells for high-intensity sound, and Dr. Bredberg's work seems to bear this out. Various figures have been given of the contribution made by the outer hair cells: Lurie (Stevens, S. S., Davis, H., and Lurie, M. H. [1935]. *J. gen. Psychol.*, **13**, 297–315) was one of the first, and he claimed that loss of outer hair cells raised the threshold by about 40 db. His technique was certainly not so refined as yours, Dr. Bredberg; I wonder if you could give us some definite value now on the basis of your observations?

Bredberg: I have not made such a study yet, but my present investigation is meant to form a background for further studies of this kind.

Bosher: One point which interests me about this loss of cells is: how is the gap closed? The hair cells form a barrier between the endolymph, which has a very high potassium content, and the fluid in the tissue spaces of the organ of Corti, which almost certainly has a very high sodium content. Normally, there are also so-called ion-tight junctions between the hair cells and the adjacent phalangeal processes of Deiters' cells preventing diffusion of potassium into the tissue spaces around the hair cells. When one of the hair cells is lost, can this diffusion be prevented? Because if not, surely that region of the organ of Corti will cease to function.

Bredberg: When a sensory cell disappears the phalangeal processes of the adjacent Deiters' cell occupy the free upper surface completely, forming what we call a phalangeal scar.

Bosher: But it must take some time for these cells to come together and to form ion-tight junctions. Do you ever find gaps where there is no cellular continuity between one phalangeal process and the next?

Bredberg: I have not seen gaps in human material, but this is not the best material for this kind of study.

Engström: We have studied this problem in animals where the cochlea has been damaged by ototoxic antibiotics and by noise. The cuticular plate disintegrates and is replaced gradually by the outgrowth of the phalanxes of the Deiters' cells. It is easy to recognize the collapsing cells. There is of course a possibility that for a short while, leakage could occur and this might cause dysfunction in the ear. The different stages of this disintegration can be seen in our book (Engström, H., Ades, H. W., and Andersson, A. [1966]. *Structural Pattern of the Organ of Corti.* Stockholm: Almqvist and Wiksell).

Fex: There might well be a potassium leak, and it would be interesting to put a microelectrode through and see what the level of the scala media potential is, in the scar areas, if it is possible to do this.

Whitfield: Schuknecht actually pushed a needle through the basilar membrane, and made a hole, and afterwards obtained behavioural discriminations from his animals (Schuknecht, H. F. [1960]. In *Neural Mechanisms of the Auditory and Vestibular Systems*, chapter 6, ed. Rasmussen, G. L., and Windle, W. F. Springfield: Thomas).

Tumarkin: Schuknecht has said that despite the fact that he perforated the Reissner's membrane and allowed the endolymph to mix with perilymph, he found surprisingly little functional loss.

Whitfield: Is there some mechanism which removes the ions as they leak through?

Lowenstein: There are the sodium-absorbing cells that do this normally; there may be a stepping-up of the pumping mechanism (Dohlman, G. F. [1967]. *Ciba Fdn Symp. Myotatic, Kinesthetic and Vestibular Mechanisms*, pp. 138–143. London: Churchill).

Davis: We may be confusing two issues here; one is the endolymph going where perilymph (or "Cortilymph") should be: the other is perilymph getting into the scala media. These might have quite different effects, the former much more serious than the latter.

Lowenstein: The endolymph is then the dangerous fluid, as far as the synaptic junctions are concerned.

Davis: I would expect it to be.

SECTION III

AFFERENTS AND EFFERENTS OF THE AUDITORY NERVE

PATTERNS OF ACTIVITY IN SINGLE AUDITORY NERVE FIBRES OF THE SQUIRREL MONKEY

J. E. Rose, J. F. Brugge, D. J. Anderson and J. E. Hind
Laboratory of Neurophysiology, University of Wisconsin, Madison, Wisconsin

We wish to consider here how information concerning the frequency of a sinusoidal stimulus is transmitted from the cochlea to the cochlear nuclear complex in the brainstem.

All the findings to be considered are based on analysis of discharges recorded by sodium chloride filled micropipettes which were inserted into the auditory nerve of a deeply anaesthetized squirrel monkey, under direct visual control.

It has been noted by many workers in the past (Galambos and Davis, 1943; Tasaki, 1954; Katsuki, Suga and Kanno, 1962; Rupert, Moushegian and Galambos, 1963; Nomoto, Suga and Katsuki, 1964; Kiang *et al.*, 1965; Rose *et al.*, 1966) that a discharge of an auditory neuron may be time-locked to the stimulus cycle. While Wever (1949) utilized observations of that kind to develop his well-known volley principle, little systematic work has been done on this subject.

It will be our assumption that, at least for low frequencies, the transfer of frequency information is accomplished by means of a *time code* which consists of discharges whose spacing is determined by the intervals between the successive deflections of the cochlear partition. Deflections in one direction only are, we believe, detected; not every one needs to be effective in eliciting a discharge and the deflections must, of course, be of sufficient amplitude to reach the threshold firing level. This assumption implies that for a single pure tone of low frequency (or certain combinations of tonal stimuli) such a code may be termed a *period time code*, since the discharges can be expected to group around integral multiples of the period(s). The utility of such a time code must, of course, remain a presumption unless it can be shown that the cadence of incoming discharges into the cochlear complex is indeed instrumental in the evaluation of stimulus frequency. We shall summarize here the experimental findings which lead us to conclude that a period code does in fact exist when a single pure tone or certain tonal combinations are presented.

(1) *Distributions of interspike intervals*

The invariable finding in our experiments which immediately suggests the existence of a period code is that every fibre which responds to a low frequency discharges spikes at intervals which tend to group around integral multiples of the stimulus period. Figs. 1 and 2 illustrate such findings

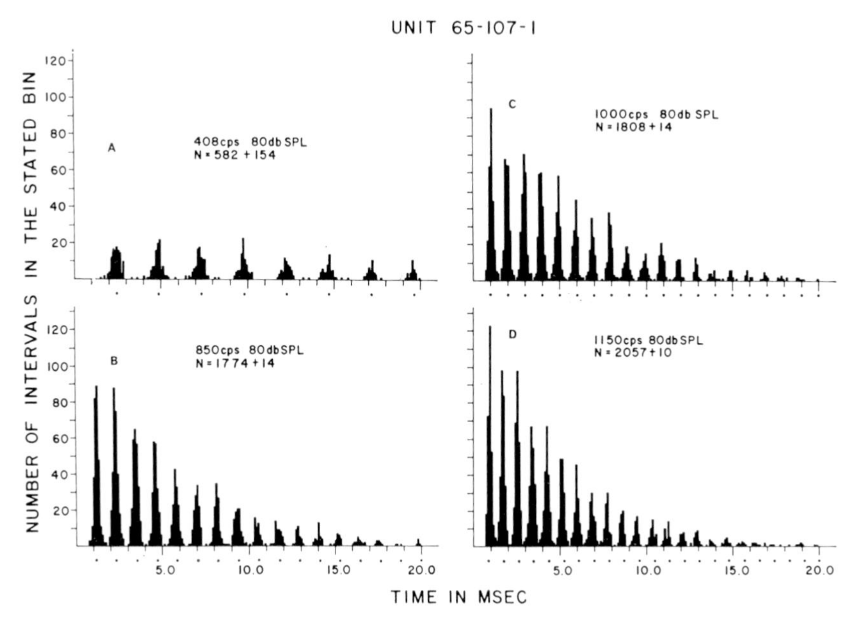

FIG. 1. Neuron 65-107-1. Periodic distributions of interspike intervals when pure tones of different frequencies activated the neuron.

Stimulus frequency in cycles per second (cps) is indicated in each graph. Intensity of all stimuli: 80 db SPL. In all figures the sound pressure level (SPL) is given in decibels (db) re 0·0002 dyne/cm². Tone duration: 1 second. Responses to 10 stimuli constitute the sample upon which each histogram is based. *Abscissa:* time in milliseconds; each bin = 100 µsec. Dots below abscissa indicate integral values of the period for each frequency employed. *Ordinate:* number of interspike intervals in the bin. N = number of interspike intervals in the sample. N is given as two numbers. The first indicates the number of plotted intervals; the second is the number of intervals whose values exceeded the maximal value of the abscissa.

for neuron 65-107-1, Fig. 3 shows its response area. We have published similar examples of responses of other fibres elsewhere (Rose *et al.*, 1967). The number of spikes discharged is—at a given intensity level—a function of the position of the stimulus within the response area of the unit and is thus highest for its best frequency. By contrast, the cadence of discharges is quite independent of the best frequency and is governed solely by the cadence of the stimulus cycles. The discharges occur preferentially during a restricted segment of the stimulus cycle and are thus phase-locked to the cycles of the applied frequency. We term such a discharge pattern a *phase-locked response*.

It is of interest to examine the shortest interspike intervals when the period of the stimulus is shorter than the refractory period of the fibre. If we assume that auditory nerve fibres do not differ in any substantial way from other mammalian nerve fibres, the problem created by refractoriness

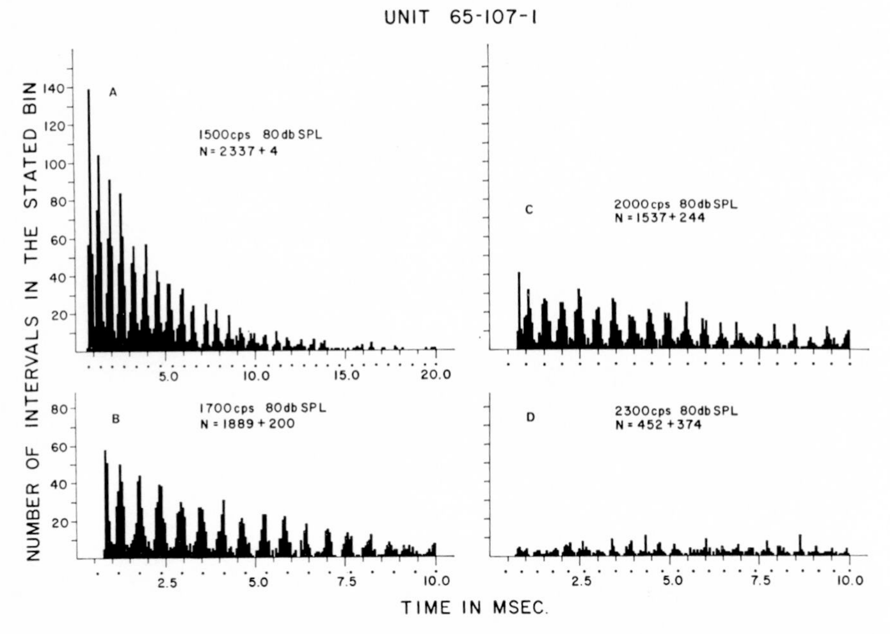

FIG. 2. Neuron 65-107-1. Periodic distributions of interspike intervals when pure tones of different frequencies activated the neuron.

Abscissa: time in milliseconds; each bin = 100 μsec. in A; 50 μsec. in B to D. Other legends as Fig. 1.

can be expected to arise for all frequencies higher than some 1,100 cyc./sec. This can be expected because the absolutely refractory period is usually about 500 to 700 μsec. and the relative refractoriness is probably some few hundred microseconds longer. It is evident from Fig. 2 what actually happens. Consider, for example, Fig. 2C, where the frequency employed was 2,000 cyc./sec. and thus the period was 500 μsec. There are no intervals grouped around this value. The first modal peak occurs around some 750 μsec.; the second peak lags slightly behind the second integral multiple of the period, while the third and subsequent modal values occur rather precisely at integral multiples of 500 μsec.

In view of such findings, which are typical, it is necessary to introduce a restriction concerning the relation of the stimulus period to the timing of the discharges. The more correct statement is that if a low frequency is employed the unit discharges at intervals which tend to group around integral multiples of the stimulus period provided that this value exceeds the refractory period of the fibre.

This statement holds true for every unit in our material and for any effective low frequency regardless of its intensity. In fact, if a unit discharges spontaneously it is a common observation, for successive increases

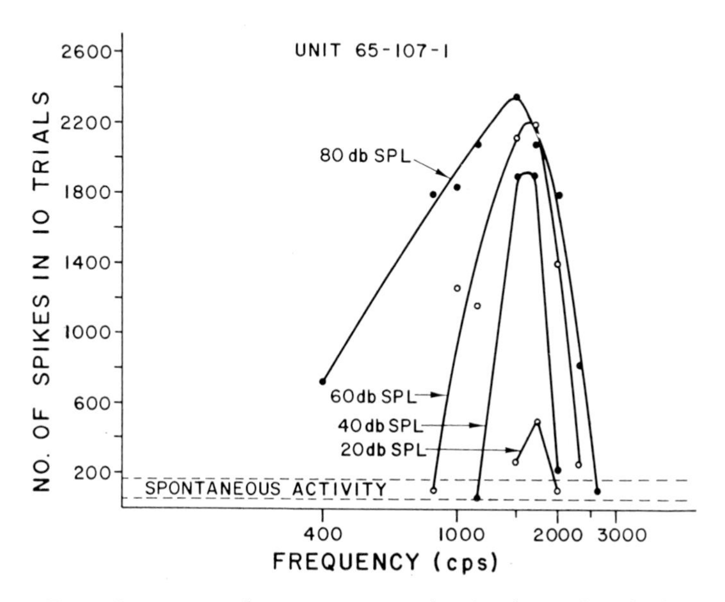

FIG. 3. Response area for neuron 65-107-1 showing the number of spikes produced at different intensities of the tone.

Abscissa: frequency in cycles per second. *Ordinate:* number of spikes in 10 trials. Duration of each trial: 1 second. Dashed lines indicate the range of spontaneous activity.

in stimulus intensity, that the discharges first become phase-locked before their rate of occurrence changes substantially from that which is spontaneous. The only exception may occur when one attempts to drive a unit which has a substantially higher best frequency, by a very low tone at high intensity. In such a situation, responses may be grouped also around the period of a harmonic of the frequency used. Fig. 4 illustrates such a case. Here the tone employed was 195 cyc./sec. The discharges, however, group not only around multiples of the period of the frequency used but also around integral multiples of the period of its third harmonic. Since such responses typically disappear as soon as the intensity of the fundamental is lowered it seems reasonable to attribute them to harmonics produced either in the ear or in the stimulus generator.

Our sample consists by now of some 250 fibres. While this figure is quite small, the consistency of the findings suggests not only that informa-

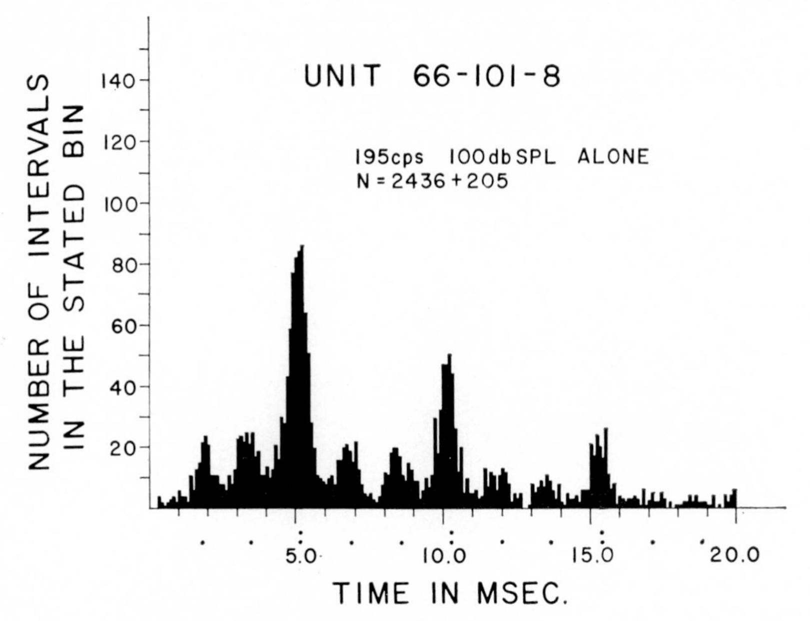

FIG. 4. Interspike interval distributions for neuron 66-101-8 when a tone of 195 cyc./sec. was presented at 100 db SPL.

Best frequency of the neuron about 1,000 cyc./sec. Note that the interspike intervals group around the integral multiples of 5,128 μsec. (period of 195 cyc./sec.) as well as around the integral multiples of 1,709 μsec. (period of 585 cyc./sec.). The histogram is based on three trials, each 8 seconds in duration. *Abscissa:* time in milliseconds; each bin = 100 μsec. *Ordinate:* number of intervals in each bin. Dots below abscissa indicate integral values of the periods of 195 cyc./sec. and 585 cyc./sec. tones. Other legends as in Fig. 1.

tion regarding low frequency is transmitted in a time code but that there need not be another mechanism by which this transfer is accomplished.

Two observations are common in the phase-locked responses to very low frequencies (Fig. 5). The first is that in addition to distributions grouped normally around integral multiples of the period there appears another skewed distribution at intervals close to one millisecond, indicating that at low frequencies a unit often discharges more than once during a stimulus cycle. The second observation of special interest is that not every cycle is effective even if the stimulus is fairly intense and its period is longer than 10 msec. The implication is that the refractory period of the fibre may be a limiting factor but is not a determinant of whether or not any given stimulus cycle will be effective in eliciting a discharge.

(2) *The periodogram and the coefficient of synchronization*

Since any effective low frequency produces a phase-locked response, it

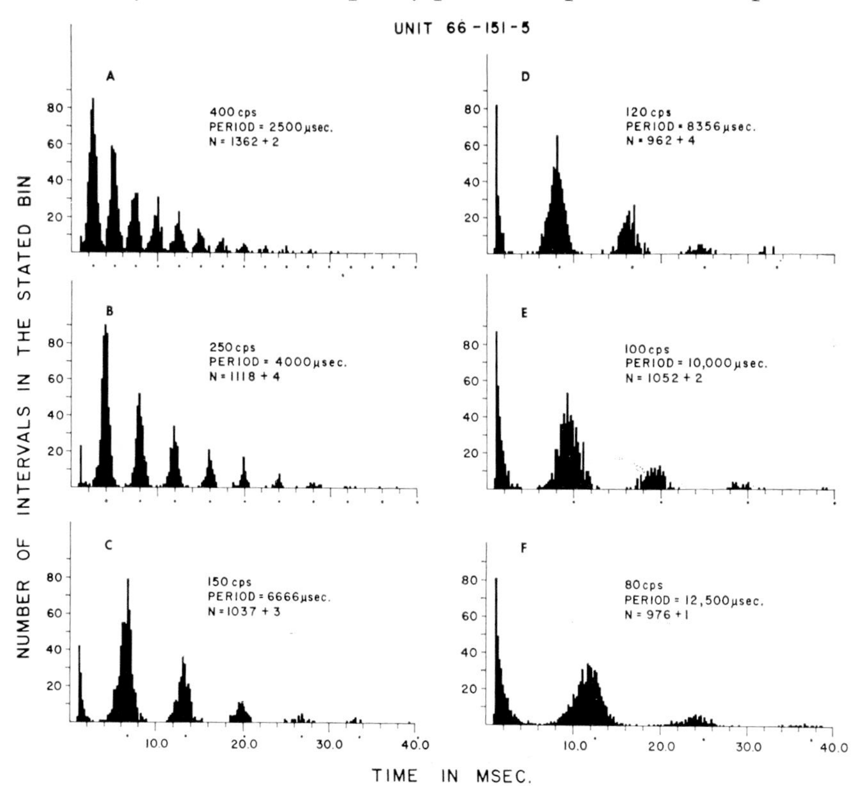

FIG. 5. Neuron 66-151-5. Periodic distributions of interspike intervals when tones of different frequencies activated the neuron.

Period obtained by the LINC. Frequency computed from the period. Each histogram is based on responses to a tone of 10 seconds duration. Intensity of all tones: 100 db SPL. *Abscissa:* time in milliseconds; each bin = 200 μsec. *Ordinate:* number of intervals in each bin. (From Rose *et al.*, 1967.)

is advantageous to evaluate the degree of locking by timing the discharges in relation to the sinusoidal stimulus itself rather than from the time of occurrence of the previous spike (as is done in determining interspike intervals).

If a fixed point on each stimulus wave-form is used to reset the timing counter then each spike will be timed in reference to a fixed point in time during one period of the stimulus cycle. In this way one can obtain "folded histograms", better termed periodograms, which reveal immediately whether or not the responses are locked to the stimulus cycle. Fig. 6 shows three such periodograms for unit 66–86–13. For all three there is an obvious locking to the stimulus cycle and it is clear that the locking is best for the 1,000 cyc./sec. tone, less for 2,000 cyc./sec. and substantially less for the 3,195 cyc./sec. stimulus. In order to express numerically the degree of locking we proceed as follows. We determine first the most effective half

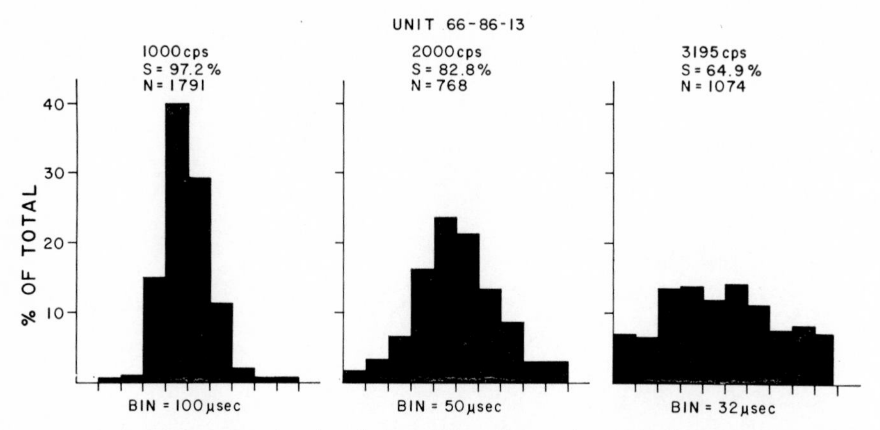

FIG. 6. Periodograms showing the relation of the spikes to the cycle of the stimulating frequency for neuron 66-86-13.

Each spike is timed in relation to a fixed point on the phase of the sine wave and its timing value cannot exceed the duration of the period of the frequency employed. Origin of each periodogram is arbitrary. *Abscissa:* time in microseconds; value for each bin is indicated in each graph. *Ordinate:* percentage of spikes in the bin. All stimuli: 90 db SPL; tone duration is different for each sample. N = number of spikes in the sample. S = coefficient of synchronization (see text).

of the stimulus cycle and term this half the "major half-cycle". We then express the number of spikes which occur during the major half-cycle as a percentage of the total number of spikes discharged and call this measure the "coefficient of synchronization". This coefficient can be as high as 100 per cent—if all spikes should occur during the major half-cycle—or as small as 50 per cent—if there is no relation at all to the cycle. In practice, the synchronization coefficient can hardly ever be 50 per cent, since this would mean that one cannot divide the period into two halves in such a way that the number of spikes during each half is different. With samples of about 1,000 spikes a coefficient of synchronization of about 56 per cent usually indicates a statistically highly significant departure from a random distribution.

(3) *Upper limit of the phase-locked response*

In Fig. 6 the coefficient of synchronization is over 97 per cent for the 1,000 cyc./sec. tone but only 65 per cent for the 3,195 cyc./sec. stimulus. A tone of 6,000 cyc./sec. was the best frequency for this unit. Fig. 7 assembles all synchronization data available for this neuron. The frequency of the stimulus is plotted against the synchronization coefficient of the respective periodograms. There is a systematic decrease in the value of this coefficient as the frequency is raised. The findings shown are typical. While units

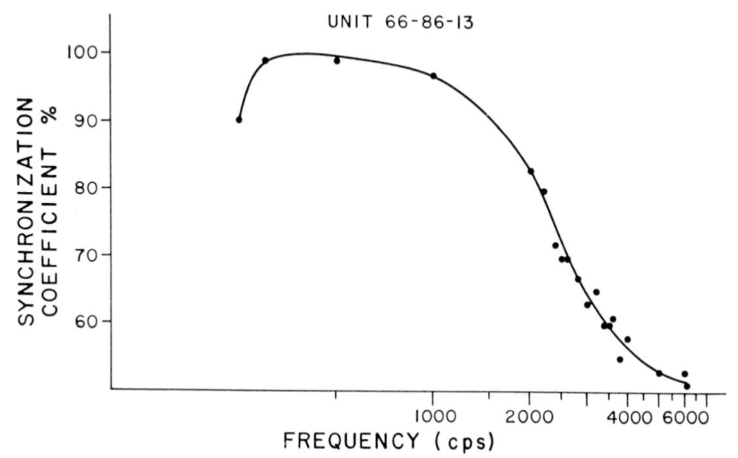

FIG. 7. Relation between the coefficient of synchronization and the stimulating frequency for neuron 66-86-13.

Three periodograms for this unit are shown in Fig. 6. All stimuli: 90 to 100 db SPL. Best frequency of the neuron: 6,000 cyc./sec. *Abscissa:* frequency of the stimulus. *Ordinate:* value of the coefficient of synchronization.

differ somewhat in this respect, one cannot usually demonstrate any significant locking to the stimulus cycle above some 4,500–5,000 cyc./sec.

Although such findings fit rather well the current concepts of most workers, we are hesitant at present to accept as established that locking to the stimulus cycle necessarily fades out at higher frequencies. We have detailed elsewhere (Rose *et al.*, 1967) some of the factors which are known or may be supposed to contribute to the dispersion of the timing values and which become rapidly more significant as the stimulus period shortens. We think it therefore prudent to leave open the question whether and to what extent there may be phase-locking for high frequencies.

(4) *Probability of stimulus cycle being effective*

It is of considerable interest to inquire whether the effectiveness of a low

TABLE I

CONDITIONAL PROBABILITY THAT A UNIT WHICH DISCHARGED AT TIME ZERO WILL DISCHARGE AGAIN IN RESPONSE TO THE 1ST TO 10TH CYCLES FOLLOWING THE TIME-ZERO DISCHARGE

Each population consists of responses to 10 stimuli each lasting 1 second. Neuron 65–107–1.

Frequency Cyc./sec.	Cycle										Number of spikes discharged	Number of cycles delivered	Spikes/Cycles
	1st	*2nd*	*3rd*	*4th*	*5th*	*6th*	*7th*	*8th*	*9th*	*10th*			
408	0·16	0·16	0·19	0·20	0·17	0·19	0·18	0·18	0·25	0·25	746	4,080	0·18
850	0·16	0·20	0·22	0·23	0·21	0·22	0·23	0·30	0·22	0·26	1,798	8,500	0·21
1,000	0·14	0·16	0·18	0·20	0·20	0·19	0·18	0·24	0·18	0·18	1,832	10,000	0·18
1,150	0·14	0·16	0·18	0·18	0·21	0·20	0·20	0·21	0·21	0·19	2,077	11,500	0·18
1,500	0·11	0·15	0·15	0·17	0·16	0·17	0·17	0·19	0·19	0·15	2,351	15,000	0·16
1,700	0·07	0·11	0·12	0·14	0·13	0·14	0·13	0·12	0·13	0·16	2,099	17,000	0·12
2,000	0·04	0·06	0·08	0·08	0·10	0·09	0·10	0·09	0·10	0·09	1,791	20,000	0·09

frequency stimulus can be related in a simple way to stimulus cycles. We approached this problem by calculating the conditional probabilities of the discharge. Table I shows such calculations for the data shown in Figs. 1 and 2. The question is: if a unit fired at time zero, what is the probability that it will fire again in response to the consecutive 1st to 10th stimulus cycles which follow the original discharge? The condition is that the second firing did not yet occur.

There are a number of observations which can be made on the basis of these figures. However, we shall consider here only three findings. The first is that the extreme left-hand probability figure in each horizontal row in Table I is the smallest in each row. Since this figure expresses the probability of firing during the next cycle after the original discharge it follows that this probability is, as could be expected, the smallest. Secondly, the probability of firing oscillates in each horizontal row around a more or less constant value and there is no systematic increase in the probability of firing after any number of ineffective cycles. Thus, except when the time elapsed is quite short, it matters little or not at all how long the fibre actually rested. We take this to mean that the refractory period of the fibre is acting as a limiting factor but is not a major determinant of whether or not a cycle will be effective. This suggests in turn that the mechanism governing the effectiveness of the cycle is likely to lie peripherally to the fibre, perhaps in nerve endings. Thirdly, the probability figures in each horizontal row are sufficiently close to each other to suggest that a fair estimate of the actual firing probabilities should be obtained if one divides the number of spikes discharged by the number of cycles delivered. The last three columns list the appropriate figures, and it is reasonable to say that this expectation is fulfilled. If this is so, it should be true that the quotient of spikes discharged divided by the number of cycles delivered should reflect the actual probability of firing regardless of the duration or strength of the stimuli. This usually holds true in our material, and Table II shows the actual and calculated probabilities when a sample consisting of responses to ten one-second tones is divided into four samples, each of 250 msec. duration, or when a tone of over 30 seconds duration is presented. The only major exception is the situation when the stimulus is so low in frequency that multiple firings during one cycle become common. In such cases it is necessary to subtract the number of multiple firings from the total number of discharges to obtain a quotient which approximates to the actual probabilities observed.

The fact that the probability of firing is approximated by the quotient mentioned provides, we believe, substantial evidence that a sinusoidal

TABLE II

ACTUAL AND CALCULATED PROBABILITIES OF DISCHARGE OF UNIT 65–107–1

The population of responses of unit 65–107–1 to 1,500 cyc./sec. stimuli (5th row in Table I) has been divided into four sub-populations. Each sub-population consists of responses which occurred during successive 250 msec. periods of the 10 stimuli presentations. The first four horizontal rows show the conditional probability calculations for these four sub-populations. The last horizontal row shows such calculations for unit 66–59–12 when a stimulus of 301 cyc./sec. at 100 db SPL was sounded for over 30 seconds.

	Cycle												
Unit 65–107–1 1,500 cyc./sec.	*1st*	*2nd*	*3rd*	*4th*	*5th*	*6th*	*7th*	*8th*	*9th*	*10th*	*Number of spikes discharged*	*Number of cycles delivered*	*Spikes/Cycles*
0–250 msec.	0·15	0·19	0·18	0·21	0·21	0·22	0·21	0·21	0·21		699	3,750	0·19
250–500 msec.	0·10	0·13	0·13	0·16	0·15	0·18	0·19	0·17	0·15	0·18	561	3,750	0·15
500–750 msec.	0·09	0·14	0·11	0·15	0·14	0·14	0·14	0·20	0·24	0·14	554	3,750	0·15
750–1,000 msec.	0·10	0·12	0·16	0·15	0·14	0·16	0·14	0·19	0·15	0·16	536	3,750	0·14
Unit 66–59–12 301 cyc./sec.	0·24	0·37	0·37	0·40	0·44	0·45	0·44	0·39			3,462	9,230	0·38

stimulus acts in general as if it consisted of as many individual stimuli as there are cycles.

(5) *Response to two tones*

We have reviewed the essential evidence which leads us to believe that it is the cadence of the stimulus cycle which determines the temporal aspects of the response if a single pure tone is presented. Whether or not one accepts that groupings of the discharges in time are essential for the transfer of frequency information is of no immediate consequence in a study of responses to a single pure tone. However, the situation is different if two tones are applied, for it is evident that if the temporal aspects of the responses are relevant it is quite easy to determine from periodograms or interspike interval distributions what frequency information is actually being transmitted. We have reported elsewhere (Hind *et al.*, 1967) some of our results on the interaction of two tones and are at present engaged in further studies on this subject. Here, we wish merely to point out that it is useful to proceed on the assumption that low frequency information is, in fact, relayed in a time code. We shall consider but one experimental situation when two tones not harmonically related are sounded simultaneously.

Fig. 8 illustrates the interspike interval distributions under various stimulus conditions for unit 66–101–12. A tone of 1,315 cyc./sec. was the best frequency for this neuron. Fig. 8A shows the distributions when the best frequency tone was presented alone at 40 db sound pressure level (SPL). Fig. 8B illustrates the distributions when a 503 cyc./sec. tone at 80 db SPL was presented together with the 1,315 cyc./sec. stimulus. The phase-locked response pertains mainly to the 503 cyc./sec. tone and we may conclude that the best-frequency stimulus has been masked. Actually the periodograms (not shown) indicate that the latter frequency was not altogether ineffective, but we need not be concerned with this fact here. When the intensity of the 503 cyc./sec. tone is lowered by 10 db it is this frequency which is now largely masked (Fig. 8C), since the responses are grouped around integral multiples of the period of the higher frequency. Fig. 8D provides evidence that a 503 cyc./sec. tone at 70 db SPL was a highly effective stimulus when presented alone. The spike counts suggest that whichever is the mode of the response to a two-tone combination, the discharge rate remains virtually unaffected. This rate, however, is here lower than that produced by the more effective stimulus acting alone (a usual finding in our experiments). We shall not elaborate here on the problem of masking but shall merely state that if two tones, not harmonically related, are presented together the phase-locked response of the neuron is in one of three modes. The neuron may respond: (1) to the first tone only; (2) to the second tone only; (3) to both tones. Which mode will actually occur is a function of the intensity of the component stimuli and their position within the response area of the neuron. The significant point to be made is that with the assumption that a time code is actually

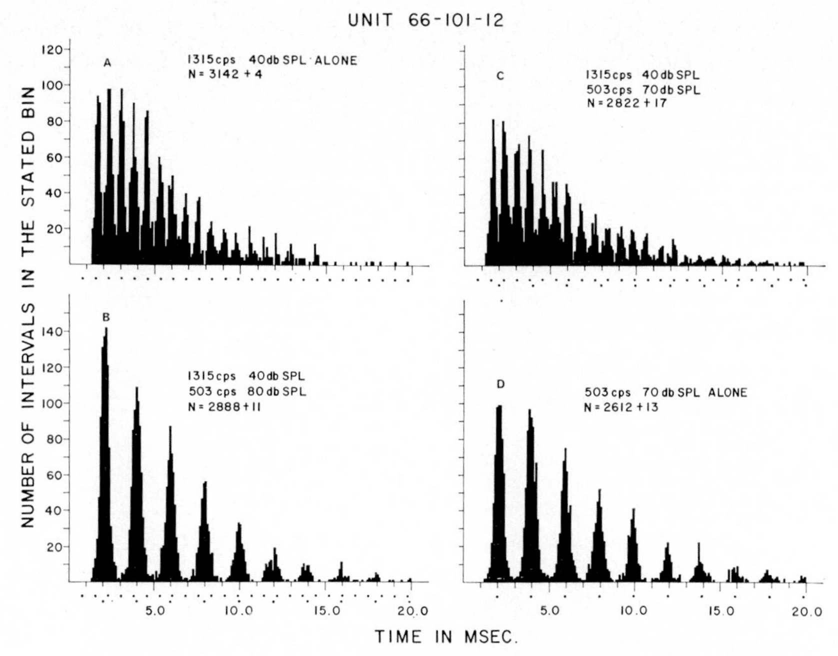

FIG. 8. Neuron 66-101-12. Distributions of interspike intervals when a tone of 1,315 cyc./sec. at 40 db SPL was sounded together with a 503 cyc./sec. tone at indicated intensities. Duration of all tones: 8 sec. A: response pattern to 1,315 cyc./sec. tone alone; B and C: response patterns when both tones were sounded together; D: response pattern to 503 cyc./sec. tone alone at 70 db SPL. *Abscissa:* time in milliseconds: each bin = 100 μsec. Dots below abscissa indicate integral multiples of the period of the component tones. *Ordinate:* number of intervals in each bin. All graphs are scaled to correspond to two tonal presentations of 8 sec. duration.

utilized for the transfer of low frequency information, some intriguing problems in psycho-acoustics become accessible to electrophysiological investigation.

SUMMARY

Responses of single nerve fibres to a pure tone of relatively long duration (1–20 sec.) were studied. The results indicate, in agreement with those of other observers, that each fibre responds only within a restricted frequency-intensity domain, and therefore for each a response area can be determined.

In a response to a tone of low frequency the discharges are spaced at intervals which group around the integral multiples of the period of the stimulating tone, regardless of the frequency or intensity of an effective stimulus, and thus regardless of the best frequency of the responding fibre. Locking of the discharges to a portion of the stimulus cycle occurs up to some 4,500–5,000 cyc./sec. but this relation becomes progressively less strict when the tone is higher than some 2,000 cyc./sec.

It is suggested that the cadence of the stimulus cycles determines the phase-locked response and thus a period-time code which is utilized by the spiral ganglion neurons to transmit to the cochlear nuclear complex information about the frequency of a low tone. It is further suggested that a fruitful way to view a sinusoidal stimulus is to consider it as consisting of as many individual stimuli as there are cycles. Some evidence to support these conclusions is presented.

REFERENCES

GALAMBOS, R., and DAVIS, H. (1943). *J. Neurophysiol.*, **6,** 39–57.

HIND, J. E., ANDERSON, D. J., BRUGGE, J. F., and ROSE, J. E. (1967). *J. Neurophysiol.*, **30,** 794–816.

KATSUKI, Y., SUGA, N., and KANNO, Y. (1962). *J. acoust. Soc. Am.*, **34,** 1396–1410.

KIANG, N. Y.-S., WATANABE, T., THOMAS, E. C., and CLARK, L. F. (1965). M.I.T. Research Monograph No. 35. Cambridge, Mass.: M.I.T. Press.

NOMOTO, M., SUGA, N., and KATSUKI, Y. (1964). *J. Neurophysiol.*, **27,** 768–787.

ROSE, J. E., GROSS, N. B., GEISLER, C. D., and HIND, J. E. (1966). *J. Neurophysiol.*, **29,** 288–314.

ROSE, J. E., BRUGGE, J. F., ANDERSON, D. J., and HIND, J. E. (1967). *J. Neurophysiol.*, **30,** 769–794.

RUPERT, A., MOUSHEGIAN, G., and GALAMBOS, R. (1963). *J. Neurophysiol.*, **26,** 449–465.

TASAKI, I. (1954). *J. Neurophysiol.*, **17,** 97–122.

WEVER, E. G. (1949). *Theory of Hearing.* New York: Wiley.

DISCUSSION

Whitfield: Professor Rose, in discussion in an earlier meeting in this series ([1966]. *Ciba Fdn Symp. Touch, Heat and Pain*, p. 162. London: Churchill) you said that you could see no evidence that the auditory system utilized periodicity of discharge in the auditory nerve for frequency coding. I gather that you have changed your view as a result of these new experiments and that you are now implying that this is a method by which information is transmitted. Of course, information transmitted along a line is related to an agreement between a sender and a receiver by which the receiver is modified to take up one of an agreed number of states; it is not something which some other kind of receiver can see in the line. So that because it is possible by using an oscilloscope or computer to see what the frequency was, as you have done, this does not mean that the cochlear nucleus can do it or that it is "information" for that nucleus. We have in your results, a necessary but not a sufficient condition for this transmission, and it seems to me that we need to know whether the cochlear nucleus or some higher level can detect the difference, not simultaneously but serially, between a period of 1,000 μsec. and a period of say, 1,050 μsec. This is the problem.

One would expect from the way in which the transducer works that there would be some kind of periodicity, and your findings throw a great deal of light on the details of the way in which the transducer does work, but they do not seem to provide any more evidence than we have already that this is a method of transmission of frequency information in the auditory system. It could be so, but if so we lack knowledge of the clock. I started off in 1950 by looking for this clock, but I failed, and as far as I know nobody has found any evidence that there is any time-base which could measure the intervals to the necessary percentage accuracy. There may be, but we have not found it.

Rose: The findings of which I spoke at the previous symposium suggested the experiments on which I reported here. We have observed in the inferior colliculus (Rose, J. E., Gross, N. B., Geisler, C. D., and Hind, J. E. [1966]. *J. Neurophysiol.*, **29**, 288–314) that some neurons there are highly sensitive to interaural time-differences. A time-difference as small as 10 to 20 μsec. can be recognized by such neurons. This fact implies that the auditory system can evaluate with considerable precision the timing of stimulus events. We observed moreover in the inferior colliculus that discharges of some (but by no means all) neurons in response to a tone of low frequency are grouped around integral multiples of the stimulus period. We thought it therefore important to study in detail the timing of discharges in the auditory nerve fibres.

I quite agree with you that periodicity of discharges as seen in the cochlear nerve fibres does not constitute a proof that such periodicity is necessarily utilized for transfer of low frequency information. However, the invariable occurrence of such periodicity makes the classic idea that the transfer of low frequency information is based on it both possible and, I think, rather likely. There are some important consequences if one entertains this view and considers that timing between discharges determines the information relating to frequency. I have presented some data on responses to a two-tone combination and I think that the results are quite compatible with psychophysical experiences on masking. Our group is now exploring responses to harmonically related tones. While I shall not detail here the results, I may state that the findings are again in agreement with an assumption that a time-code is being utilized in the transfer of frequency information.

I wish to stress emphatically that the suggestion of the utilization of such a time-code applies only to the cochlear nerve fibres. I have no suggestion to make about transformations which may and indeed must occur at different synaptic levels. One could only speculate on transformations which probably occur already in the different nuclei of the cochlear complex.

Erulkar: It seems a shame that such a beautiful analysis has to be relayed to the central nervous system only to be messed up! The problem is that once these impulses reach the cochlear nucleus and one records the patterns of discharges of the spikes, there are few cells that appear to respond to the cycle or will lock in with the cycle at high frequencies. At the inferior colliculus, you find something else. Dr. P. G. Nelson and I found a unit whose response was time-locked at about 280 cyc./sec. ([1964]. *J. Neurophysiol.*, **26**, 908–923), while Dr. Rose found the same for a unit at about 500 cyc./sec., but the problem is that at higher levels, this type of locking-in does not occur at frequencies much higher than this. Something else strange does occur. The multiple discharge recorded in response to a tonal stimulus may now be stopped by a hyperpolarizing potential (Nelson, P. G., and Erulkar, S. D. [1964]. *Loc. cit.*). At the inferior colliculus half the cells respond with just a single or double-spike discharge; at the cortex the majority of units respond with this short discharge. The point is that the time-locking information which reaches the cochlear nucleus is then changed to some other form of coding, until at the cortex there is just the single-spike discharge of the cell which provides information on frequency. What one wonders is whether this single-spike discharge is also related temporally to the stimulus periodicity. At any rate, at the present time it seems that we have to think of two different mechanisms for analysis at the auditory nerve of tonal frequencies—one in which the impulses are time-locked for frequencies below 5,000 cyc./sec., and the other which is not time-locked for frequencies above this.

Two other points: first, it is nice to know that there are still a few afferent fibres left in the auditory nerve for Professor Rose to record from, especially after the earlier papers! Secondly, I notice that you cut the vestibular nerve, presumably cutting the efferent supply to the cochlea. You have a beautiful coding system here without the efferent system intact.

Davis: Professor Rose, you have provided a picture of an excitatory half of the cycle producing time-locked responses, and everything works out very beautifully. Is the other half of the cycle inhibitory? I ask because there is spontaneous activity which is the background among which the time-locked responses appear, and yet in many of your distributions at low frequencies there is a notable absence of any such spontaneous discharge in the intervals between the large groups of time-locked responses.

Rose: I do not know the answer to this. There is, I think, little doubt for the neurons in the inferior colliculus that a stimulus cycle may be excitatory–inhibitory or inhibitory–excitatory even under conditions of deep anaesthesia. The situation is less clear for the auditory nerve fibres under the conditions stated (deep anaesthesia and cutting of some branches of the vestibular nerve). If one pairs a highly efficient stimulus (say the best frequency tone, which alone produces some 2,000 spikes in 10 seconds) with a marginal frequency (which alone produces but 50 to 100 spikes in 10 seconds) the number of spikes to a two-tone combination may be any number between these two extremes, depending on the intensities of the component tones. However, if the number of discharged spikes approaches the higher figure the discharges will be locked only to the more efficient frequency; they will be locked only to the less efficient frequency if the number of discharges approaches the lower figure. It is also possible to find a marginal frequency such that it alone produces but a few spikes. If such a frequency is paired (at a suitable intensity level) with a highly efficient frequency the number of discharges to a tonal combination may approach zero. However, even then the few discharges which occur are locked to the less efficient frequency. One hesitates therefore to ascribe at least this class of phenomena to inhibitory events.

Following a highly effective stimulus there is as a rule a cessation or diminution of spontaneous activity for some time. This is, of course, a common finding at many synaptic levels. I do not know whether one should attribute it to inhibition.

Finally, if a unit is highly active spontaneously an efficient tonal stimulus near threshold will always cause locking of the discharges to stimulus cycles before the discharge rate increases above the level of spontaneous activity.

Schwartzkopff: I have no evidence of this but would like to present a hypothesis. Fig. 1 (see p. 86) shows the response of a unit in the superior olive to clicks. The time of occurrence of each spike discharge following the click is plotted for different click intensities. The response is periodic, with a period corresponding to 560 cyc./sec., and the characteristic frequency of the unit to continuous tonal stimulation was 550–620 cyc./sec. This is, however, not a common type of response in the superior olive, and represents only 5–10 per cent of our units. Using various species of birds and mammals, we have found about 12 units of this type but we have never found a unit in which the characteristic frequency was essentially different from the corresponding click-evoked period. I have used this kind of coincidence to build up a hypothesis, to which there are many objections, however. It suggests how the central nervous system would evaluate "time" or "period" information presented to it.

There are two ways in which this could be done. The first is based on an idea that J. C. R. Licklider has promoted ([1956]. In *Third London Symposium on Information Theory*, pp. 253–258, ed. Cherry, C. London: Butterworth) where a time delay device would produce a differentiation of period. However, we ourselves suggest the following. There exist in the central nervous system neurons which respond preferentially with a periodic discharge, the period corresponding to their characteristic frequency. This periodic discharge will be evoked by a single "impulsive" input, and in the manner of a resonant system will be evoked preferentially by repetitive inputs of the characteristic frequency. Inputs of frequency different from the natural or resonant frequency of the neural system will be ineffective in evoking a response. This would enable the central nervous system to analyse the period or time information entering it. This is very important in the lower vertebrates since they have no real peripheral analysing mechanism. (See Schwartzkopff, J. [1962]. *Grundlagenstudien aus Kybernetik und Geisteswissenschaft*, **3**, no. 4, 97–109.)

Lowenstein: In the face of the evidence that there is pitch discrimination up to 800 cyc./sec. in fish, we must assume that some such mechanism operates (Stetter, H. [1929]. *Z. vergl. Physiol.*, **9**, 339–447; Dijkgraaf, S., and Verheijen, F. J. [1950]. *Z. vergl. Physiol.*, **32**, 248–256).

Johnstone: In the lizard *Trachysaurus rugosus*, we have not been able to find any phase-locking at any frequency in the auditory nerve (Johnstone, J. R., and Johnstone, B. M. [1967]. *Aust. J. exp. Biol. Med.*, **65**, P24). The discharge patterns fall into two groups, one spontaneous and the other non-spontaneous. Did you find any non-spontaneous fibres, Professor Rose?

Rose: This depends on the length of time you wish to wait in order to meet the criterion. In the auditory nerve of the squirrel monkey there are units which discharge from less than 10 impulses per second to something over 150 per second.

Johnstone: In the lizard, information which we would think should be almost completely coded on a volley system, because the analyser looks poor, is in fact not so coded and appears to be coded on a place system. This is curious, because the analyser does not appear to be capable of it.

Lowenstein: Nevertheless there is the beginning of the presence of an analyser.

Johnstone: Yes, and it apparently can code at the frequencies we have tried, which are from about 400 cyc./sec. up to about 4 kcyc./sec. At the lower frequencies there is a 1:1 ratio but it is irregular and once you go to higher frequencies you do not get phase-locking.

Lowenstein: Would you say that at about roughly 100–120 cyc./sec. a difference appears, and below that you get the 1:1 ratio?

Johnstone: I think so, but I did not go so low as that myself.

Whitfield: Did not Dr. Enger suggest that there is some place analysis in fish?

Lowenstein: Yes, but it is difficult to suggest the mechanism.

Davis: When I introduce my students to the operation of the nervous system I tell them that although the axons do operate according to the all-or-none principle and give discrete all-or-none discharges, the nervous system as a whole is organized to disguise this fact as completely as possible! If we go to the cortex, we see the disguise begin to appear with the loss of the beautiful orginal time-locking described by Professor Rose. The moral is that if we are looking for the place where frequency information is extracted and put into some more stable form, we have to look very close to the periphery.

Rose: I do not wish to revive the old controversy in which place or periodicity of impulses are alternatives for frequency coding. I do not think that we have to choose between these alternatives. Both are likely to be employed at some levels.

Davis: They are both employed, certainly; but the question is, to what extent?

Neff: If you use a discontinuous sound, such as 1,000 clicks per second, and make the same analysis as you do with a continuous sound at 1,000 cyc./sec., what does your picture look like? Can you distinguish between the responses to these two stimuli?

Rose: We did not try clicks.

Erulkar: Did you use any modulated tones, Professor Rose?

Rose: No.

Fex: There was a beautiful study by A. Rupert, G. Moushegian and R. Galambos ([1963]. *J. Neurophysiol.*, **26**, 449–465) on conscious cats with implanted microelectrodes. They recorded from single primary auditory fibres and found a group of neurons which were not phase-locked but which responded with increased firing to an increase in the loudness of a tone.

Evans: May I take up Professor Schwartzkopff's hypothesis that there exists a resonant neural mechanism capable of selectively responding to a periodic input? It seems to me that this hypothesis is unnecessary and that there is yet no evidence for it. It seems unnecessary to invoke a *neural* mechanism for the kinds of patterns that Professor Schwartzkopff found in cells of the superior olive, because N. Y.-S. Kiang ([1965]. *Discharge Patterns of Single Fibers in the Cat's Auditory Nerve.* Cambridge, Mass.: M.I.T. Press) has shown clearly that responses of this kind are very prominent at the level of the primary neuron. One therefore does not have to postulate a *cellular* mechanism to give such repetitive responses; it is most likely that they reflect mechanical events in the cochlea. Secondly, for his hypothesis to be valid he must demonstrate that his units respond to a narrower band of tonal frequencies than those presented to the input of the cell; in other words, he must show that these cells possess a "Q factor". To show that they respond to an impulsive input with a periodic discharge is not enough. As I have just said, it seems more likely that Professor Schwartzkopff's neurons are responding passively to, or "following", a periodic discharge pattern in the primary neuron.

Schwartzkopff: When I originally proposed this hypothesis I was aware that this kind of selective mechanism may be built into the first-order neuron already. I believe that this is so in fish. This was the original hypothesis, but it fits, of course, with Dr. Enger's work—he has given us evidence that auditory nerve fibres in fish have tuning curves. There is no proof, however, that these tuning curves result from a peripheral mechanism in the sense that the hair cells are differently organized in respect to position.

Spoendlin: Professor Rose, you find in all fibres a best frequency; on what do you think this sensitivity of each fibre to a special frequency relies—on the topographical arrangement of that fibre, or on some special structure of that fibre, or what?

Rose: I think that there is a fairly general agreement, usually implicit rather than explicit, on what "best" frequency implies. One assumes that the best frequency of a neuron will be that frequency which causes optimal deflection of that particular spot of the cochlear partition which supplies directly or indirectly the innervation of the neuron in question. I use the word "spot" vaguely without giving it any real dimensions except that one expects it to be in length but a fraction of the cochlear partition.

I have a question myself. You spoke (p. 107) of the distribution of fibre endings over a length of some 0·6 mm. This would be a rather restricted distribution for the usual response area. One could, however, attempt to estimate from such anatomical facts how large a "spot" of the cochlear partition is actually likely to be.

Spoendlin: You accept a tonotopical localization, then. I must emphasize that 0·6 mm. for the outer spiral fibre's spiral extension is an average value. The dendrites might differ considerably in their lengths, so that the longest ones are possibly longer than 0·6 mm.

Evans: Professor Rose mentioned (p. 158) data from the inferior colliculus as possible evidence for a neural "clock". It should be noted that these data concerned binaural interaction and not monaural interaction, and there would seem to be a great difference here. There certainly are brainstem mechanisms which are specifically organized to measure very small time-differences between inputs from the two ears but I am not sure that it has been shown that the auditory system contains mechanisms which could operate on inputs from *one* ear with this resolution.

My next point is that one of the problems with the translation of a period code at the periphery or at the lowest parts of the central nervous system into some other code at the upper levels, is that if one records from single cells in the upper levels one never finds cells that are narrowly "tuned" to low frequencies. They all have very wide response areas, and in fact their sensitivity as a function of frequency is what one might predict on the basis of place in the cochlea, and not on the basis of transformation from a period code into some other code, such as place, which J. C. R. Licklider ([1956]. In *Third London Symposium on Information Theory*, pp. 253–268, ed. Cherry, C. London: Butterworth) has suggested.

My last point is a question. If the nervous system is to utilize the kind of information that Professor Rose has demonstrated so beautifully, then it surely must integrate over space (that is, integrate the activity over many fibres) and not over time, as Professor Rose was able to do using the LINC computer, for whereas Rose's measurements result from integration over many seconds, pitch discrimination can be made in a few milliseconds. For the central nervous system to effect spatial integration however, it would seem necessary for the discharges of input fibres of similar characteristic frequency to be phase-locked to approximately the same phase of the stimulus cycle. If this were not so, it would be an impossible task for a neuron higher up in the pathway to use the period code information from *many* cells. My question to Professor Rose is, then, has he noticed whether the discharges of cells of neighbouring characteristic frequencies are phase-locked to the same phase of the stimulus cycle?

Rose: In my judgment no major differences in the width of the response areas have been so far demonstrated convincingly between the different synaptic levels, despite some claims to this effect. One finds broader and narrower response areas for auditory nerve fibres, in the cochlear complex, in the inferior colliculus, in the medial geniculate and in the auditory cortex. I do not deny that significant differences may exist but one has to demonstrate them convincingly.

I am not certain that I understand the question you raise about the phase. I am talking about responses in a steady state to tones of long duration. It is the same stimulus cycle that activates the different units which are capable of responding to it and the discharges group around absolute values in real time.

Schwartzkopff: I have a record from first-order auditory fibres synapsing with the secondary neurons in the bird's medulla, using a microelectrode which was built to record from cells, not fibres, and so it was recording from about ten fibres at a time. This exactly answers Dr. Evans' question. The recording shows neuronal activity of the same frequency as the stimulus. This means that the separate elements involved are all locked to the same part of the phase, at about 1,000 cyc./sec. These 10 or 12 elements, all close together, are synchronized with about the same variation in time that you see if you study a single unit.

Davis: There may also be differences in phase, as Tasaki demonstrated in his original study of the first-order neurons in our laboratory (Tasaki, I. [1954]. *J. Neurophysiol.*, **16**, 97–122). Different neurons turned out to have exactly opposite phase relations. We suggested that these fibres came from different positions on the basilar membrane and that the phase differences expressed the delay of the Békésy travelling wave. We can feel confident that there is this distribution, but neighbouring fibres, with nearly the same characteristic frequencies, should be expected to behave exactly as Professor Schwartzkopff has indicated. It is a very interesting question, as you say, how the central nervous system handles this situation. The delay of the Békésy travelling wave can amount to as much as a full cycle. So the central nervous system would have all phases to choose from!

Whitfield: On the question of response areas, my colleague H. F. Ross has shown that if you transform frequency on a square-root basis you can obtain a "normalized" threshold/frequency curve which is the same for all frequencies. Furthermore, if you calculate its standard deviation, there is no significant difference between neural levels. In fact the greatest variation is between the results of different workers at the same level! So this appears to be a completely uniform system in which there is no need to invoke a break between one range and another.

This question which Dr. Evans put forward is also a crucial one; at some point there has to be conversion from periodicity into a probability of response. This transformation has been shown to occur for pulses in two fibres converging, but nothing equivalent has been shown for pulses of different interval in the same system at different times. There might be an experiment which would resolve this: if one could destroy all the fibres from the cochlea responding to a particular frequency, say 1,000 cyc./sec., using an animal trained to respond to this frequency, and could then stimulate electrically the remaining fibres with 1,000 pulses/sec., a response would be evidence that this kind of pattern could be processed by the nervous system.

Neff: We have done experiments in the cat something like those about which Dr. Whitfield is asking. We have trained cats to respond to a train of clicks presented through a loudspeaker and have recorded the evoked response to this stimulus from implanted gross electrodes in the inferior colliculus, in the cortex, and at other locations in the auditory nervous system. After being trained to respond to a train of clicks, animals transferred immediately to a train of shocks given at the same rate through the implanted electrode. The inference might be made that the electric shocks were producing the same sensation to the animal as clicks heard over the loudspeaker. We have also done the experiment using the reverse procedure, first training cats on the electric shocks and then testing them with a sound stimulus (Neff, W. D., Nieder, P. C., and Oesterreich, R. E. [1959]. *Fedn Proc. Fedn Am. Socs. exp. Biol.*, **18**, 112 [abstract]; Nieder, P. C., and Neff, W. D. [1961]. *Science*, **133**, 1010–1011).

We have also stimulated electrically at lower frequencies, training an animal to respond when we changed from 1 shock/sec. to 10 shocks/sec. We can do this with electrodes in both inferior colliculi; if the cats are trained to respond to shocks on one side, they will transfer to the other. If they are trained to respond to electrical stimulation of the primary cochlear nucleus they will transfer to stimulation of the inferior colliculus (W. D. Neff, unpublished data).

Whitfield: To support the theory that periodicity alone can convey pitch, you would have to destroy all the auditory nerve fibres which would normally be stimulated by the 1,000-cycle tone and stimulate some others periodically. As long as any fibres which would be stimulated by the 1,000-cycle tone remain to be stimulated, you cannot answer the question. Clearly this is a difficult experiment to do, because you have selectively to destroy fibres in the auditory nerve.

Rose: I would think that you would have to destroy almost the entire cochlea.

Whitfield: Yes, but there are nevertheless only two kinds of evidence which can be used to support this theory. One is an experiment of this type which may be impossible to carry out; the other is an experiment of the "Hall" type in which you can show that somewhere there is a unit whose probability of response depends on whether the impulses coming up the auditory nerve are spaced 1,000 μsec. apart or 1,050 μsec. apart.

Davis: We (Davis, H., Silverman, S. R., and McAuliffe, D. R. [1951]. *J. acoust. Soc. Am.*, **23**, 40–42) made a fair approximation to Dr. Whitfield's requirement in psycho-acoustic experiments in which human beings were stimulated with high-frequency (2,000 Hz) pips, which (in animals at least) do not produce any mechanical disturbance that is reflected in the cochlear microphonic in the apical region. The periodicity of such a stimulus train in the range from about 150 to 400 pips/sec. gives a subjective experience of pitch with which it is easy to recognize a tune. Here is the separation in a psycho-acoustic experiment of periodicity pitch from the more musical-sounding "place pitch".

Whitfield: You are saying that these pips produce no responses in the apical fibres?

Davis: We cannot prove this absolutely, but I am relying strongly on the animal experiment in which 2,000 cyc./sec. pips did not register at the apical turn (Tasaki, I., Davis, H., and Legouix, J.-P. [1952]. *J. acoust. Soc. Am.*, **24**, 502–519). This is the pattern of the Békésy travelling wave which goes just so far and then dies out. The low frequency (150 to 400 pips/sec.) was not present as a sinusoidal component in the acoustic stimulus, and this is a fair approximation to what you are demanding from the animal experiments. Here one is giving verbal instructions to a human subject rather than training an animal.

Whitfield: I am not convinced that these are not phenomena which belong to the same category as the failure to see one's blind spot. When you set up a series of harmonics which could have a difference tone (the "missing fundamental" experiment), psycho-acoustically the nervous system tends to assign to the combination some situation with which it is familiar, and the situation with which it is familiar is the existence of a fundamental at the difference frequency. This is again an experiment which is almost impossible to do, because one would need, in order to test the theory, a naïve animal which had never heard a complex tone.

Davis: This would even be impossible to design in principle!

Schwartzkopff: There have been a few experiments in human beings who had lost the entire external ear. Russian workers have applied electric current to the root of the nerve, and it was possible to obtain a very rough discrimination of pitch.

Davis: The best job by far of electrical stimulation in the human being has been done by F. Blair Simmons ([1966]. *Archs Otolar.*, **84**, 2–54). His patient was studied psycho-acoustically at the Bell Telephone Laboratory and by the Simmons group independently. The results fit well with the concept of a certain degree of rather crude pitch perception up to roughly 1,500 cyc./sec., enough to recognize the tone of voice or even to discriminate words chosen from a small vocabulary but insufficient for complete understanding of speech.

Erulkar: We have been dealing mainly with time-locked responses at lower frequencies (e.g. 2,000 cyc./sec.), and Professor Rose said that he felt that the nerves did respond in a similar way to higher frequencies. What is your feeling on what happens in a cat that hears a sound at 30 or 40 kcyc./sec.?

Rose: The actual experimental results indicate that the periodicity fades out usually at about 5 kilocycles. However, this limit is likely to be too low since there is at least one major timing error in our set-up which contributes to the dispersion of timing values. Let me elaborate on this point. Our counter is accurate to 2 μsec. The problem is the choice of a point in time from which we wish to start the timing. This point in time is given at the moment when the voltage of the spike reaches a predetermined discrimination level. This measure will be a consistent measure of the time of occurrence of the spike only if the amplitudes of the spikes remain constant and if there are no fluctuations in the baseline. Since neither is usually strictly true, the timing of the spike becomes a function of its rise-time. Now, the rise-time in our set-up is about 200 μsec. There is thus some uncertainty as to the precise time at which the spike actually occurred. How large this uncertainty may be is difficult to assess. If one should guess that it may be as much as some 10 to 20 per cent of the rise-time, an error of some 20 to 40 μsec. would be introduced. Such an error is, of course, quite trivial for lower frequencies. It becomes, however, increasingly more important as the frequency rises and an error of, say, 40 μsec. would tend to blur almost completely a locking to 5·0 kilocycles even if such a relation existed. In addition to this known source of error there are some other considerations which suggest that timing problems may possibly be introduced. Cochlear fibres are quite small and it seems reasonable to assume that they are usually injured even by the finest probing electrode. If an injury should cause some uncertainty in timing this again would tend to obscure the relation to cycles at higher frequencies.

I do not wish to suggest that a relation to cycle exists for higher frequencies. I am merely justifying the view that at the present time it appears proper to view the negative findings with considerable caution. It is, I think, certainly true that some cells in the inferior colliculus can in fact discern a time-difference of some 10 to 20 μsec. The question thus is not whether the auditory system can evaluate a very small time-difference; it certainly can. The question is, I believe, whether relaying on this capacity a time-code is utilized in the transfer of higher frequency information in the auditory nerve.

Whitfield: This is between two fibres, and not within the same fibre, and that is surely the crucial point.

Rose: I am not certain that I see your point. The discharges in a single fibre are spaced comfortably apart often by a number of integral values of the period. I suppose evaluation of any minute interval value must be done by means of a converging input of a number of fibres.

Whitfield: The ability to resolve time-differences of 10–50 μsec., which we all agree the nervous system can do, is not relevant. What has to be done is to distinguish an interval of the order of a millisecond from an interval one or two per cent longer or shorter. We can all distinguish an interval between two taps on the bench of half a second, but how many, without a watch, can distinguish an interval of 50 seconds from one of 51 seconds when these are presented successively?

Schwartzkopff: From a comparative point of view, you do not find sound discrimination above 2,000–3,000 cyc./sec. in species without basilar membranes. W. H. Dudok van Heel ([1956]. *Experientia*, **12**, 75–77) has shown that minnows are able to discriminate up to about 1,200 cyc./sec., the upper limit of frequency discrimination being dependent upon temperature. I would say that high frequency discrimination is correlated with the presence of a basilar membrane, which means that it is related in some degree to a pure place system, while the lower frequencies always use the frequency-locked mechanism, even if a basilar membrane is present.

Lowenstein: We should clarify our position on the necessity of frequency-synchronized transmission in lower vertebrates that have pitch discrimination. Dr. Enger, have you any reason to assume there is a place-representation mechanism operating in animals without a basilar membrane?

Enger: No. In teleost fish at least, there is no structural difference along the saccule or in the sacculus contrasted with the lagena. From recordings of single auditory nerve fibres in the sculpin (*Cottus scorpius*) there is no indication that the sacculus and the lagena cover different frequency ranges.

Lowenstein: All you found was that there are some fibres that seem to be tuned to certain frequency ranges?

Enger: Yes; but I should mention that a student of mine in recording saccular microphonics from the sculpin has found that high frequencies—400–500 Hz—can be obtained more readily from the anterior than from the posterior portion of the sacculus. Low frequency responses were recorded along the full length of the sacculus.

EFFERENT INHIBITION IN THE COCHLEA BY THE OLIVO-COCHLEAR BUNDLE

JÖRGEN FEX

Laboratory of Neurobiology, National Institute of Mental Health, Bethesda, Maryland

THIS paper is concerned with three main aspects of efferent inhibition in the cochlea: site of action, transmitter substance, and function. New experimental findings on transmitter substance are presented.

SITE OF ACTION

It will be argued that the site of action of the crossed olivo-cochlear fibres (Rasmussen, 1946, 1953, 1960) in the cat cochlea is the membrane of outer hair cells. The uncrossed olivo-cochlear efferents (Rasmussen, 1960) probably act directly on auditory afferents, perhaps only underneath inner hair cells.

The crossed olivo-cochlear fibres

Recent evidence for an action on the outer hair cells by the crossed efferents came from studies (Fex, 1967*a*,*b*) on intra-cochlear potentials evoked by the crossed olivo-cochlear fibres. The following techniques were used.

The round window of the cat's cochlea was exposed and part of the round window membrane was removed to open the scala tympani. Perilymph was sucked away from the scala tympani to expose the basilar membrane of the organ of Corti. A microelectrode for recording was advanced under visual control through the round window until the tip reached the basilar membrane. Then, either there was a pause until the scala tympani had spontaneously refilled, or it was refilled by artificial perilymph. In some experiments the artificial perilymph contained strychnine or an increased content of calcium chloride. The ear was stimulated by sound, the crossed olivo-cochlear fibres were electrically stimulated in the floor of the fourth ventricle (Fig. 1, arrows 1 and 3) and responses were recorded. This was done with the electrode tip in the scala tympani, in the organ of Corti, and in the scala media (Fig. 2).

The time-course of the centrifugally evoked potentials was the same in the three regions, which led to the conclusion (Fex, 1967*b*) that the potentials mirrored the same event. The evoked potentials were larger in the scala media than in the scala tympani by a factor of from 4 to 10, reaching an amplitude of 3 mv. The latency of the potential ranged between 12 and 40 milliseconds in 14 experiments when natural perilymph was used. The latency was shortened (Fig. 3) with an increase of the calcium content

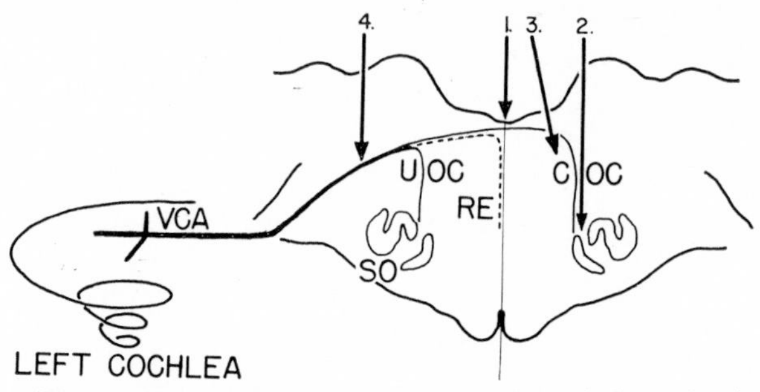

FIG. 1. Diagram illustrating the placement of electrodes for stimulating efferent cochlear fibres.

COC, crossed olivo-cochlear fibres; UOC, uncrossed olivo-cochlear fibres; RE, reticular cochlear and vestibular efferents, which are present in rodents but, in all probability, not in the cat (see text); SO, superior olivary complex, VCA, vestibular cochlear anastomosis. At, and just below, the point of arrow 1 is the region most commonly used for electrical stimulation of the crossed olivo-cochlear fibres. Arrow 2 indicates where crossed olivo-cochlear fibres have been stimulated by stereotactically placed electrodes. Uncrossed olivo-cochlear fibres have been stimulated by electrodes placed slightly to the right of arrow 2. Arrow 3 points slightly below a region in which crossed olivo-cochlear fibres have been stimulated. Arrow 4 points approximately to a region which Desmedt and LaGrutta, and also Sohmer, have used for stimulation of uncrossed olivo-cochlear fibres. See text for references. (Reproduced from Fex, 1967*b*, by permission of the *Journal of the Acoustical Society of America*.)

in the scala tympani. From this it was suggested that release of transmitter substance at the inhibitory synapse is dependent upon calcium (Fex, 1967*a*). Also the latency with which the augmentation of the cochlear microphonics (CM) appeared with efferent stimulation was shortened with an increase of calcium in the scala tympani (Fex, 1967*b*).

It was concluded that the evoked potential reflected postsynaptic activity of crossed efferents (Fex, 1967*b*). This follows from the finding that the potential could outlast stimulation for hundreds of milliseconds. There are no interneurons of the crossed efferents (Rasmussen, 1953) to account for this.

The evoked potentials in all probability reflect events in hair cells. An amplitude of 3 mv is reached by sound-evoked potentials only when strong sound stimuli are used. Centrifugally evoked postsynaptic potentials of this size are more likely to reflect current generated in membranes of rows of parallel hair cells than current in thin unmyelinated afferent terminals with no regular spatial orientation. No other structures are likely to receive efferent synapses of olivo-cochlear efferents in the organ

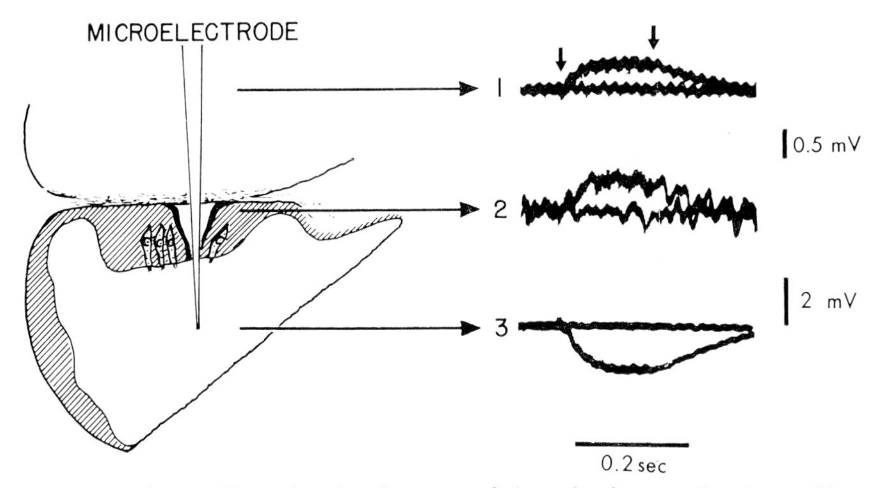

FIG. 2. Diagram illustrating the placement of electrodes for recording intracochlear potentials evoked by stimulation of the crossed olivo-cochlear fibres.

It should be noted that it was not known whether or not the electrode passed through the tunnel of Corti on its passage through the cochlear partition. In 1, 2 and 3 are shown potentials which were recorded in the scala tympani, in the organ of Corti, and in the scala media, during one experiment. Each record shows a potential evoked by repetitive electrical stimulation of the crossed olivo-cochlear fibres; the three records were obtained with the same parameters of stimulation. The duration of stimulation is indicated by the arrows in the first record; each potential is superimposed on a baseline representing the resting condition. Note that 1 and 2 have a common voltage calibration, different from that of 3. D.c. recording was used. Negativity is downward. (Reproduced from Fex, 1967*b*, by permission of the *Journal of the Acoustical Society of America*.)

of Corti of the cat. That these postsynaptic events take place in the hair cell membrane is also indicated by the change of sign of the evoked potential, from positive in the organ of Corti to negative in the scala media (Fig. 2).

Histological evidence indicates that potentials evoked by stimulation of the crossed efferents would be generated by outer hair cells. Crossed efferents innervate only outer hair cells in the rat cochlea, according to preliminary findings by Iurato (1964). The cat would probably be similar to the rat in this respect. In the cat, in the basal turn of the cochlea, there are large efferent endings on the outer hair cells of all three rows (Spoendlin and Gacek, 1963), and we are concerned here only with activity in the basal turn, because of the electrode placements. Many of these efferent endings would not be close enough to an afferent to form efferent–afferent synapses, to judge from the paper by Spoendlin and Gacek (1963). Furthermore, no such synapses were found by Rodríguez Echandía (1967) who looked specifically for synapses in the outer hair cell region of cat, using

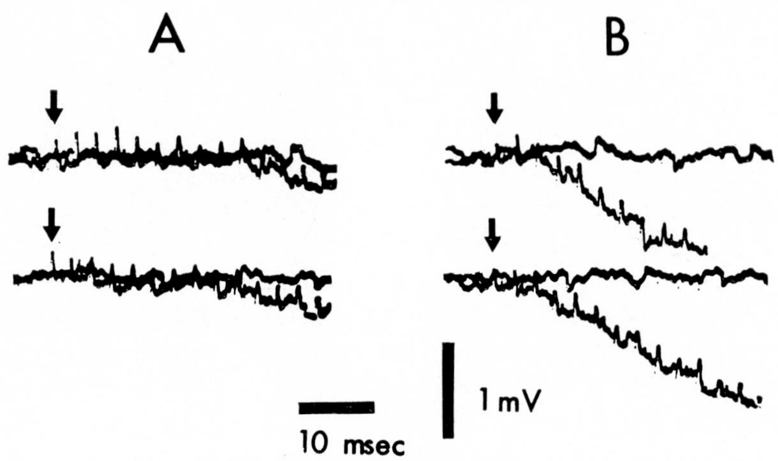

FIG. 3. Records illustrating the action of calcium on centrifugally evoked potentials. The initial time-course is shown of potentials that were evoked by electrical stimulation of crossed olivo-cochlear fibres. Each potential is superimposed on a baseline representing the resting state. All the records were taken with the microelectrode in one position in the scala media. The start of stimulation is indicated by an arrow. A. Control with natural perilymph in the scala tympani. B. The natural perilymph had been partly replaced by artificial perilymph, containing 5·0 mM-calcium chloride. This increase of calcium content was the significant change; see text. D.c. recording was used. Negativity is downward.

electron microscopy. Neither is there any other evidence that efferent and afferent structures that are close together underneath outer hair cells actually form functioning efferent–afferent synapses.

It was concluded that the centrifugally evoked potentials reflected postsynaptic activity generated in outer hair cells by crossed efferent fibres and that the crossed efferents of the cat inhibit by their action on outer hair cells (Fex, 1967*b*).

It was pointed out that all findings on inhibition by the crossed efferents in the cat cochlea can be explained from an action on outer hair cells. A full discussion of this would include the following arguments:

(1) Crossed cochlear efferents in the rat innervate only outer hair cells (Iurato, 1964).

(2) Three to five thousand auditory afferent neurons innervate outer hair cells only. The remaining afferents, about 45,000, innervate inner hair cells only (Spoendlin, 1966).

(3) Crossed efferents inhibited acoustically evoked activity in 52 out of 56 auditory primary afferents in one study on single nerve fibres (Fex, 1962) and in all of 141 afferents in another such study (M. L. Wiederhold, personal communication).

These arguments are mutually incompatible, which suggests the following comments:

ad (1) Future histological studies may show that crossed efferents in the cat also impinge on auditory afferents for inner hair cells.

ad (2) Spoendlin's figures of 3,000 to 5,000 primary auditory neurons for the outer hair cells could be much too small (cf. Eldredge, 1967). Also, Spoendlin leaves unconsidered the possibility of an intricate system of afferent collaterals formed centrally to the habenula perforata. Thus, all, or some, of the outer hair cells may share spiral ganglion cells with inner hair cells through an afferent innervation by collaterals.

ad (3) There may be as yet unrecognized differences, for instance in size, between auditory nerve fibres for outer and inner hair cells. Thus a bias causing an over-representation of outer hair cells may inadvertently have been introduced in Fex's and in Wiederhold's studies. Also, in Wiederhold's study several or many auditory fibres were very little suppressed by efferent stimulation. Such weak suppression might have been due to indirect effects of the efferent stimulation.

Because of the evidence for the hair cell membrane as the site of action of crossed olivo-cochlear fibres, the CM change with crossed efferent stimulation (Fex, 1959) is not surprising. The change is actually an increase and has been called paradoxical, since the CM is hypothesized (Davis, 1965) to be a causal agent in the generation of activity in the auditory nerve. However, there is no paradox here if the reasonable assumption is granted that d.c. events in the hair cell membrane may be as important as the CM in regulating activity in the auditory nerve. Then it may be concluded that an increase in excitation, parallelling the increase of the CM, is more than cancelled by the inhibitory components of the efferent action on the hair cells.

Desmedt and Delwaide (1965) have suggested that the hair cells of the cat's cochlea are of no significance for the mechanism of centrifugal inhibition but that in the pigeon, efferents act significantly on hair cells. They had found the pigeon's crossed cochlear inhibition to be weaker than that in the cat cochlea, but the increase of the CM was more marked in the pigeon than in the cat. They also referred to Cordier's (1964) histological findings as supporting evidence that cochlear efferents probably end only on hair cells in the pigeon cochlea. But as argued above, in the cat cochlea also there may well be no synapses between crossed efferent neurons and afferent auditory neurons, particularly not in the outer hair cell region. More important for this argument may be that there are large differences between the mammalian and the avian cochlea on the afferent side (Katsuki, 1965; Schmidt and Fernández, 1962). These differences could be the reason for the differences in efferent actions. Thus, my concept of efferent inhibition in the mammalian cochlea is not in opposition to the findings in the avian cochlea.

The uncrossed cochlear efferents

Repetitive electrical stimulation of the uncrossed olivo-cochlear fibres in the brainstem of the cat also suppresses the click-evoked action potential (AP) of the auditory nerve, recorded with gross electrodes at the round window (Desmedt and LaGrutta, 1963). However, no change of the CM with such stimulation was reported by Desmedt and LaGrutta, neither could Sohmer (1966) nor I (Fex, 1967*b*) see any such change. This suggests that these uncrossed efferents do not act on hair cells.

The pertinent histological evidence for where the site of action of the uncrossed efferents may be is difficult to interpret. Thus, there may be species differences between cats and rodents. Rodents have uncrossed reticulo-cochlear efferents (Rossi and Cortesina, 1962, 1965; Rasmussen, 1965, personal communication to G. Rossi), which cats have not (cf. Rasmussen, 1960). Also in the guinea pig, in the outer hair cell region, there are the *2a* endings of Smith and Sjöstrand (1961) which may form synapses with afferent fibres, perhaps with hair cells. The *2a* endings are not found in the basal turn (Smith and Sjöstrand, 1961), have not been determined to be efferent, and similar endings have not been identified in the cat. The *2a* endings were not explicitly mentioned in Iurato's study (1964) of the differential innervation in the rat cochlea by crossed and uncrossed efferents. Iurato claimed that uncrossed fibres participate in the efferent innervation of outer hair cells. It may be suggested here, partly from lack of knowledge, that this innervation in the rat corresponds to the *2a* endings in the guinea pig and to reticulo-cochlear fibres in rodents in general, and that the cat lacks *2a* endings.

Iurato claimed further (1964), as mentioned above, that the inner hair cell region in the rat cochlea receives its total efferent innervation by uncrossed fibres. This may be true also for the cat. In this region in the cat there are probably efferent synapses with auditory afferents but perhaps none with hair cells (cf. Spoendlin and Gacek, 1963).

TRANSMITTER SUBSTANCE

Gisselsson (1960) applied acetylcholine (ACh) electrophoretically to the vicinity of hair cells in the organ of Corti in the guinea pig and found an increase of CM. Gisselsson used relatively weak electrophoretic currents and tried to use the sound stimuli that would optimally stimulate the regions to which ACh was applied. In these respects his approach was different from that of Katsuki and co-workers (Katsuki, Tanaka and Miyoshi, 1965; Tanaka and Katsuki, 1966), which may explain the differences in results between the two studies. Katsuki and co-workers saw only a decrease of CM with application of ACh. Gisselsson (1960) suggested that ACh is the transmitter substance of the crossed olivo-cochlear fibres. For supporting evidence he referred to previous findings (Churchill and Schuknecht, 1959) that the heavy staining for acetylcholinesterase (AChE) in the organ of Corti depends on the presence of olivo-cochlear fibres and that electrical stimulation of the crossed olivo-cochlear efferents causes an augmentation of CM (Fex, 1959). Gisselsson (1960) mentioned unpublished observations of a decrease of the AP of the auditory nerve of the guinea pig after intra-arterial injection of ACh, but he considered the evidence too indirect to permit any suggestion concerning nervous transmission in the cochlea.

Since then many experiments have been done in different laboratories in attempts to define the transmitter substance, or substances, of the cochlear efferent inhibition (see also for references Fex, Fuxe and Lennerstrand, 1965; Tanaka and Katsuki, 1966). In many such experiments pharmacological agents were injected somewhere outside the cochlea with negative results, which may have been caused by diffusion barriers in the organ of Corti. It is also useful to recall that we know neither that nerve fibres staining heavily for AChE by necessity are cholinergic (cf. Lewis, Shute and Silver, 1967), nor that blocking by strychnine of a synaptic transmission by necessity excludes that such transmission has a cholinergic link (cf. McKinstry and Koelle, 1967).

Using my new techniques (Fex, 1967*a*,*b*) I have applied *d*-tubocurarine to the organ of Corti of the cat. In four adult cats artificial perilymph containg 1·3 mM (millimoles/l.) calcium chloride was used to approximate normal conditions, as in previous studies (Fex, 1967*a*,*b*). The centrifugally

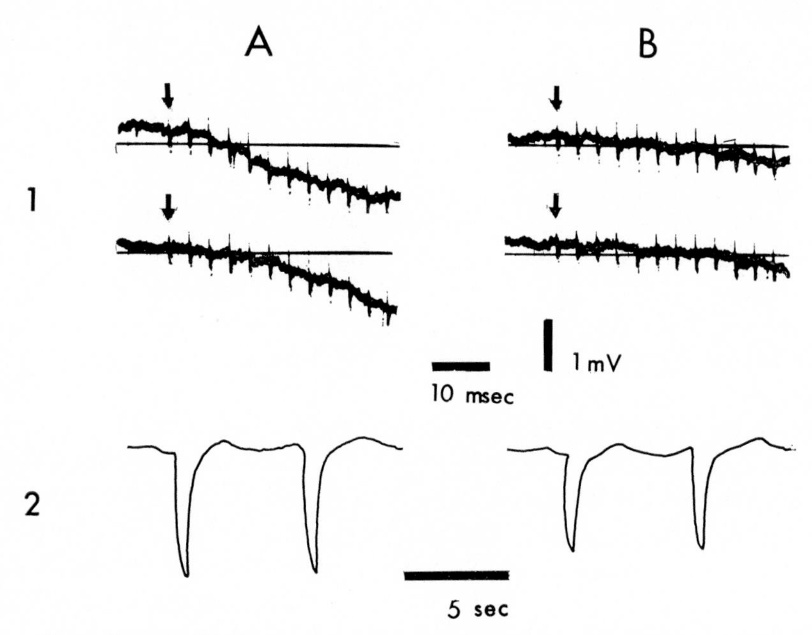

FIG. 4. Records illustrating the action of *d*-tubocurarine on centrifugally evoked intra-cochlear potentials.

Row 1 shows the initial time-course of potentials that were evoked by repetitive electrical stimulation of crossed olivo-cochlear fibres. Row 2 shows the full time-course of the potentials of row 1. All the records were taken with the microelectrode in one position in the scala media. The start of stimulation is indicated by an arrow in row 1. A. Control, artificial perilymph in the scala tympani. B. 12·5 minutes later than A. Artificial perilymph containing 1·0 μmole ($0{\cdot}7 \times 10^{-6}$ g./ml.) *d*-tubocurarine had, 5–8·5 minutes later than A, partly replaced the solution without *d*-tubocurarine. Note that after the application of *d*-tubocurarine, the time of rise and the latency of the potentials were prolonged and the amplitude decreased.

Note that voltage calibration is common to all the records, while time calibration is different for the two rows. The ink-writer that produced the records of row 2 had a rise time constant of 0·15 seconds; d.c. recordings were used for all records. Negativity is downward.

evoked postsynaptic potential was recorded in the scala media, as was the resting potential in the scala media.

Then such artificial perilymph, to which 1·0 μM ($0{\cdot}7 \times 10^{-6}$ g./ml.) of *d*-tubocurarine chloride had been added, was applied to the scala tympani. As a result the centrifugally evoked potential was changed (Fig. 4). Both the time of rise and the latency were prolonged, and the amplitude decreased. These changes could be at least partly reversed with artificial perilymph containing 5 mM-calcium chloride and no tubocurarine, at a time when perilymph with 1·3 mM-calcium chloride had little or no such reversing effect.

There was no change in the resting potential in the scala media larger than 1 mv that could safely be ascribed to *d*-tubocurarine. Further studies are needed to determine the changes of the resting potential that *d*-tubocurarine might cause.

The action of *d*-tubocurarine observed here was in all probability due to one or several of the following mechanisms:

(1) Conduction block in the unmyelinated portion of the olivo-cochlear fibres.

(2) Changes in the conductivity of the hair cell membrane, beyond the subsynaptic membrane opposite the efferent endings.

(3) Interference with the transmitter mechanism of the crossed cochlear efferents, presynaptically or postsynaptically or both.

ad (1) Tubocurarine chloride at a concentration of 10^{-2} g./ml. does not change the evoked C-potential recorded from the desheathed vagus nerve of the rabbit (Arnett and Ritchie, 1961). This implies that the tubocurarine did not block conduction in the unmyelinated fibres of the vagus nerve. It therefore is unlikely that 1·0 μM ($0{\cdot}7 \times 10^{-6}$ g./ml.) *d*-tubocurarine chloride would impair conduction in olivo-cochlear fibres in the cat.

ad (2) The resting potential in the scala media was probably changed less than 1 mv by the application of *d*-tubocurarine. It therefore seems unlikely that *d*-tubocurarine would change the conductivity in the hair cell membrane, beyond the subsynaptic area of the crossed efferents, so that the centrifugally evoked potential would consequently change.

ad (3) Tanaka and Katsuki (1966) could depress the AP at the round window of the cat by electrical stimulation of the crossed efferents after electrophoretic application of *d*-tubocurarine to the organ of Corti. They therefore concluded that the crossed olivo-cochlear fibres are non-cholinergic. However, their description of the *d*-tubocurarine experiments is not detailed, which makes an analysis of the results difficult. It also leaves us room for statements that are contrary to their conclusions from their negative findings. Thus the present new positive findings may safely be taken as evidence that ACh is a transmitter substance of the crossed olivo-cochlear fibres.

In an attempt to obtain stronger evidence that ACh is an inhibitory transmitter substance in the cochlea a different approach to the problem is now being tried in my laboratory. The problem may be complicated by the presence of adrenergic fibres in the cochlea that do not innervate blood vessels (Vinnikov *et al.*, 1966; Spoendlin and Lichtensteiger, 1966; Lichtensteiger and Spoendlin, 1967).

FUNCTION

We cannot yet put our pieces of information together to determine the function of the olivo-cochlear systems. The following pieces of information will have to fit in.

Galambos (1956) found that repetitive electrical stimulation of the crossed olivo-cochlear fibres depresses the AP at the round window. Recordings from primary auditory neurons during such stimulation showed that the crossed efferent fibres are inhibitory (Fex, 1962), as suggested by Galambos (1956). The additional findings that the crossed efferents are activated by sound and respond selectively to different tone frequencies led to the conclusion that these efferents form part of an auditory feedback mechanism (Fex, 1962). It has also been found that the crossed cochlear efferents are best activated from the ear which they innervate (Fex, 1963). The uncrossed olivo-cochlear fibres are also part of an auditory feedback mechanism (Fex, 1965), but conclusive evidence that they exclusively inhibit is lacking.

Several investigators have tried to change acoustically evoked potentials at the round window by natural activation of the cochlear efferent fibres. Thus, the click-evoked AP at the round window has been found to decrease slightly during habituation (Al'tman, 1960; Burgeat, Andrianjatovo and Burgeat-Menguy, 1963; Veselý, 1963) and during the development of an orientation reaction (Maruseva, 1961). The effect of habituation on the AP was inhibited through cortical ablation (Burgeat, Andrianjatovo and Burgeat-Menguy, 1963) or through spreading cortical depression (Veselý, 1963). In the experimental animals of these studies the middle ear muscles were not functioning.

In a more recent study, changes of both AP and CM at the round window of the guinea pig were found during habituation, distraction and conditioning to acoustic stimuli (Buño *et al.*, 1966). No such changes were observed when the crossed efferents had been sectioned in the brainstem.

In two very recent studies changes of behaviour have been the criterion for effects of the crossed cochlear efferents (Capps, 1967; Trahiotis, 1967).

Capps (1967) used four squirrel monkeys in her study. After transection of the crossed olivo-cochlear fibres the animals showed deficit in frequency discrimination, tested at 1,000 cyc./sec. and 4,000 cyc./sec. Capps concluded that the changes were due to transection of the crossed olivo-cochlear fibres. She suggested that the efferent auditory system operates to sharpen the frequency-resolving power of the ear.

Trahiotis (1967) used six cats with the right cochlea destroyed in his behavioural study. In three cats, the crossed olivo-cochlear fibres were cut; the other three cats were sham-operated. There was an increased amount of masking at 1,000 cyc./sec. and 2,000 cyc./sec. with transection of the crossed efferents.

Thus, what we now know about the function of the cochlear efferents can be summarized as follows. The cochlear efferent systems form part of a complex auditory feedback mechanism. The cochlear efferent activity may be under cortical and subcortical control. The crossed olivo-cochlear bundle may sharpen the ear's frequency discrimination (Capps, 1967) and may reduce the masking effect of noise (Trahiotis, 1967). What the other cochlear efferents might do is an open question.

SUMMARY

The crossed olivo-cochlear fibres in the cat's cochlea act on the membrane of outer hair cells and inhibit activity in primary auditory neurons. The inhibition is thus presynaptic relative to the synapse between hair cells and afferent terminals.

Repetitive electrical stimulation of the crossed olivo-cochlear fibres evokes postsynaptic potentials in the cochlea, which are generated by the membrane of the outer hair cells. The new findings that these postsynaptic potentials are depressed by *d*-tubocurarine, at a concentration of *d*-tubocurarine chloride of 1·0 μmole (0·7 $\times 10^{-6}$ g./ml.) in artificial perilymph in the scala tympani, suggest that acetylcholine is a transmitter substance of efferent inhibition in the cochlea.

The uncrossed olivo-cochlear fibres may also be inhibitory, although conclusive evidence for this is lacking. These efferents may act on afferent endings, perhaps only underneath inner hair cells.

Uncrossed reticulo-cochlear efferents are present in rodents, but not in cats. These neurons may participate in the efferent innervation of the outer hair cell region of the rodent cochlea, with the *2a* endings of Smith and Sjöstrand, while the cat cochlea would have no *2a* endings.

Adrenergic neurons may innervate the inner hair cell region of the organ of Corti.

The functions of the different cochlear efferents are not yet well understood. However, recent evidence indicates that the crossed olivo-cochlear bundle may sharpen the ear's frequency discrimination and may reduce the masking effect of noise.

Acknowledgements

The *d*-tubocurarine experiments and the manuscript of this paper were completed at the Laboratory of Neurobiology, National Institute of Mental Health, Bethesda, Maryland, U.S.A. The friendly encouragement of the Chief of the Laboratory of Neurobiology, Dr. Ichiji Tasaki, has made my work here even more pleasant than it would otherwise have been.

The cost of my family's travel from Stockholm to Washington, D.C., was partly covered by a grant from the Swedish Medical Research Council.

REFERENCES

AL'TMAN, IA. A. (1960). *Sechenov physiol. J. USSR*, **46**, 617–629. (*Fiziol. Zh. SSSR*, **46**, 526–536.)

ARNETT, C. J., and RITCHIE, J. M. (1961). *J. Physiol., Lond.*, **155**, 372–384.

BUÑO, W., Jr., VELLUTI, R., HANDLER, P., and GARCIA AUSTT, E. (1966). *Physiology and Behavior*, **1**, 23–35.

BURGEAT, M., ANDRIANJATOVO, J., and BURGEAT-MENGUY, C. (1963). *Annls Oto-lar.*, **80**, 575–580.

CAPPS, M. J. (1967). Thesis, University of Illinois. (Capps, M. J., and Ades, H. W. (1968). In preparation.)

CHURCHILL, J. A., and SCHUKNECHT, H. F. (1959). *Henry Ford Hosp. med. Bull.*, **7**, 202–212.

CORDIER, R. (1964). *Bull. Acad. r. Méd. Belg.*, **4**, 729–753.

DAVIS, H. (1965). *Cold Spring Harb. Symp. quant. Biol.*, **30**, 181–190.

DESMEDT, J. E., and DELWAIDE, P. J. (1965). *Expl Neurol.*, **11**, 1–26.

DESMEDT, J. E., and LAGRUTTA, V. (1963). *Nature, Lond.*, **200**, 472–474.

ELDREDGE, D. H. (1967). *J. acoust. Soc. Am.*, **41**, 1386–1388.

FEX, J. (1959). *Acta oto-lar.*, **50**, 540–541.

FEX, J. (1962). *Acta physiol. scand.*, **55**, Suppl. 189.

FEX, J. (1963). *Acta physiol. scand.*, **59**, Suppl. 213, 41.

FEX, J. (1965). *Acta physiol. scand.*, **64**, 43–57.

FEX, J. (1967*a*). *Nature, Lond.*, **213**, 1233–1234.

FEX, J. (1967*b*). *J. acoust. Soc. Am.*, **41**, 666–675.

FEX, J., FUXE, K., and LENNERSTRAND, G. (1965). *Acta physiol. scand.*, **64**, 259–262.

GALAMBOS, R. (1956). *J. Neurophysiol.*, **19**, 424–437.

GISSELSSON, L. (1960). *Acta oto-lar.*, **51**, 636–637.

IURATO, S. (1964). *Atti Soc. ital. Anat.*, **72**, 60.

KATSUKI, Y. (1965). *Physiol. Rev.*, **45**, 380–423.

KATSUKI, Y., TANAKA, Y., and MIYOSHI, T. (1965). *Nature, Lond.*, **207**, 32–34.

LEWIS, P. R., SHUTE, C. C. C., and SILVER, A. (1967). *J. Physiol., Lond.*, **191**, 215–223.

LICHTENSTEIGER, W., and SPOENDLIN, H. (1967). *Life Sci.*, **6**, 1639–1645.

MARUSEVA, A. M. (1961). *Sechenov physiol. J. USSR*, **47**, 599–608. (*Fiziol. Zh. SSSR*, **47**, 542–550.)

MCKINSTRY, D. N., and KOELLE, G. B. (1967). *J. Pharmac. exp. Ther.*, **157**, 328–336.

RASMUSSEN, G. L. (1946). *J. comp. Neurol.*, **84**, 141–219.

RASMUSSEN, G. L. (1953). *J. comp. Neurol.*, **99**, 61–74.

RASMUSSEN, G. L. (1960). In *Neural Mechanisms of the Auditory and Vestibular Systems*, pp. 105–115, ed. Rasmussen, G. L., and Windle, W. F. Springfield: Thomas.

RASMUSSEN, G. L. (1965). Personal communication to G. Rossi. In *Le Basi Morfo-Funcionali del Controllo della Sensazioni Acustiche*, pp. 41–116, Filogamo, G., Candiollo, L., and Rossi, G. Turin: Societá Italiana di Laringologia, Otologia e Rinologia.
RODRÍGUEZ ECHANDÍA, E. L. (1967). *Z. Zellforsch. mikrosk. Anat.*, **78,** 30–46.
ROSSI, G., and CORTESINA, G. (1962). *Minerva otorinolar.*, **12,** 1–63.
ROSSI, G., and CORTESINA, G. (1965). *Acta anat.*, **60,** 362–381.
SCHMIDT, R. S., and FERNÁNDEZ, C. (1962). *J. cell. comp. Physiol.*, **59,** 311–322.
SMITH, C. A., and SJÖSTRAND, F. (1961). *J. Ultrastruct. Res.*, **5,** 523–556.
SOHMER, H. (1966). *Acta oto-lar.*, **62,** 74–87.
SPOENDLIN, H. (1966). *The Organization of the Cochlear Receptor. Advances in Oto-Rhino-Laryngology*, vol. 13. Basel and New York: Karger.
SPOENDLIN, H. H., and GACEK, R. R. (1963). *Ann. Otol. Rhinol. Lar.*, **72,** 660–686.
SPOENDLIN, H., and LICHTENSTEIGER, W. (1966). *Acta oto-lar.*, **61,** 423–434.
TANAKA, Y., and KATSUKI, Y. (1966). *J. Neurophysiol.*, **29,** 94–108.
TRAHIOTIS, C. (1967). Thesis, Wayne State University. Paper in preparation.
VESELÝ, C. (1963). *Supplementum Sborníku vědeckých prací Lékařshé fakulty KU v Hradci Králové*, **6,** 247–267.
VINNIKOV, IA. A., GOVYRIN, V. A., LEONT'EVA, G. P., and ANICHIN, V. F. (1966). *Dokl. Akad. nauk SSSR*, **171,** 484–486 (in Russian).

DISCUSSION

Lowenstein: You mentioned the difference in the results obtained by Katsuki and Gisselsson in applying acetylcholine to the hair cells; did they use the same concentration, because with significantly different concentrations, I take it one can have opposite results?

Fex: Neither Gisselsson nor Katsuki mentioned the concentrations they used. But more critical is the electrophoretic current they used and the time for which they applied it. Gisselsson used a very much weaker current (2×10^{-8} A) than did Katsuki (8×10^{-6} A) and applied it for a short time (although he does not state his time). The amount of acetylcholine applied by Gisselsson could therefore have been lower by a factor of 1,000, or more, than the amount that Katsuki applied.

Lowenstein: If you have too much acetylcholine, you can get the opposite effect?

Fex: Yes.

Erulkar: I would agree with that, but it would seem that the experiment which is necessary to see whether there is any transmitter release (especially if the transmitter is acetylcholine), is to add magnesium, as Katz and Miledi did at the neuromuscular junction (Katz, B., and Miledi, R. [1965]. *Proc. R. Soc. B*, **161**, 496–503). Magnesium will still allow the action potential to be relayed in the nerve but it will block any release of transmitter. I was not quite sure of the significance of your calcium experiments, but at the neuromuscular junction calcium is essential for the release of transmitter substance. With your preparation, magnesium would be a critical substance to try.

Fex: No, it is not critical. This is clear from the following argument. Calcium facilitates the release of acetylcholine at the neuromuscular synapse (Katz, B., and Miledi, R. [1965]. *Proc. R. Soc. B*, **161**, 496–503). Calcium regulates the release of gamma-amino-butyric acid (GABA) at inhibitory synapses in the claw-opener muscles of the lobster (Otsuka, M., Iversen, L. L., Hall, Z. W., and Kravitz, E. A. [1966]. *Proc. natn. Acad. Sci. U.S.A.*, **56**, 1,110–1,115) and there is evidence that calcium also regulates the release of noradrenaline by sympathetic nerve stimulation (Boullin, D. J. [1966]. *J. Physiol., Lond.*, **183**, 76P). Thus calcium is not specific for the release of one particular kind of synaptic transmitter substance. Concerning magnesium, less is known. It has been shown that magnesium and calcium have strictly antagonistic effects at the neuromuscular synapse—that is, that magnesium blocks the release of acetylcholine at this synapse (del Castillo, J., and Engbaek, L. [1954]. *J. Physiol., Lond.*, **124**, 370–384). We do not yet know whether magnesium blocks transmitter release at all chemical synapses in vertebrates or, less generally, at all synapses in vertebrates at which calcium facilitates transmitter substance release. Magnesium does perhaps block the release of acetylcholine at all different kinds of cholinergic synapses in vertebrates and it would be surprising if exceptions to this were to appear. However, such exceptions are certainly still conceivable. Therefore, if local application of magnesium at the cochlear, efferent, inhibiting synapses were to leave the action of these synapses unchanged this would not conclusively rule out acetylcholine as a transmitter substance for these synapses. Neither would a positive finding establish that these synapses are cholinergic, but would no more than suggest that the release of an inhibitory transmitter substance, whatever its nature might be, is inversely related to the magnesium concentration. Similarly, my calcium experiments ([1967]. *Nature, Lond.*, **213**, 1233–1234) were interpreted to suggest that the release of an inhibitory transmitter substance at the outer hair cells depends upon calcium.

Also, other pieces of evidence presented in my paper here together indicate strongly that there *is* a transmitter release. In particular, my results with an amount of *d*-tubocurarine which is of the same order as is commonly used for blocking the effects of release of acetylcholine, indicate that acetylcholine has a physiological action at this efferent synapse. The relevant experiment to do now is to see if the inhibition *per se* becomes less when *d*-tubocurarine has been applied to a concentration of one μmole/l. in the scala tympani. Also, one could try to pick up acetylcholine from the perilymph after stimulating the crossed efferents.

Johnstone: One thing we are certain of is that the inhibitory efferent fibres are blocked by strychnine and by brucine very specifically and not by picrotoxin. This is in my opinion good evidence against a cholinergic mechanism. Secondly, I have perfused perilymph with a variety of drugs including various amounts of *d*-tubocurarine and also physostigmine (eserine) and acetylcholine. If this were a cholinergic mechanism, I would expect physostigmine to give at least some activation of the efferents and so cause a depression of the N_1 (afferent) action potential, but it does not (Johnstone, B. M. [1965]. *Proc. XXIII Int. physiol. Congr.*, abstract no. 867).

The published experiments of Gisselsson were related to a phase change in cochlear microphonics. I believe that somebody has recently repeated these experiments.

Fex: Concerning the significance of the blocking of inhibition by strychnine we may note the findings by D. N. McKinstry and G. B. Koelle ([1967]. *J. Pharmac. exp. Ther.*, **157**, 328–336) that very small doses (1·0 to 100 μg./ml.) of strychnine sulphate cause a significant reduction of acetylcholine release by preganglionic stimulation. They have related these findings to a suggested mechanism in which acetylcholine is released by the crossed efferent fibres in the cochlea ([1967]. *Nature, Lond.*, **213**, 505–506). On the one hand I believe their story, and on the other hand there are no results on strychnine that do rule out the possibility of acetylcholine as a transmitter substance of the crossed cochlear efferents.

Secondly, since in your perfusion experiments you did not stimulate the crossed efferents, I am not sure that you can say much about these. It is easy to miss the efferent effects if experimental conditions are unfavourable or if one has no criterion of efferent effects to begin with. Thus, did you control the absolute d.c. level of the scala media? If you did not, how did you know that your guinea pigs were in an optimal state? If my cats are not in an optimal state I do not see very much of an effect by the crossed cochlear efferents. It is probably important how the hair cells are biased when they are hit with the efferent barrage. If they are slightly depolarized or injured you might get no efferent inhibition at all. It is also well known that if one tries to see efferent effects on sound-evoked potentials at strong sound pressures, little will be seen.

Johnstone: The animals were in good condition and gave normal responses. I agree that it would be best to stimulate the efferents and to pick up the effect on the N_1 action potential. I did not do this as my experiments were primarily designed to reveal the afferent transmitter. But if acetylcholine is the inhibitory transmitter, why does not putting a fair quantity around the hair cell, either by perfusing with acetylcholine or by applying physostigmine, mimic the action of stimulating the efferents? As far as I know it does in all other systems.

Fex: But you did report a transient blocking of the afferent action potential by Prostigmine (neostigmine) at a concentration of 10 μmole/l. ([1965]. *Proc. XIII Int. physiol. Congr.*, abstract no. 867). Surely this blocking could have been mediated by cochlear efferent synapses; you did not prove the opposite. Also, Y. Tanaka and Y. Katsuki ([1966]. *J. Neurophysiol.*, **29**, 94–108) did observe the disappearance of sound-evoked activity in single auditory nerve fibres with intracochlear application of acetylcholine, but the mechanism of this effect is not well understood.

I tried using physostigmine but realized very quickly that, with extracellular recording and therefore necessarily repetitive stimulation of the efferent fibres, an effect of physostigmine might be too complicated to interpret. One might get an effect as if acetylcholine had been applied or an opposite effect, as if too much acetylcholine had been applied. Such a reversal of effect was mentioned earlier in the discussion as a possible explanation of the results when acetylcholine had been applied in the cochlea and changes of the cochlear microphonics were observed. If one wants to try to mimic a transmitter action of the cochlear efferents with experiments that lend themselves to simple interpretations and strong conclusions, one may be limited to intracellular recordings from hair cells, looking at the centrifugally evoked postsynaptic potentials.

As for the paper by L. Gisselsson that I mentioned in my paper ([1960]. *Acta oto-lar.*, **51**, 636–637) this is preliminary work that he did in S. Thesleff's laboratory—preliminary, but well controlled. You were thinking of his thesis in 1950 (Gisselsson, L. [1950]. *Acta oto-lar.*, Suppl. 82). There he looked at the effect of physostigmine on the cochlear microphonics and found a phase shift. This phase shift has later been shown to have been due possibly to changes of body temperature (Gannon, R. P., Laszlo, C. A., and Moscovitch, D. H. [1966]. *Acta oto-lar.*, **61**, 536–546).

Whitfield: We have in one of the centrifugal pathways to the cochlear nucleus an apparently cholinergic system; the pathway is blocked by gallamine, by atropine and by dihydro-β-erythroidine, but strychnine has no effect on it. Such behaviour is substantially different from that of the olivo-cochlear bundle. On the other hand we did some pilot experiments on perfusing the guinea pig cochlea and stimulating the crossed olivo-cochlear bundle, and found a positive result for acetylcholine, tested on leech muscle, although admittedly, it was right down at the limits of sensitivity of the preparation. If the results are correct they certainly have not clarified the problem!

Davis: It is extremely difficult to draw clear conclusions from these varied pieces of information when there are so many factors to consider. We should recall that the action potential, which is used as one measure of activity, is probably highly dominated by the inner hair cells, if we accept Dr. Spoendlin's counts. On the other hand, the cochlear microphonic, particularly in the guinea pig in the first turn, is strongly dominated by the outer hair cells; two or three lines of evidence suggest that the microphonic of the inner hair cells contributes very little indeed. So that if there is a clear difference in distribution between crossed and uncrossed efferents relative to inner and outer hair cells, we might find that the crossed and uncrossed efferents produce selective, different effects on the action potential and cochlear microphonic respectively. Perhaps this would help to resolve some of the apparently contradictory findings.

Fex: Yes. We could first then accept as an alternative the suggestion in my paper that the crossed efferents work with the two kinds of synapses, the one acting directly on outer hair cells and the other on afferent endings underneath the inner hair cells. M. L. Wiederhold's (personal communication) and my findings ([1962]. *Acta physiol. scand.*, **55**, Suppl. 189) that almost all auditory nerve fibres are inhibited by crossed efferent stimulation would then be compatible with Dr. Spoendlin's counts. Also, at these two different kinds of crossed efferent synapses, neuropharmacologically active agents might have selectively different actions, either because of actual differences in structure between the synapses (cf. Tauc, L. [1967]. *Physiol. Rev.*, **47**, 521–593) or because of differences in accessibility for applied substances between the two kinds of synapses. If we then accept Dr. Davis' highly interesting suggestion that the action potential be dominated by the inner hair cells, still another apparent contradiction, also mentioned here before, might be resolved. Thus the crossed efferent inhibition of the action potential that Y. Tanaka and Y. Katsuki ([1966]. *J. Neurophysiol.*, **29**, 94–108) could see after intracochlear application of *d*-tubocurarine could be due to an action of crossed efferents on afferent endings beneath the inner hair cells. This action could be relatively resistant to such application of *d*-tubocurarine, which would explain their findings. On the other hand, the action of the crossed efferents on the outer hair cells could be relatively sensitive to local application of *d*-tubocurarine, which would explain my findings as I have described them in my paper here.

Whitfield: The cochlear potentials appear also to be affected by past history. The resolution of the summating potential following the termination of a stimulus seems to depend on the amplitude and duration of the preceding stimulus (Stopp, P. E. [1967]. *Nature, Lond.*, **215**, 1400). Change of stimulus duration may affect not only its amplitude but even its polarity.

Johnstone: Dr. Fex, when you stimulated the crossed efferent fibres and got this very nice d.c. change, did you stimulate the uncrossed efferents and look for a d.c. change, particularly in the scala media? Because if what Dr. Spoendlin says is true and the uncrossed efferents go to the inner hair cell region, and not to the cells but to the afferent nerves from them, then on stimulating these uncrossed fibres you would not expect to pick up such a change in the scala media. This could be an additional confirmation of the histological picture.

Fex: I have not stimulated the uncrossed efferents in this kind of experiment but a rather equivalent negative finding is the one I mentioned in my paper that the cochlear microphonics are not changed by electrical stimulation of the uncrossed efferent fibres.

Johnstone: Incidentally, I was very happy to see your relative amplitudes of the d.c. potentials. You pick up about four to five times as much potential in the scala media as you do in the scala tympani. This would fit in very nicely with stimulation of the crossed fibres causing a change in resistance in the hair cells (Johnstone, B. M., Johnstone, J. R., and Pugsley, I. D. [1966]. *J. acoust. Soc. Am.*, **40**, 1398–1404).

Tumarkin: Dr. Fex, as I understand it, you assume that having applied a chemical, it reaches hair cells in one way or another. But I have the impression that in the normal cochlea the flow of perilymph goes from the scala tympani round through the helicotrema into the scala vestibuli and then across Reissner's membrane, and is absorbed by the stria vascularis. This is a very circuitous route which may be dominated and influenced by all sorts of factors which may vary from one species to another, and so it must be difficult to know exactly when the drug reaches the hair cells?

Fex: There is an effect very quickly, as fast as I can record it; it is seen within 10 seconds when calcium chloride, for instance, is applied in the scala tympani. Experiments done in Dr. Davis' laboratory, also with Dr. Tasaki (Tasaki, I. [1957]. *Ann. Rev. Physiol.*, **19**, 417–434) indicate that ions pass between the scala tympani and the organ of Corti through the basilar membrane. Surely 10 seconds is what you could expect for this? Also, the round window has been taken away in my experiments, one can see the scala tympani filling up quickly with natural perilymph after the suction and there would thus be a *stream* against the flow of applied artificial perilymph; it would have to work by diffusion and 10 seconds for that passage, round the helicotrema, would be too short.

Tumarkin: That is very much to the point, I quite agree.

[For further discussion on the cochlear efferent system, see p. 298.]

Note added in proof by Dr. J. Fex

Artificial perilymph in the cat's scala tympani has been assayed for acetylcholine in my laboratory. The new results from these experiments strongly suggest that acetylcholine is released in the inner ear by the crossed olivo-cochlear fibres, when these fibres are electrically stimulated in the brainstem.

SECTION IV

SPATIAL LOCALIZATION

ORIENTATION THROUGH SOUND IN FISHES

H. Kleerekoper★ and T. Malar★

Department of Biology, McMaster University, Hamilton, Ontario, Canada

INTRODUCTION

The problem of the localization of a sound source by fish has received but little attention in the past. Recently, however, interest in this important aspect of hearing in fish has increased (van Bergeijk, 1964; Dijkgraaf, 1964; Moulton and Dixon, 1967). Evidence that an ostariophysan species could localize a source of sound was brought by Kleerekoper and Chagnon (1954), while Reinhardt (1935) and von Frisch and Dijkgraaf (1935) concluded that fish cannot achieve such localization. Recent theoretical studies (Harris and van Bergeijk, 1962; Harris, 1964; van Bergeijk, 1964) have demonstrated that an underwater sound source produces, in addition to a propagated pressure wave, displacement of water particles. While the amplitude of the pressure wave decreases as $1/r$ (where r is distance), that of the displacement diminishes as the square of the distance. At a certain distance from the source (about one wavelength/π) both amplitudes are equal. Distances closer to the source are defined by van Bergeijk (1964) as "near-field" and greater distances as "far-field". All early observations on the role of sound in the orientation of fish have been carried out in tanks whose size was such that the fish were, entirely or to a great extent, within the theoretical near-field boundaries of the sound sources. Although in the near-field there are both displacements of particles and pressure waves, the relative amplitude of the latter becomes very small as the sound source is approached. Thus, responses of fish in the near-field are likely to be elicited by the displacement of water particles, to which lateral line organs are responsive (Harris and van Bergeijk, 1962), rather than by pressure waves. These and additional considerations regarding certain theoretical limitations of the available sensory equipment led van Bergeijk (1964) to conclude that fish do not have the capability of localizing a single sound source in the far-field, although the presence of the sound may be detected.

In recent years, however, it has been demonstrated that the production of sound is a fairly widespread capability among fish species and evidence has been accumulating which indicates that such sound, often produced

★ Now at Institute of Life Science, Texas A and M University, College Station, Texas.

through special anatomical adaptations, may play a significant role in adaptive behaviour in a wide variety of situations, which include feeding, defence, migration, schooling, reproductive and other behavioural aspects. (The literature on the behavioural implications of sound produced by fish was summarized by Winn, 1964.) In view of this evidence, the assumption that fish cannot extract directional information from a sound field ("far-field") is, at least for the biologist and behaviourist, a most unsatisfactory one, particularly since it is based on the failure to find a sensory mechanism required for the acquisition of such information. Rather than placing emphasis on our inability to identify an appropriate directional mechanism, we should be giving attention to the execution of experiments designed to test the capability of fish to extract directional information from sound in the far-field. Other than a few observations in the field (e.g. Wisby *et al.*, 1964; Busnel, 1959; Moulton and Backus, 1966; Moorhouse, 1933) there are no quantitative data available to elucidate this problem.

Recently, in the laboratory of the senior author, a system was completed which is capable of monitoring the locomotion of fish under controlled conditions and of extracting and processing many of the data pertinent for a quantitative description of locomotor patterns, in a tank whose dimensions allowed for observations in the acoustic far-field. In the present study, these new facilities were used to ascertain whether the locomotor pattern of carp is affected by sound in the far-field and to quantify these changes. A small monitor system provided additional information on the effects of sound on the locomotion of sunfish (*Lepomis*).

METHODS

Naïve, adult, single individuals of carp (*Cyprinus carpio*) and sunfish (*Lepomis gibbosus*), obtained from natural lakes, were used for these observations. Prior to the experiments, the sunfish had been kept in holding tanks for several months, and the carp for 2 to 3 weeks. During the experiments, feeding was interrupted. Locomotor patterns of a single animal were monitored before and during the generation of sound by one of two methods, depending on the species.

Carp. Locomotor patterns of this species were monitored in a tank measuring 500 × 500 cm. and 60 cm. high, by means of a matrix of 50 × 50 photoconductive cells placed at 10 cm. centres below the transparent floor of the tank and activated by a collimated light field. The latter covers, from overhead, the whole horizontal area of the tank at low, uniform intensity. Water is supplied through perforations uniformly distributed over the entire area of one of the lateral walls of the tank under near laminar conditions of flow at the rate of 50 litres per minute and leaves the tank, through perforations, equally distributed, over the entire area of the lateral wall on the opposite side of the tank (Kleerekoper, 1967*b*). Photocells, when affected by the presence of the fish, provide pulses to an electronic logic system which allows for the determination of the address of the photocell in the matrix. This address is punched on tape together with the time of

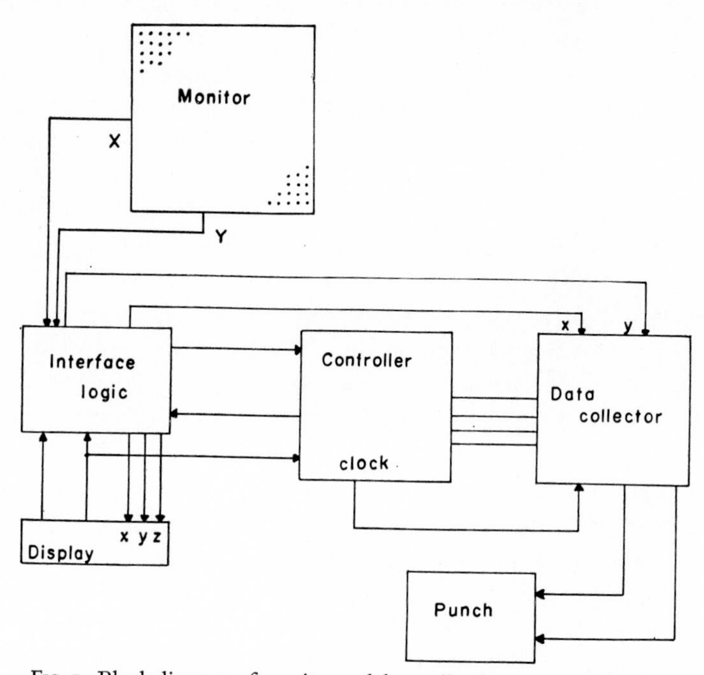

FIG. 1. Block diagram of monitor and data collection system. The display shows the displacements of the fish on a matrix background.

the event in fractions of a second and is displayed on the screen of an oscilloscope for direct observation of the locomotor patterns described by the fish. The data on the tape are transposed on to IBM cards for data analysis as well as for controlling an incremental plotter, which duplicates the animal's movements on paper (Fig. 1). In addition to the locomotor patterns themselves, processing of the data provided information on the speed and direction of locomotion, the frequency distribution of the angles of turn, the ratio of left- and right-hand turns, the relationships between speed of locomotion and angles of turns, either left- or right-handed, and the lengths of straight pathways between turns.

Sound in this tank was produced by a 20-w driver loudspeaker placed 20 cm. above and directed normally to the water surface, midway along the "outflow wall". The gross pattern of the sound pressure produced by an audio-oscillator and amplifier at 2,000 cyc./sec. is shown in Table I.

Table I

PATTERN OF SOUND PRESSURES IN TANK

X Axis

	0·0	2·6	12·8	6·4	15·3	3·8	12·8	1·3	10·2	2·6
	5·1	14·0	15·3	2·6	5·1	7·7	2·6	14·0	12·8	11·5
	6·4	5·1	11·5	3·8	20·4	5·1	15·3	6·4	7·7	3·8
	7·7	23·0	21·7	20·4	7·7	2·6	10·2	17·9	12·8	6·4
	5·1	7·7	23·0	7·7	7·7	6·4	10·2	7·7	10·2	2·6
Y Axis Speaker	20·4	14·0	19·1	17·9	11·5	6·4	15·3	17·9	7·7	5·1
	6·4	10·2	12·8	10·2	19·1	3·8	8·9	5·1	7·7	6·4
	12·8	12·8	8·9	2·6	16·6	10·2	12·8	3·8	12·8	7·7
½ m.	5·1	5·1	17·9	3·8	6·4	5·1	5·1	3·8	12·8	6·4
	1·3	3·8	7·7	5·1	14·0	6·4	6·4	2·6	10·2	5·1

½ m.

dynes/cm.²

At this frequency, in a tank 500 × 500 cm. in size, with the sound source placed outside the water, near-field effects (Harris and van Bergeijk, 1962; van Bergeijk, 1964) play no role.

Locomotion was monitored before and, without interruption, during stimulation with sound.

Sunfish. Locomotion of animals of this species was tracked in a cylindrical tank, 210 cm. in diameter and 30 cm. high, described elsewhere (Kleerekoper, 1967*a*). In this tank, hollow walls, 50 cm. long, placed radially, divide the space equally into 16 vertical compartments which communicate only centrally. Water is introduced peripherally into each compartment, at equal rates, through a glass tube originating from a glass manifold overhead. The outlet consists of a central standing pipe acting as an overflow. The hollow walls, alternately, contain banks of photoconductive

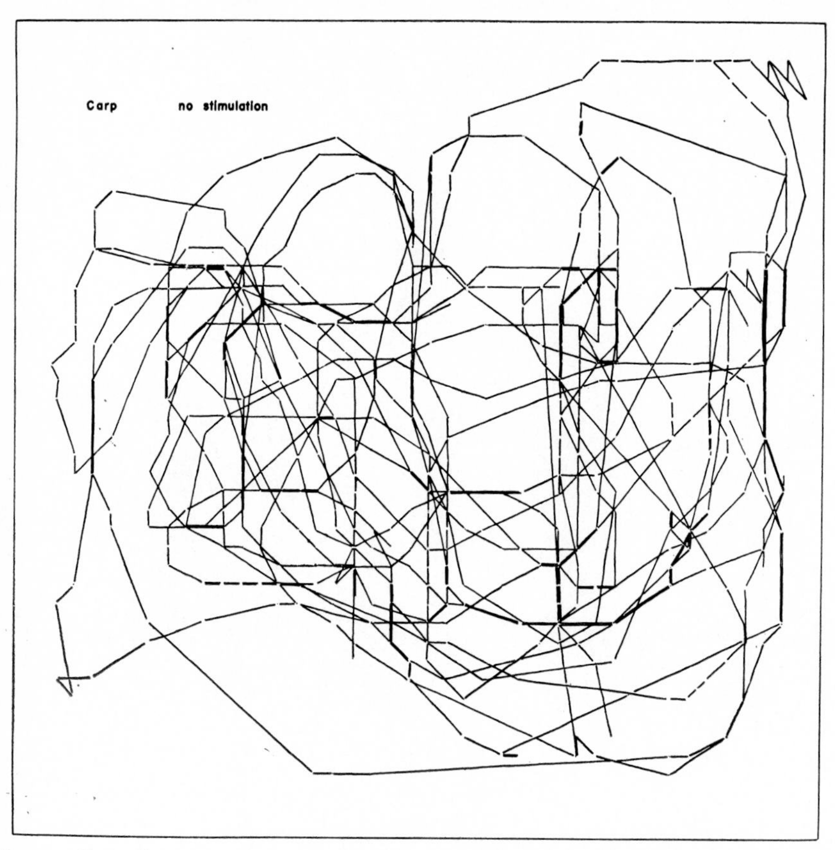

FIG. 2. Carp. August 29, 1967. Locomotor pattern with silent loudspeaker. Dimensions of tank: 500 × 500 cm.

cells and light sources. The responses of the photocells to locomotion of the animal are monitored in an adjoining instrument room where the displacements are recorded as to direction, left- and right-hand turns and angle of turn. Single headphones in plastic bags were placed peripherally in four of the compartments, directed toward the centre of the tank and connected to an audio-oscillator. In each of these experiments, the locomotor pattern of a single sunfish was monitored during periods of from

12 to 24 hours before sound stimulation. At the end of this period, without interruption in the recording, a single speaker was driven at 700 cyc./sec. to stimulate this non-ostariophysan fish. Although hollow dividing walls must have greatly restricted the near-field effect, the latter must have been present in the compartment containing the sound source.

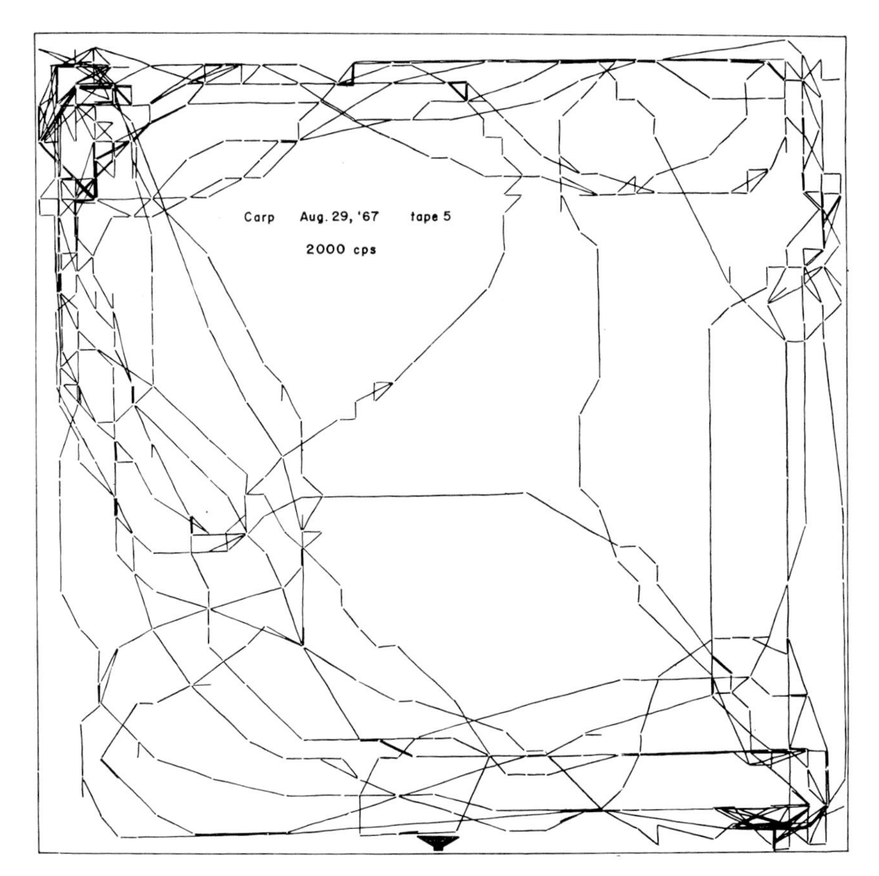

Fig. 3. Carp. (Continuation of August 29, 1967.) Locomotor pattern during transmission of sound at 2,000 cyc./sec. Position of loudspeaker is shown at centre bottom of plot.

RESULTS

Carp. The effects of a sound field containing a main sinusoidal frequency of 2,000 cyc./sec. on the locomotor patterns of carp is shown in Figs. 2 to 5, representing the results of two experiments with different individuals. The patterns described by the naïve animals in the absence of the sound field (Figs. 2 and 4) were recorded during two hours immediately following the introduction of the animal into the tank. Generation of the sound field produced the patterns of Figs. 3 and 5.

In both experiments, generation of the sound resulted in a marked change in the locomotor pattern. The movements, which, by visual inspection, had been fairly evenly distributed throughout the tank, became largely

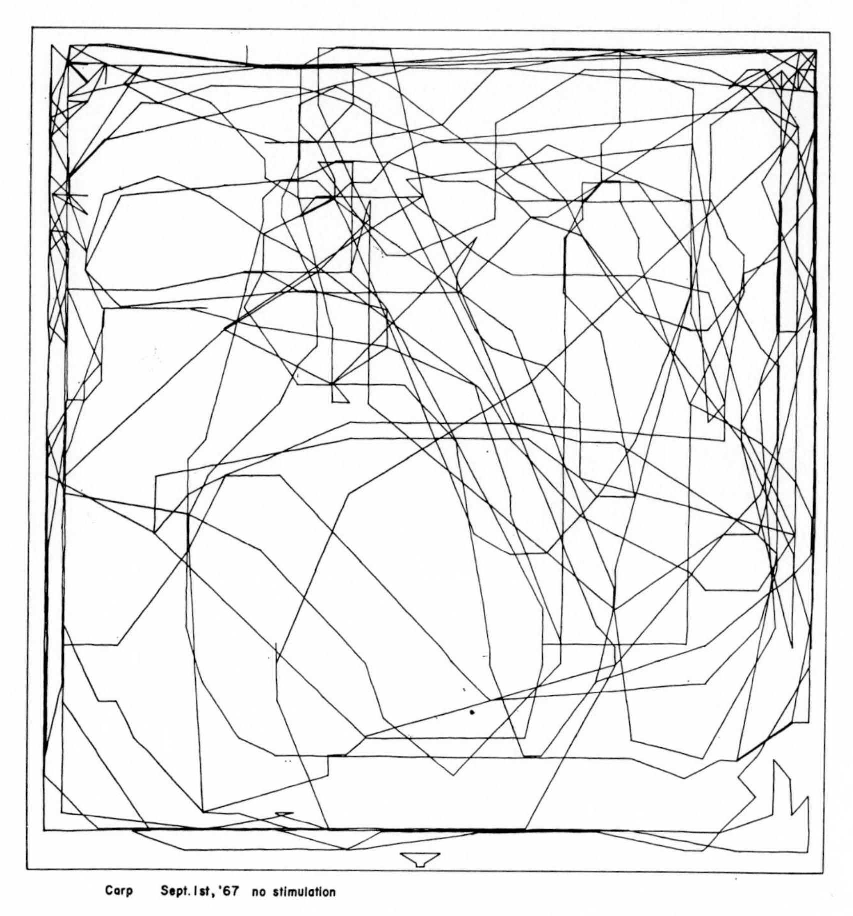

FIG. 4. Carp. September 1, 1967. Locomotor pattern with silent loudspeaker.

restricted to well-defined, small areas as soon as sound was generated. These areas coincide with those of lowest sound pressure, while areas of highest sound pressure are crossed by very few pathways.

Since locomotor patterns generally are largely determined by the magnitude of the angles of turns, the frequency distribution of these angles was computed. Fig. 6 represents this distribution in classes of 10 degrees for the period before (solid line) and during (broken line) the generation of the sound field for the experiment of September 1st. In this histogram, all the turns made by the animal are included and are separated into left- and right-handed turns. The distribution of angles before stimulation with

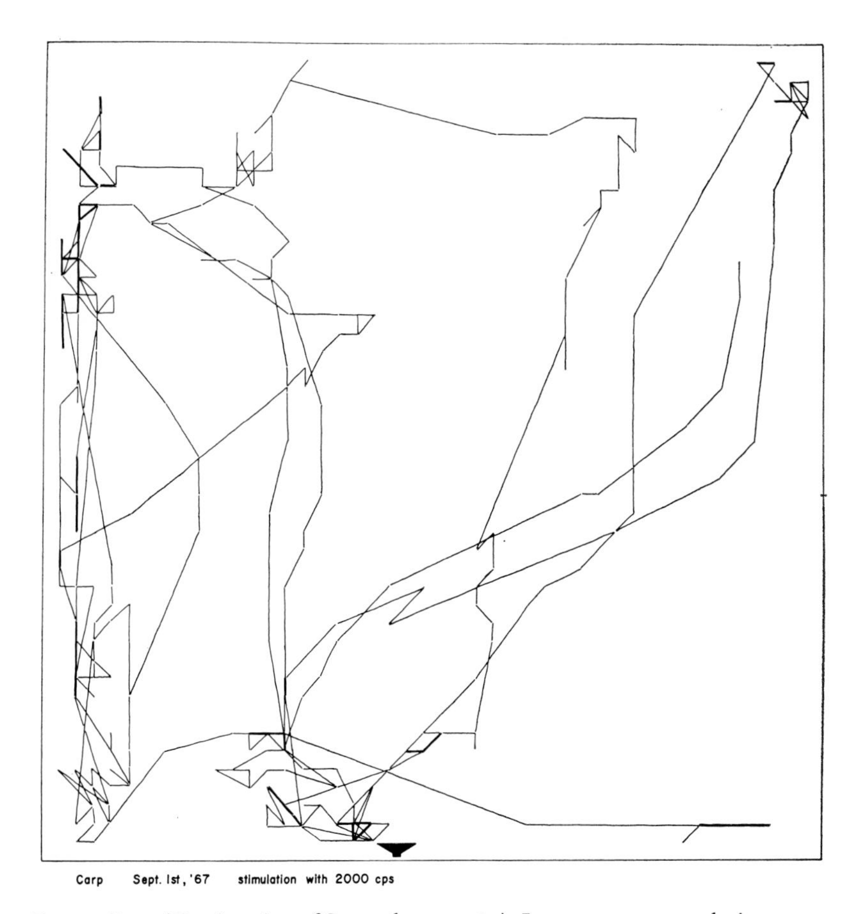

FIG. 5. Carp. (Continuation of September 1, 1967.) Locomotor pattern during transmission of sound at 2,000 cyc./sec.

sound is normal, with a peak at 10–30° left. Generation of the sound produced a pronounced shift towards sharper turns, both left- and right-handed. This shift is particularly clear when the distribution of left- and right-handed turns are combined, as in Fig. 7.

In order to ascertain whether, in the above shift, all turns participated equally or the shift was due mainly to the first turns whenever the fish

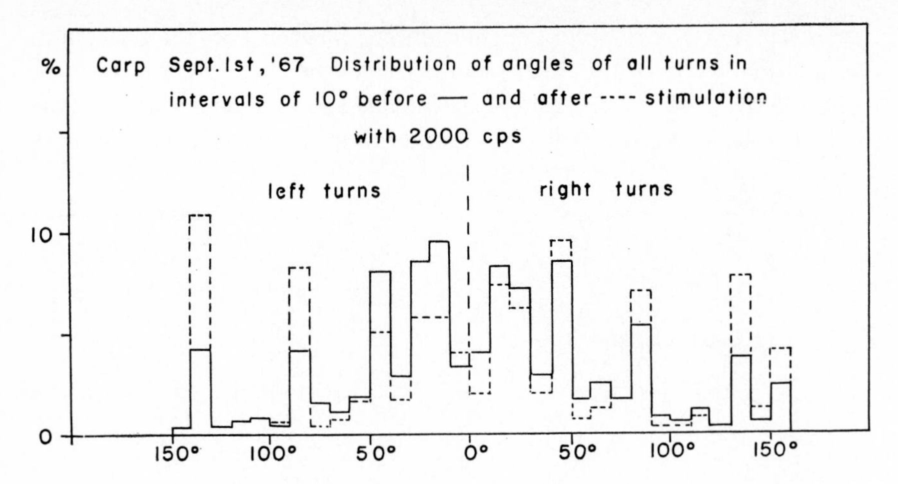

FIG. 6. Carp. September 1, 1967. The distribution of the angles of all turns in intervals of 10° before and after stimulation with sound at 2,000 cyc./sec. Left- and right-hand turns shown separately.

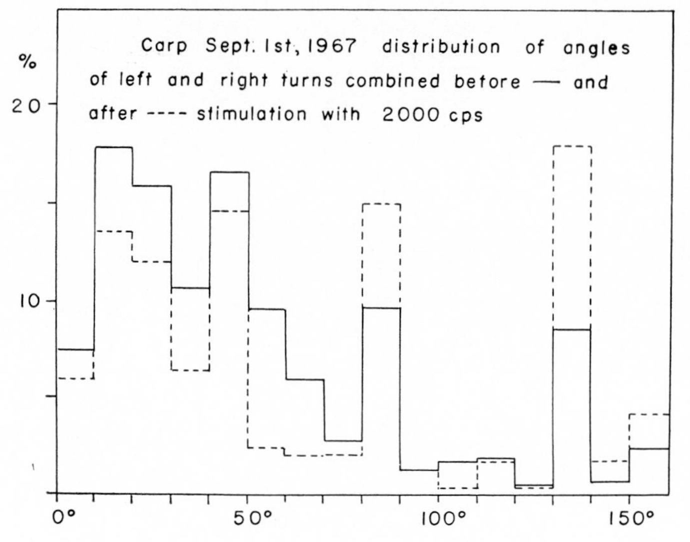

FIG. 7. Carp. September 1, 1967. The same data as in Fig. 6, but combined for left- and right-hand turns.

departed from a straight pathway, the latter turns were computed separately, with the results presented in Fig. 8. The shift in the frequency distribution is already exhibited by the first turns following a straight pathway of at least 20 cm. and its magnitude is not significantly different from that displayed by subsequent turns.

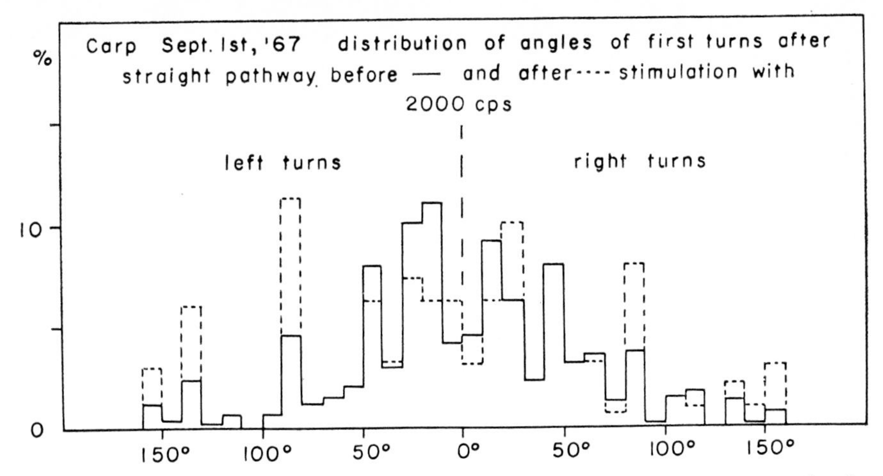

FIG. 8. Carp. September 1, 1967. The distribution of the angles of the first turns after the animal had followed a straight pathway of at least 20 cm.

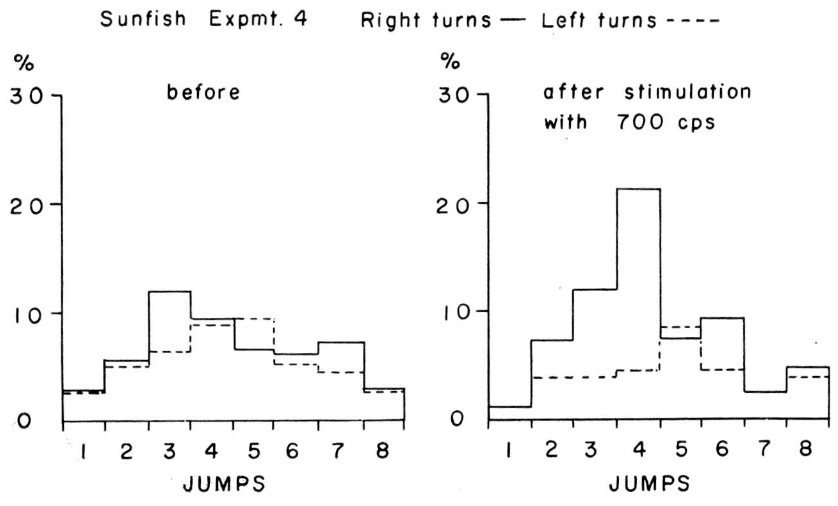

FIG. 9. Sunfish. The effect of sound (700 cyc./sec.) on the distribution of the angles of turns. See text.

In analysing the possible relationships between the above shifts in angle distributions and the exposure of the animal to sound, consideration was given to a possible interdependence of the angle of turn and the speed of locomotion of the fish. The speed of all displacements was computed. No correlation could be found between the distribution of angles and speed classes and no significant differences in speed were recorded in the periods before and during sound generation.

Sunfish. Fig. 9 represents the effects of a sinusoidal sound of 700 cyc./sec. on the locomotor behaviour of a sunfish as recorded in the cylindrical tank (see Methods). On leaving a compartment, the animal typically changes the direction of locomotion. The angle of this change determines how many compartments will be by-passed before the animal enters another compartment. Thus, the number of compartments by-passed ("jumps" in the histogram) provides information on the angles of turn and their distribution (Kleerekoper, 1967*a*,*b*). The data of Fig. 9 are representative of a series of similar experiments and indicate a pronounced change in the frequency distribution of the angles of turn when the fish is exposed to sound.

DISCUSSION

The results demonstrate a pronounced modification of the locomotor pattern of the carp when sound is transmitted into the tank. The fish cross the field of highest intensity only a few times and tend to move into the areas of lower sound pressure, whereas before the transmission of the sound the distribution of movement is much more even. This finding is in contrast with the recordings of movements of menhaden (*Breevoortia tyrannus* Latrobe, a clupeid fish) obtained by direct observation by Moulton (1963). These fish, held in a wire cage in open water, when subjected to sound waves of 15 kilocycles and 1–20 kilocycles, moved into the areas of highest sound pressure. However, unconditioned goldfish (Moulton and Dixon, 1967) moved away from a sound source. This difference in behaviour between species is, of course, unrelated to the problem at hand. It should be remembered that the fish used in our experiments were naïve animals, not only in the sense that they were unconditioned but, also, that they had never been exposed to the acoustic experience in question and were entirely unacquainted with the tank.

Examination of the locomotor pattern during sound transmission does not reveal any characteristic of a "constrained random walk" which van Bergeijk (1964) suspected to have occurred in the experiments on localization of sound by minnows (Kleerekoper and Chagnon, 1954).

The change in pattern is associated with an increased frequency of larger angles (sharper turns). The fact that this shift in angle distribution occurs at the first turn, whenever the animal departs from a straight course, suggests that turning may, itself, be the response to changes in the sound field. It is of interest that the frequency of turning is highest in those areas of the tank with steepest pressure gradients. The increase in angle of turn is also found in the sunfish, under entirely different experimental conditions. Furthermore, the absence of correlation between speed of locomotion and magnitude of angle of turn demonstrates that the shift in angle is a direct response to the characteristics of the sound field rather than a secondary effect of changes in speed of locomotion.

On the strength of the above evidence, the hypothesis is presented that orientation in a sound field results from the "adjustment" or "modulation" of the angle of turn in response to amplitude. Such an orientation mechanism would require independent, quantitative response to sound pressure (probably, indirectly through response to particle displacement) by the two receptor organs. The close proximity of the ears would require response to differences in the time of arrival of the sound wave and (or) to differences in sound pressure at the two ears of such small magnitudes that the existence of such a mechanism for orientation seems highly doubtful. Recently, however, Moulton and Dixon (1967) demonstrated that in unconditioned, restrained goldfish, in the near-field, sound directed on to one side of the head produces a tail flip to the opposite side. Such a movement, in a free-swimming fish, brings about a turning of the head away from the sound source. In conditioned fish, with sound as the conditional stimulus for the proffering of food, the tail flip occurs at the same side as the sound source, thus producing turning towards the source. These responses were found to be mediated by the Mauthner's cells, whose function makes it possible for the fish to establish directional orientation toward a unilateral sound source in spite of the incomplete acoustic isolation between the ears.

The results presented in this paper and those of Moulton's elegant experiments seem to complement each other and to provide strong evidence both for the ability of fish to respond directionally to a sound field and for the proposed mechanism by which such orientation can occur.

If, as is here proposed, the change in angle of turn in the sound field results from changes in sound pressure in the pathway of the fish, the latter should also make a larger number of turns in a sound field compared to a "silent" environment. In other words, the lengths of the straight pathways should become shorter. This hypothesis was tested by computing the lengths of all the straight pathways followed by the carp, before and during sound transmission, in three independent experiments, with three different, naïve animals. As the unit of length the interval between successive photocells on the same axis of the matrix was taken. By plotting the number of units covered by straight pathways against frequency on log-log paper, straight lines were obtained. The mean value λ for the length of the straight pathways was calculated ($N=N_o e^{-x/\lambda}$), as shown in Table II.

TABLE II

COMPARISON OF LENGTHS OF STRAIGHT PATHWAYS IN SOUND FIELD AND "SILENT" ENVIRONMENT

Experiment	*λ in cell units*	
	No sound	*Sound*
	(number of events in brackets)	
August 23, 1967	2·3 (1,608)	1·3 (4,224)
August 29, 1967	0·95 (13,380)	0·5 (13,248)
September 1, 1967	2·1 (2,820)	1·4 (1,488)

These results lend further support to the hypothesis that, in fish, orientation in a sound field results from adjustment of the angle of turning in response to pressure variations in the field and that the locomotor response is mediated by the Mauthner's cells, as described by Moulton and Dixon (1967).

SUMMARY

The locomotor pattern of single, naïve individuals of carp changed abruptly when the animal was exposed to sinusoidal sound at 2,000 cyc./sec. under conditions which precluded the near-field effect. The modified pattern showed relatively small, well-defined areas to which most of the animal's movements were confined and which were coincident with pockets of lowest sound pressure. Computer analysis of many parameters of the movements, monitored by a matrix of photo-conductive cells associated with a data-processing system and plotter, demonstrated that the modified locomotor pattern resulted from an increase in the angles and frequency of turning. The latter was highest in zones with steepest sound pressure gradients. Speed of locomotion was not affected by the exposure of the animal to the sound.

The authors postulate that, in these fish, orientation in a sound field results from modulation of the angle of turning, mediated by the function of the Mauthner's cells, in response to pressure variations. Corroborating evidence was derived from recordings of the locomotor pattern of naïve sunfish (*Lepomis gibbosus*) in which the angles of turning also increased in response to sound of 700 cyc./sec.

Acknowledgements

The authors are indebted to Mr. H. Lycklama of the Department of Physics of McMaster University for the assembly of the computer programmes required in the analysis of the data and to Miss V. Anderson, Mr. A. Timms and Mr. G. Westlake, graduate students in the Department of Biology, for their participation and assistance in the experimental work and the elaboration of many of the data.

REFERENCES

BUSNEL, R. G. (1959). *Bull. Inst. fr. Afr. noire*, **21** (A), 346–360.

DIJKGRAAF, S. (1964). *Experientia*, **20,** 586–587.

HARRIS, G. G. (1964). In *Marine Bio-Acoustics*, pp. 233–247, ed. Tavolga, W. N. Oxford: Pergamon.

HARRIS, G. G., and VAN BERGEIJK, W. A. (1962). *J. acoust. Soc. Am.*, **34,** 1831–1841.

KLEEREKOPER, H. (1967*a*). In *Olfaction and Taste II (Proceedings of the Second International Symposium, Tokyo*, 1965), pp. 625–645, ed. Tavolga, W. N. Oxford and New York: Pergamon.

KLEEREKOPER, H. (1967*b*). *Am. Zool.*, **7(3),** 385–395.

KLEEREKOPER, H., and CHAGNON, E. C. (1954). *J. Fish. Res. Bd Can.*, **11(2),** 130–152.

MOORHOUSE, V. H. K. (1933). *Contr. Can. Biol. Fish., N.S.*, **7,** 465–475.

MOULTON, J. M. (1963). *Ergebn. Biol.*, **26,** 27–29.

MOULTON, J. M., and BACKUS, R. H. (1955). *Maine Dept. of Sea and Shore Fisheries Circular*, **17,** 1–7.

MOULTON, J. M., and DIXON, R. H. (1967). In *Second Symposium on Marine Bio-Acoustics*, pp. 187–232, ed. Tavolga, W. N. Oxford: Pergamon.

REINHARDT, F. (1935). *Z. vergl. Physiol.*, **22,** 570–603.

VAN BERGEIJK, W. A. (1964). In *Marine Bio-Acoustics*, pp. 281–299, ed. Tavolga, W. N. Oxford: Pergamon.

VON FRISCH, K., and DIJKGRAAF, S. (1935). *Z. vergl. Physiol.*, **22,** 641–655.

WINN, H. E. (1964). In *Marine Bio-Acoustics*, pp. 213–231, ed. Tavolga, W. N. Oxford: Pergamon.

WISBY, W. J., RICHARD, J. D., NELSON, D. R., and GRUBER, S. H. (1964). In *Marine Bio-Acoustics*, pp. 255–268, ed. Tavolga, W. N. Oxford: Pergamon.

DISCUSSION

Johnstone: Professor Kleerekoper, is there any long-term drift of behaviour in the unstimulated animal?

Kleerekoper: There are changes in respect to intensity of movement—frequency of movement per time unit—but not in respect to the angle–straight pathway relationships which make up the pattern. The animals do become quieter as time goes on and finally stop moving if they are left too long in the same tank in the absence of other individuals of the same species.

Lowenstein: Am I right in thinking that the influence of sound caused the fish to swim nearer the wall of the tank? If this is always so, one would think that the sound brings about an increased thigmotactic response.

Kleerekoper: It happens that in these experiments many movements were near the walls, but this is not always so. I am not ruling out thigmotaxis as a factor in these locomotor patterns in a sound field. One of our plans is to put fences of different shapes inside the tank in order to see what influence dimension and shape of free swimming areas have on locomotor patterns.

J. D. Pye: Some fish, such as surface-feeding Poecilids, move quite differently when away from the surface. Are there any changes in these locomotor patterns according to the depth at which the fish are swimming—for example, do they go to the bottom and then swim in a different way when the sound is present?

Kleerekoper: I have only two records on this; in one we lowered the level of the water to 8 in., which was a reasonable depth for the fish in question, and in the other the level of the water was 2 ft.; however, there is no obvious difference in the locomotor pattern in this species (carp). But depth may be an important factor in other species.

Tumarkin: Professor Kleerekoper says he does not believe that fish follow an intensity gradient. I must confess I have always assumed that they do, and that is why the swim bladder can act as a direction finder. The difference between a fish and a terrestrial vertebrate is that the latter is relatively stationary; consequently it localizes sound by using two ears to enable it to sample the sound field at two different points simultaneously. The fish, being almost constantly on the move, prefers to use one detector to sample the sound field at two (or rather many) points successively. In either case, however, mere location of a distant source is only half the story. The creature can only act purposively if it has a memory store against which it can compare the incoming sound and decide whether it is benign or malign. Once the computer reports "benign", the fish will seek out the source, and the obvious way would be to follow up the gradient; and conversely with a malign source. We do not know what your signals meant to the fish but clearly they did mean something, since they induced purposeful movements. It looks as if the fish were scanning the tank, remembering from moment to moment what it was like and so choosing certain areas in preference to others.

Kleerekoper: I agree with you that intensity is the signal, but the response to change in intensity is not, according to my present understanding of the situation, to the *gradient* of intensity. The data support the conclusion that a *change* in intensity brings about a *change* in the angle of turn. This may, in itself, constitute the orientation mechanism.

Tumarkin: I still think that the fish will be influenced by memory at two different levels. It will use its long-term memory bank to decide whether the sound is malign, benign or indifferent, and it will use its short-term memory to decide which way it is going relative to the sound. If, at a given moment, it receives a malign sound and a moment later the sound is louder it will turn tail, but in response to exactly the same sound it will continue on course if a moment later the sound is fainter. It could only do this if it could recognize an intensity gradient on the basis of short-term memory, and then take action on the basis of its long-term memory.

Kleerekoper: No; this is where hearing and smell have something in common, in that the animal cannot directly locate the source of either sound or smell as it can do with a light source. The animal could directly locate the source only if it were able to follow a sound gradient. We could not find a mechanism for such direct localization; actually there is no evidence that the animal requires such a mechanism, if I am correct in assuming that for each change in intensity level there is a change in the locomotory pattern.

Tumarkin: If you hold a lighted candle I can locate it with my eyes closed by using my finger as a single sensor of thermal gradient. For that purpose I need three things. My proprioceptors tell me the path of my finger tip in relation to my body as a frame of reference. My short-term memory enables me to compare the temperature at any moment with what it was previously, and my long-term memory tells me that to find the candle I must climb up the temperature gradient. I see no reason why the fish should not be able to do something similar, albeit on a more rudimentary plane.

Lowenstein: But that is a more complex mechanism than that postulated by Professor Kleerekoper. It invokes memory and also invokes comparison between two situations, on a higher level. We know that in lower animals there are guiding mechanisms which involve no such comparison or memory, which are directly steered by an innate response to, say, intensity—like klinokinesis, although I gather that Professor Kleerekoper does not think we deal with klinokinesis in his case.

Kleerekoper: My interpretation of the data differs from Mr. Tumarkin's only in that I believe that the fish does not require this rather difficult absolute "memory", since at each particular point of sound-pressure perception, a corresponding, behaviourally innate change in direction occurs, so that from point A to point B there is no need for the fish to remember what has occurred at point A because if the situation is different at B, there will be a change in locomotor pattern.

Schwartzkopff: What Mr. Tumarkin has described is exactly what *Paramecium* does in approaching a certain point or avoiding another point, according to a phobic orientation. *Paramecium*, without a brain, can do this, and it is of course based on some very primitive kind of memory.

Lowenstein: I understand that sharks have now been claimed to do this too (Kritzler, H., and Wood, L. [1961]. *Science*, **133**, 1480–1482; Nelson, D. R., and Gruber, S. H. [1963]. *Science*, **142**, 975–977).

Batteau: One point that prejudiced me towards a molecular model of function for sensory systems (see. p. 237) was that it would apply to a single-celled organism, which has no nervous system.

Kleerekoper: I do not see why in the case of *Paramecium* you require memory at all. It may well work along the lines I have suggested for fish. At any particular level—in the case of my experiments with fish, it is of the intensity of sound—you have a corresponding value for the angle of turn; if you move into a new level you get a different turn; you do not need to postulate a memory because the situation is new and the organism acts accordingly.

Whitfield: But how does the fish know that it is new, without having some reference point? You must either build into the fish a reference which is an absolute standard of intensity, just as a person may have absolute pitch, or have a temporary standard of intensity (that is, what happened to the fish so many seconds or milliseconds ago), which is a memory. There must be a reference for you to be able to say that the situation is new—or in other words, that it is different from what it was.

Lowenstein: It is not necessarily so; think of steam pressure in a piston in a steam engine. With great steam pressure the piston moves fast while with low steam pressure it moves slowly; does it remember that it moved faster before, because it moves more slowly? If you have a very simple mechanical system, as Professor Kleerekoper assumes to exist in fish, you do not need any memory or any comparison, although sensory adaptation may be involved (Fraenkel, G S., and Gunn, D. L. [1940]. *The Orientation of Animals.* Oxford: The Clarendon Press).

Whitfield: You do have some standard of reference, in that the same steam pressure acting on a different piston, which has different characteristics of friction or inertia, will cause it to behave differently; in the same way you may have a standard of reference of frequency because you have a metal fork which has certain mechanical properties with a certain resonant frequency; you are still providing a standard within the system.

Lowenstein: But is that standard comparable to memory? That is the question. Obviously the physical constants play a part, but does that mean memory?

Whitfield: For this purpose I was merely defining memory as a short-term standard. But you must have one or the other.

Engström: Professor Kleerekoper has said that the fish went to the quietest part of the tank; must that not mean that it had a choice, that it chooses its position and finds the best place for itself?

Kleerekoper: This is the behavioural aspect: I am thinking of ways by which it achieves that position.

Engström: But it must compare the sound with something else, and that comparison must be in favour of the quietest part.

Neff: I don't think it has to make a comparison. If the fish moves more slowly and makes a smaller angle of turn to the lowest intensity, it will remain longer in the place of lowest intensity without ever comparing it with any other place. You could build a mechanical device that would perform in this manner.

Tumarkin: This is so if we are assuming that a fish hears one sound and has one reaction to the sound, which is to get to the noisiest place or to the quietest place. But the fish is a living organism with presumably a spectrum of different sounds to which it attaches certain values. It has the problems of finding a mate, of escaping from predators, and of catching prey, and must have various codes of behaviour: I presume it would want to follow the increasing intensity of the signal of the female and decreasing intensity of the signal of the predator, and so on. That surely implies that each meaningful signal must involve a memory for it.

Enger: From the title of Professor Kleerekoper's paper, "Orientation Through Sound in Fish", I had expected evidence on whether fish could determine the *direction* of a sound source; while your fish *may* be able to determine the direction of a sound source, your experiments did not show that. As you say, they can be interpreted to mean that fish are able to find the place in the tank of the lowest sound pressure. I feel that if I were a fish exposed to 2,000 Hz for 2 hours, whether I could determine the *direction* of the sound or not, I would eventually find the areas of lowest sound pressure! You used a carp exposed to a sound pressure of 40 to 60 db above the threshold at that frequency; this is not very loud and certainly is not painful, but for as long as 2 hours—I would hide! With regard to your push-button system, I do not at all disagree with you, but I would like to cause you to give better evidence that if you change the external conditions by sound at 2,000 Hz, you change the motor pattern of the fish. You have not really shown that.

Lowenstein: There is an unresolved discrepancy between von Frisch and Dijkgraaf's failure to demonstrate localization of a sound source in minnows by conditioning and other methods (Frisch, K. von, and Dijkgraaf, S. [1935]. *Z. vergl. Physiol.*, **22**, 641–653) and recent claims that long-distance sound direction-finding is possible in sharks (*loc. cit.*, see p. 203), in the face of the fact that we have generally assumed that the head of a fish is transparent to sound, the ears being equally exposed on both sides, so that in theory the fish should find this difficult to do. It can orientate by its lateral line, but this is a near-field effect.

Enger: At the Institute of Marine Science in Miami experiments have been done on the ability of sharks to determine the direction of a sound source, and it is claimed that sharks are able to define the direction of a sound source within 600 feet (Nelson, D. R. [1966]. *Diss Abstr.*, **27**, no. 1). However, very low frequencies (20–60 Hz) were used, so it may be a near-field effect for part of the range, and also the sound source was rather powerful. The wavelength for a 10-Hz sound in water is 150 metres, which is not very much less than 600 feet, which was the distance claimed. I certainly find it difficult to understand how these sharks should be able to detect the direction of a sound source, because even a big shark is small compared with the wavelength of the sound, so it cannot operate on a displacement gradient from the tail to the head, for example.

Kleerekoper: First, the meaning of "orientation" is a matter of definition, and I have interpreted it to mean that an animal can locate a place of different acoustic intensity, whether it is towards or away from it, which I feel is immaterial. Secondly, Dr. Enger brought up the question of exposure to sound for 2 hours. The results are similar during much shorter exposures to sound. I gave examples of long-term records to make sure that from a statistical point of view we are not dealing with a freak situation taken out of context but a long-term event; essentially similar results are obtained in the first half hour.

As to the behavioural significance of what the animal does, this does not really fit into the context of this paper. *Why* the fish should go to the lowest intensity under these circumstances is a different problem altogether. If I had conditioned it to go to the strong field, I am now confident that it would have gone to it. The experiments in the small tank, particularly, show that the change in prevalent angle of turn is immediate.

LOCALIZATION AND LATERALIZATION OF SOUND IN SPACE

WILLIAM D. NEFF

Center for Neural Sciences, Indiana University, Bloomington, Indiana

THE binaural perception of sound was a phenomenon that early attracted the attention and investigative interests of experimental psychologists, physiologists and physicists. In the late 1800's and early 1900's many experiments were done that defined reasonably well the physical conditions that enabled man, and presumably lower animals with similar auditory systems, to recognize the direction of a sound-producing source in the space surrounding him or, as apparatus was designed for laboratory experiments, to lateralize sound led to the two ears by conductors from one or more sound-producing instruments.

It is only in relatively recent years (fifteen or twenty) that we have, through experimentation, acquired sufficient knowledge that we can begin to give an account of how the auditory system—the receptor and its neural connexions—receives and processes the information that makes possible the localization of sounds in space or the lateralization of sounds presented by two separate channels to the two ears.

I shall not, in this paper, attempt to review and summarize the evidence from the, now, large number of anatomical, electrophysiological and psychological experiments that have dealt with localization or lateralization of sound. And I shall make only brief reference to clinical studies based upon damage to the auditory system of man.

My objectives are: (1) To examine the results of experiments in which localization or lateralization of sound has been measured by behavioural training and testing methods after destruction of the auditory receptors, transection of auditory pathways or ablation of auditory centres in the central nervous system; (2) To point out what parts of the system are most important for localization and lateralization; and (3) To speculate on how the system may operate in handling the information that makes possible localization and lateralization of sound.

For details of methods and procedures, I shall refer to published reports, Summaries of results are given in Tables I–IX. Brief reviews of results are given in the following sections, beginning with the peripheral end-organ and proceeding to pathways and centres higher in the auditory system. Additional sections are concerned with commissural pathways and reflex orientation to sound.

Finally, an attempt is made to formulate some generalizations that may be a start in explaining the results now at hand and in pointing the way to future experiments.

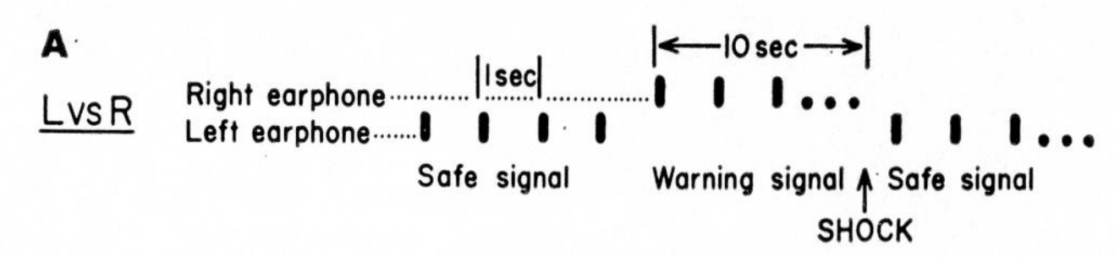

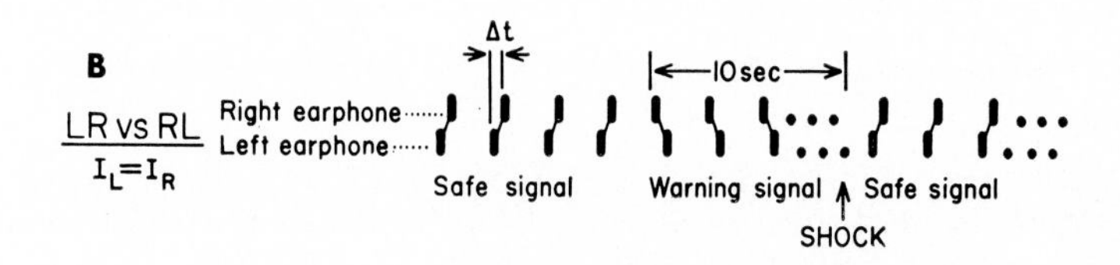

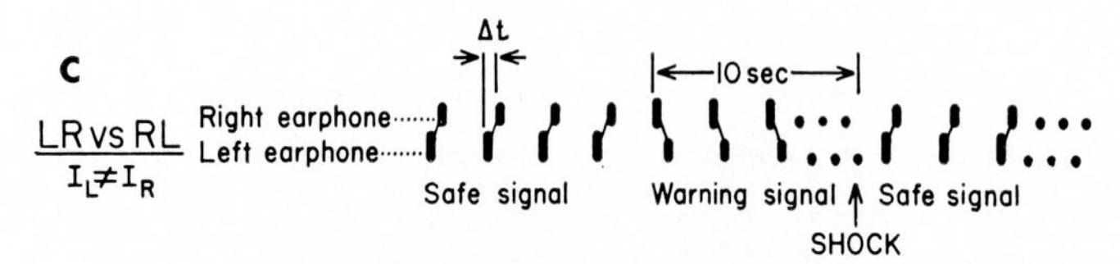

FIG. 1. Schematic representation of the temporal sequence of clicks comprising the three discriminations.

A: left vs. right (L vs. R); B: left-right vs. right left (LR vs. RL); C: graded left-right vs. right-left post-operative retraining discrimination. In C the unequal heights of the bars representing the clicks are intended to illustrate unequal intensity ($I_L \neq I_R$). In both B and C the value of *t* is 0·5 msec. (From Masterton and Diamond, 1964.)

METHODS

Most of the experiments referred to in Tables I–IX were done in the laboratory of Diamond and his colleagues (Diamond, Masterton, Jane) or in our own laboratory (Neff, Fisher, Diamond, Yela, Arnott, Aase, Wegener, Nauman, Strominger, Colavita, Chen, P. Neff). In the experiments of Diamond and his colleagues, earphones were held in place on the cat by a helmet so that signals could be presented to one ear only or to both ears with the time of arrival and intensity at the two ears controlled. Animals were trained in a double-grill box to avoid shock by moving from one compartment of the box to the other when a change in the acoustic signal occurred. The usual procedure was first, to train the animal to respond when a period of silence was followed by the presentation of a train of clicks to one ear (silence vs. R in tables). Next, the animal was trained to remain still when a train of clicks was presented to one ear and to respond when the acoustic signal was shifted to the other ear (L vs. R in tables).

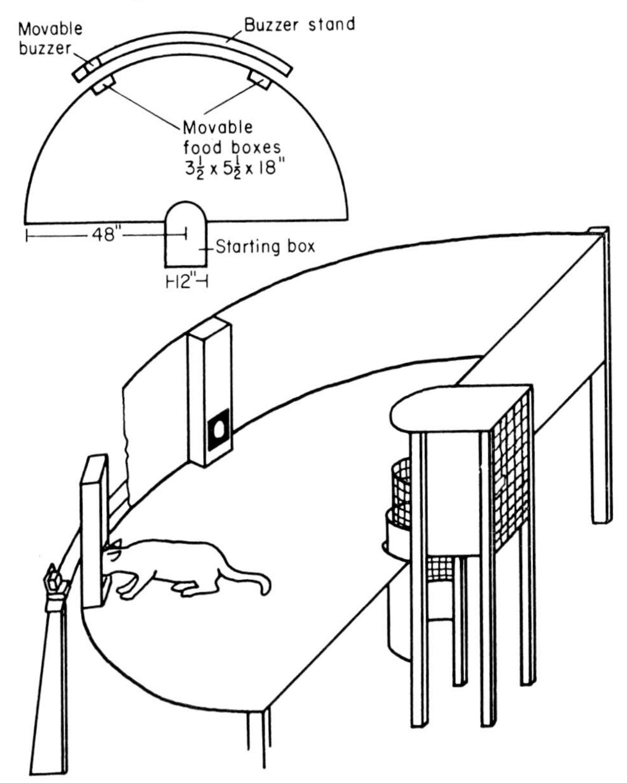

Fig. 2. Apparatus for testing localization of sound in space.

Finally, tests for lateralization were made by presenting trains of clicks to both ears but with each click in one ear preceding that in the other by 0·5 msec. (LR vs. RL in tables) (see Fig. 1). Further details of this procedure may be found in the published reports of Masterton and Diamond (1964) and Masterton, Jane and Diamond (1967).

In our laboratory, we have trained animals to localize sound in space by approaching and opening a door to obtain food from a container behind which a sound signal has been given. The distance between the food boxes and between the position of the sound signal on different trials can be changed so that a measure may be obtained of the smallest angle that the animal can discriminate between successive positions of the sound source. The animal is allowed to move in the starting cage during presentation of the sound signal. After initial training, the signal is stopped before the animal is released from the starting cage. (See Fig. 2 and published reports of Neff *et al.*, 1956; Neff, 1960, 1961*a*.)

PERIPHERAL END-ORGAN

Localizing the source of sound in space is a completely different task for an animal with two ears and an animal with one; this has long been known from psychophysical studies. The first uses the cues of difference in time of arrival, phase and intensity of sound at the two ears; the second, by scanning head movements, brings about differences in intensity and quality at the single receptor. This difference, in listening with one or two ears, is immediately apparent to the human observer and is also equally apparent in the behaviour of experimental animals. An animal that has been trained to localize with both ears intact behaves as if learning a new problem (which he is) after one ear has been destroyed. A fairly severe deficit in accuracy of localization also occurs (see Table I).

In an experiment with human subjects, in which one ear was covered by an ear-muff such as to render the peripheral sound source audible to only the uncovered ear, it was found that when head movements were allowed accuracy of localization was, at first, poor but quickly improved. Subjects reported that they aimed the uncovered ear at the place where the sound appeared loudest. One subject, who had been almost totally deaf in one ear for several years, performed better on initial tests than normally hearing subjects with one ear covered, but he also showed considerable improvement with practice. To the subject with normal hearing in both ears and with experience in localization experiments, loudness of the stimulus (wide-band noise) was, perhaps, the first cue used, but quality (the "hiss" of higher frequencies) was reported as the cue that seemed to provide more exact localization (Aase, 1962; see Table I).

MEDULLA

Current evidence from both anatomical and electrophysiological experiments indicates that the superior olivary complex (Fig. 3) is the centre of the auditory nervous system in which the first interaction of nerve impulses from the two ears takes place (Rasmussen, 1946; Galambos, Schwartzkopff and Rupert, 1959; Hilali and Whitfield, 1952; Stotler, 1953; Rosenzweig and Amon, 1955; Rosenzweig and Sutton, 1958; Nelson and Erulkar,

TABLE I

Experimenters	Animal		Lesion	Task	PERFORMANCE			Comment
	Number tested*	Identification†			Pre-operative	Post-operative	Change	
Nauman (1958)	(2)*	Cat	Unilateral destruction of cochlea	Localization of sound in space (buzzer)	5·0° 4·3°	14·6° 15·8°	−9·6° −11·5°	Head movement allowed
Masterton, Jane and Diamond (1967)	(1)*	Cat	Unilateral destruction of cochlea	Silence vs. R ‡; L vs. R; LR vs. RL (lateralization; clicks)	Not tested	Silence vs. R (+); L vs. R (+); LR vs. RL (−)§	Unable to lateralize	Control for method: dichotic stimulation via headphones
Aase (1962)	(1)*	Man	Unilateral deafness	Localization (broad-band noise)	Not tested	11° before training; 4° after training		Head movements allowed

* If only one number is given in parentheses, it is the number of animals tested; identification numbers are not always given in the tables (see Table II).
† Identification number of animals.
‡ L= sound presented to left ear via earphone; R= sound presented to right ear via earphone; LR= sound to left ear precedes sound to right ear; RL= sound to right ear precedes sound to left ear (See Fig. 1).
§ (+)= animal was able to learn discrimination. (−)= animal was not able to learn discrimination.

1963; Moushegian, Rupert and Whitcomb, 1964*a*,*b*; Hall, 1964; Rupert, Moushegian and Whitcomb, 1966). Results of behavioural studies support this finding (see Table II). In fact, the first results of experiments in which an attempt was made to section the trapezoid body suggested that, for the most accurate localization of sound, it was necessary that interaction of impulses from the two ears occur at a lower centre of the auditory system (Neff and W. Blau, unpublished data, 1954; Nauman, 1958).

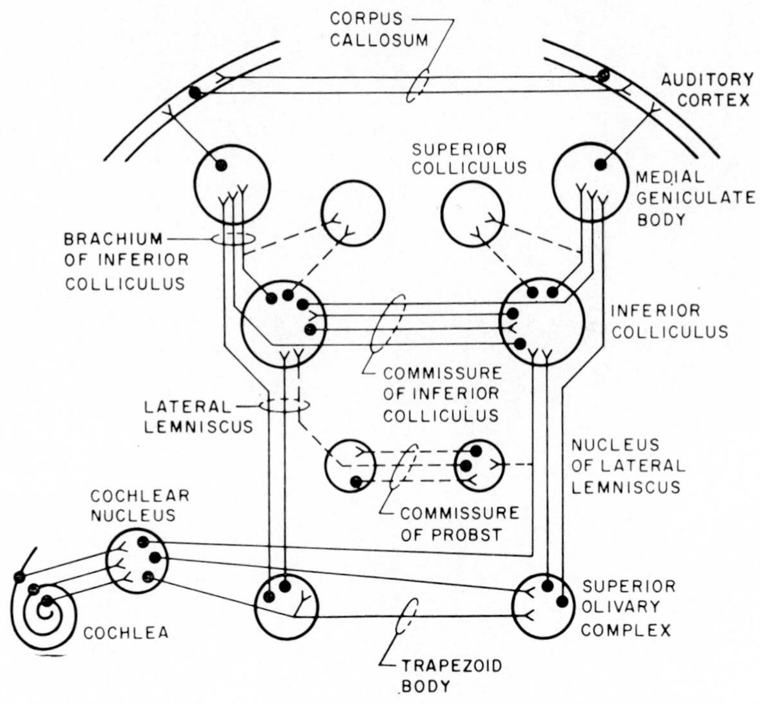

FIG. 3. Diagram of neural pathways of the auditory nervous system.

Masterton, Jane and Diamond (1967) conclude from their experiments in which they sectioned the trapezoid body in the cat and tested lateralization (see Cats 329 and 330, Table II) that "one specific function of the trapezoid body is to allow the analysis of natural time disparities". Partial section of the trapezoid body in the study of Masterton, Jane and Diamond resulted in a deficit that differed only in degree from that produced by complete section; the time-difference between arrival of sound at the two ears (Δt) had to be increased in order for the LR vs. RL discrimination (lateralization) to be made; the larger the trapezoid body lesion, the greater the increase in Δt for lateralization.

TABLE II

Experimenters	Animal: Number tested	Animal: Identification	Lesion	Task	PERFORMANCE: Pre-operative	PERFORMANCE: Post-operative	PERFORMANCE: Change	Comment
Nauman (1958)	(1)	Cat (588)	Transection of caudal third of trapezoid body	Localization (buzzer)	10·0°	11·8°		Good evoked* R's from cortex on both right and left sides to stimulation of right or left ear
Nauman (1958)	(1)	Cat (517)	Nearly complete transection of trapezoid body	Localization (buzzer)	4·5°	8·8°	4·3°	No retraining needed. Good evoked R's from contralateral cortex; poor from ipsilateral
Masterton, Jane and Diamond (1967)	(2)	Cat (329) (330)	Complete transection of trapezoid body	Silence vs. R; L vs. R; LR vs. RL; (lateralization)	Not tested, but normal cats transfer from L vs. R to LR vs. RL	Silence vs. R (+); L vs. R (+); LR vs. RL (learned with increase of Δt)†	Failure to transfer from L vs. R to LR vs. RL, and increase in Δt	Δt intervals necessary for relearning greater than those for natural sounds in space. *Note:* Cat 329 had considerable damage to left superior olive; Cat 330 had 2 transections of the trapezoid body, one incomplete
Masterton, Jane and Diamond (1967)	(4)	Cat (241) (342) (382) (330)	Partial transection of trapezoid body	Same as above	Same as above	Silence vs. R (+); L vs. R (+); transfer to LR vs. RL	Increase in Δt threshold	
Wegener (1954)	(4)	Cat (317) (324) (325) (327)	Ablation of auditory and visual projection areas of cerebellum	Localization (buzzer)	Reached criterion (27 of 30 at 5°)	Reached same criterion	None	Learned habit retained

* evoked R's = evoked responses.
† Δt = difference between time of arrival of sound at two ears.

Results of unilateral or bilateral destruction of the superior olive are inconclusive because too few animals with appropriate lesions have been tested. There is some indication from both Nauman's study (1958) and that of Masterton, Jane and Diamond (1967) that nearly complete unilateral destruction of one superior olivary nucleus does not produce any severe deficit in localizing or lateralizing behaviour. This conclusion must, however, be held as tentative in view of the severe disability found in some animals after unilateral transection of the lateral lemniscus.

The report by Masterton, Jane and Diamond (1967) that cats with nearly complete unilateral transection of the lateral lemniscus failed to transfer from a left vs. right (L–R) discrimination to LR vs. RL (Table III) led us to re-examine the protocols of animals with similar lesions that were trained and tested in the Nauman study. Cat 459 (Table III), which had a large lesion of the trapezoid body as well as complete or nearly complete transection of the left lateral lemniscus, not only showed a decrease in accuracy of localizing post-operatively but also had great difficulty in relearning to localize. After the lesion, the cat was unable to relearn for several months; more than 500 trials were required before its performance was above the chance level when the sound source on successive trials was shifted by 180°. Having learned to discriminate at 180°, it quickly learned to perform accurately at much smaller angles.*

The above results of a post-operative deficit in relearning after unilateral transection of the lateral lemniscus suggest that the surgical lesion has disrupted the auditory system in such a fashion that the animal no longer perceives the sound presented at the two ears as it did before surgery; it must, therefore, learn a discrimination based upon new cues. Masterton, Jane and Diamond concluded that "subtotal lesions of the lateral lemniscus result in a loss of the equivalence between normally fused LR and L, and normally fused RL and R. This change in perception of the clicks is not accompanied by any change in the capacity to perform the several routine discriminations nor any clear change in Δt acuity". They state further, "Therefore, even though the discrimination of sounds from two different directions is still possible after subtotal damage to the lateral lemniscus, the quality of directionality may be lost".

Unlike the effects that unilateral transection of the lateral lemniscus has on learned localization or lateralization discriminations, transection of the trapezoid body, in the experiments done so far, has not resulted in a loss

* Another explanation for the post-operative behaviour of Cat 459 is that transection of the trapezoid body and of the lateral lemniscus on one side left it essentially a one-eared animal and it therefore had to learn to use the cues available for localizing with one ear.

TABLE III

Experimenters	*Animal*		*Lesion*	*Task*	PERFORMANCE			*Comment*
	Number tested	*Identification*			*Pre-operative*	*Post-operative*	*Change*	
Nauman (1958)	(2)	Cat (442) (462)	Unilateral section of LL* and part of trapezoid body	Localization (buzzer)	4·0° 3·7°	19·0° 6·8°	−15·0° −3·1°	Some retraining required for smaller angles. Electrophysiological evidence of some intact fibres on side of lesion in Cat 462
Nauman (1958)	(1)	Cat (559)	Unilateral destruction of SO† and severe damage to trapezoid body	Localization (buzzer)	4·6°	3·9°	None	Evoked R's from right cortex only to stimulation of either ear. No retraining required
Nauman (1958)	(1)	Cat (459)	Complete transection of trapezoid body and of left LL; also destruction of left SO	Localization (buzzer)	3·5°	10·0°	−6·5°	After transection of trapezoid body and lateral lemniscus long retraining period necessary
Masterton, Jane and Diamond (1967)	(2)	Cat (215) (355)	Nearly complete transection of LL (unilateral)	Silence vs. R; L vs. R; LR vs. RL (lateralization)	Not tested	Silence vs. R (+); L vs. R (+); LR vs. RL (+ after retraining)	Lack of transfer from L–R to LR–RL	After partial transection of lateral lemniscus LR vs. RL is no longer equivalent to L vs. R

* LL = lateral lemniscus
† SO = superior olive

of the learned habit and the need for a long relearning period. This difference may be important because it raises the question: does the trapezoid body supply all or most of the input of binaural information from the superior olivary complex to the contralateral lateral lemniscus?

CEREBELLUM

Wegener (1964) tested cats in a localization situation such as shown in Fig. 2 and described above (p. 209). The animals showed no loss in accuracy of performance and retained the learned habit after ablation of the auditory and visual areas of the cerebellum, as defined by Snider and Stowell (1944) (see Table II).

TECTUM

As may be noted in Table IV, Masterton, Jane and Diamond (1968) found that cats with bilateral ablation of the dorsal two-thirds of the inferior colliculi performed about as well as normal animals on the left–right and lateralization tests that they used. Larger ablations that included most of the inferior colliculi and invaded adjacent structures produced a complete loss of the capacity to discriminate between stimuli applied in succession to the left and then the right ear and in the LR vs. RL lateralization task. Bilateral transection of the brachium of the inferior colliculus (BIC) produced complete inability to learn LR vs. RL and a deficit in performance on L vs. R.

In our own experiments (N. L. Strominger and Neff, unpublished observations, 1961), we have obtained variable results in performance on the localizing task after unilateral transection of the BIC (see Table IV). The severity of the deficit in performance appears to bear some relation to the extent of the transection. Not shown in the table are the results for one animal that failed completely to relearn the localizing discrimination after a large unilateral transection that included not only the BIC but adjacent pathways.

The loss of the learned habit of lateralizing or localizing after unilateral section of BIC may have an explanation similar to that for the deficiency in performance and necessity for relearning noted in some animals after unilateral lesions of the lateral lemniscus (see under Generalizations, p. 225).

COMMISSURAL PATHWAYS

Accuracy of localizing sound in space is not affected in the cat by transection of the corpus callosum (CC), the commissure of the inferior colliculus (CIC), or both (Nauman, 1958) (see Table V). As noted under

TABLE IV

Experimenters	Animal: Number tested	Animal: Identification	Lesion	Task	Performance: Pre-operative	Performance: Post-operative	Performance: Change	Comment
Strominger and Neff (1961)‡	(2)	Cat (816) (882)	Bilateral transection of BIC*	Localization (buzzer)	11·4° 8·0°		Complete inability to localize	
Strominger and Neff (1961)	(1)	Cat (778)	Incomplete bilateral transection of BIC	Localization (buzzer)	4·2°	16·8°	−12·6°	No relearning necessary
Strominger and Neff (1961)	(3)	Cat (882) (816) (872)	Unilateral transection of BIC	Localization (buzzer)	8·0° 11·4° 9·6°	12·5° 17·7° 42·3°	−4·5° −6·3° −32·7°	Lesion in 872 more extensive than in 882 and 816
Masterton, Jane and Diamond (1968)	(2)	Cat (301) (381)	Bilateral transection of BIC	Silence vs. R; L vs. R; LR vs. RL	Silence vs. R (+); L vs. R (+); transfers to LR vs. RL	Silence vs. R (+); L vs. R (?); LR vs. RL (−)		Unstable performance on L vs. R; no transfer to LR vs. RL; inability to learn LR vs. RL
Masterton, Jane and Diamond (1968)	(2)	Cat (243) (232)	Bilateral ablation of dorsal 2/3 of IC†	Same as above	Same as above	Silence vs. R (+); L vs. R (+); transfers to LR vs. RL	None	Performance approximately same as that of normal cat
Masterton, Jane and Diamond (1968)	(2)	Cat (R–150) (R–151)	Deep bilateral ablation of IC plus adjacent tissue	Same as above	Same as above	Silence vs. R (+); L vs. R (−); LR vs. RL(−)		Cannot discriminate L from R nor LR from RL

* BIC = brachium of inferior colliculus; † IC = inferior colliculus; ‡ Unpublished results.

TABLE V

Experimenters	Animal: Number tested	Animal: Identification	Lesion	Task	Performance: Pre-operative	Performance: Post-operative	Performance: Change	Comment
Nauman (1958)	(3)	Cat (462) (517) (588)	Transection of CC*	Localization (buzzer)	3·5° 4·0° 8·8°	3·7° 4·5° 10·7°	None	
Nauman (1958)	(3)	Cat (459) (521) (559)	Transection of CC as 2nd or 3rd in series of commissural transections	Localization (buzzer)	10·5° 4·8° 7·7°	10° 4·0° 5·0°	None	Transection of CC, CIC or both results in little or no deficit in localization of sound in space
Nauman (1958)	(2)	Cat (521) (588)	Transection of CC and CIC†	Localization (buzzer)	3·0° 8·8°	4·0° 10·0°	None	

* CC=corpus callosum
† CIC=commissure of inferior colliculus

Medulla (p. 212), transection of the trapezoid body does produce a deficit in both localizing and lateralizing behaviour. The deficit is one of accuracy in localizing a single sound source in space and, in the case of lateralization, the necessity for a larger time-difference between the sounds arriving at the two ears (see Table II). Cat 517 in Table II also had CC and CIC as well as the trapezoid body transected in the experiment with no greater change in threshold than that shown in the table.

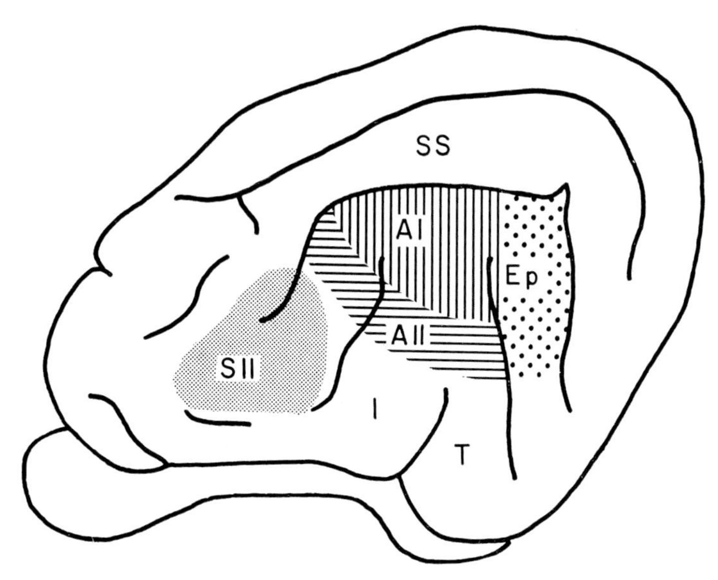

FIG. 4. Auditory projection areas of the cerebral cortex.
AI = auditory area I; AII = auditory area II; Ep = posterior ectosylvian area; I-T = insular-temporal cortex; SII = somatic area II; SS = suprasylvian gyrus.

CEREBRAL CORTEX

Most of the results of experiments listed in Tables VI, VII and VIII have been summarized in a number of earlier papers (Neff, 1960, 1961*b*, 1962, 1964, 1967, 1968; Neff and Diamond, 1958). Briefly they are as follows:

(1) After bilateral ablation in the cat of all of the cortex that receives projection from known thalamic auditory centres (Fig. 4), there is a complete loss of ability to localize sound in space (Strominger and Neff, unpublished observations, 1961) (see Table VII). In the lateralization experiment (Masterton and Diamond, 1964), the animal can relearn L vs. R but does not transfer from L vs. R to LR vs. RL. Furthermore, it has great difficulty relearning LR vs. RL and does so only after being given training in which

TABLE VI

Experimenters	Animal: Number tested	Animal: Identification	Lesion	Task	Performance: Pre-operative	Performance: Post-operative	Performance: Change	Comment
Neff *et al.* (1956)	(4)	Cat (Y-2) (Y-3) (F-4) (F-5)	Bilateral ablation of AI, AII, Ep* (approx.)	Localization (angles of 25°, 50°, 130°) (buzzer)	Learned to high level of performance	25° (−) 50° (−) 130° (+)	Failed to relearn at smaller angles. Relearned with difficulty at 130°; level of performance at 130° was not stable	No attempt to measure threshold. Ablations did not include all projection areas of medial geniculate body
Arnott (1953)	(4)	Cat (Group A)	Bilateral ablation of AI, AII, Ep, and part of I-T	Localization (buzzer)	Learned to criterion of 75% at 5°	Relearned to criterion of 75% at 40°	Severe shift in threshold angle discriminated	Performance, even above threshold, was unstable
Arnott (1953)	(4)	Cat (Group B)	Bilateral ablation of AI, AII, Ep, and part of I-T	Localization (buzzer)	Not trained	Learned to criterion of 75% at 40°	Threshold much higher than that of normal cat	Performance above threshold unstable but better than for Group A

* Auditory areas (see Fig. 4).

L in LR is greater in intensity than R, and likewise for R in RL. As long as L and R are of equal intensity, performance remains more erratic than that of the normal animal (see Table VIII).

(2) After bilateral ablations that include nearly all of the auditory projection areas of the cat or monkey, there is almost a complete loss of capacity to localize sound in space (Strominger and Neff, unpublished observations, 1961; Wegener, 1964) (see Table VII). After a one-stage bilateral ablation the learned habit is lost, and relearning is slow, usually requiring more time and trials than initial, pre-operative learning.

(3) After bilateral ablation of AI, AII, Ep and some adjacent cortex in I–T (see Fig. 4), there is a severe deficit in capacity to localize (Neff *et al.*, 1956). Again, the learned habit is lost and must be relearned. The threshold shifts from approximately 5° to 40° (Arnott, 1953; Arnott and Neff, 1950) (Table VI).

(4) After bilateral ablation of AI or after bilateral ablations that include AI on one side and a considerable portion of auditory cortex on the other, a deficit in accuracy of localization and/or slow post-operative learning have been observed (Strominger and Neff, unpublished observations, 1961) (Table VII). In tests of lateralizing behaviour, bilateral ablation of AI produces a loss in the transfer of the correct response from the learned L vs. R habit to LR vs. RL (Masterton and Diamond, 1964) (Table VIII).

(5) Accuracy of localizing and the learned habit are unaffected by all other bilateral ablations of auditory sub-areas that have been tested in the cat (Strominger and Neff, unpublished observations, 1961) (Table VII).

(6) Unilateral ablations of auditory projection areas (AI, AII, Ep, I–T, and, in some cases, parts of SII and SS) have produced variable results that are not readily interpretable in terms of extent of ablation or retrograde degeneration of thalamic centres. In 5 out of 6 animals, some decrease in accuracy of localizing was found (Strominger and Neff, unpublished observations, 1961) (Table VII).

One case of a man with right hemispherectomy is listed in Table VII for comparison with the results of unilateral ablation in animals. This patient was tested under carefully controlled conditions using a broad-band noise source. The subject, seated in a chair with head-rest, was asked to manipulate a lever that moved a spot of light on an opaque screen to the position at which he localized the sound source. The angle of error between the position of the light spot and the actual position of the sound source was

TABLE VII

Experimenters	Animal Number tested	Animal Identification	Lesion	Task	PERFORMANCE Pre-operative	Post-operative	Change	Comment
Strominger and Neff (1961)*	(4)	Cat (645) (862) (664) (663)	Bilateral ablation of AI, AII, Ep, I-T, part of SII	Localization (buzzer)	6·1° 12·4° 4·8° 8·0°	62% at 90° 62% at 90° 67% at 90° 57% at 90°	Large shift in threshold, almost complete inability to localize	Animals trained and tested in dark to control for deficits due to damage to visual system
Strominger and Neff (1961)	(1)	Cat (784)	Bilateral ablation of AI, AII, Ep, I-T, SII, anterior and mid SS	Localization (buzzer)	6·9°	49% at 90°	Complete loss; chance performance	
Wegener (1964)	(9)	Monkey	Bilateral primary auditory cortex	Localization (buzzer)	Less than 10°	No animal better than 60% at 90°	Gross deficit at all angles from 5° to 180°	Post-operative relearning required more trials than pre-operative
Strominger and Neff (1961)	(2)	Cat (827) (846)	Bilateral ablation of AI	Localization (buzzer)	7·4° 4·2°	8·3° 15·0°	−10.8°	Although 827 showed no significant shift in threshold it did show considerable post-operative deficit in relearning
Strominger and Neff (1961)	(1)	Cat (833)	Left AI, AII, Ep, I-T, SII, right AI	Localization (buzzer)	3·7°	27·6°	−23·9°	

Strominger and Neff (1961)	(2)	Cat (902) (847)	Bilateral AII, at least, or AII and Ep	Localization (buzzer)	8·4° 14·2°	8·4° 9·6°	None None	
Strominger and Neff (1961)	(1)	Cat (859)	Bilateral ablation of I-T	Localization (buzzer)	6·7°	5·3°	None	
Strominger and Neff (1961)	(1)	Cat (807)	Left I-T, right I-T, AI, AII, Ep	Localization (buzzer)	3·9°	9·3°	−5·4°	
Strominger and Neff (1961)	(2)	Cat (838) (861)	Bilateral ablation of anterior SS	Localization (buzzer)	12·4° 9·3°	12·0° 9·6°	None None	
Strominger and Neff (1961)	(2)	Cat (838) (861)	Bilateral ablation of anterior SS + Ep	Localization (buzzer)	12·4° 9·3°	11·0° 7·6°	None None	
Strominger and Neff (1961)	(6)	Cat	Unilateral ablation of AI, AII, Ep, I-T, and in some cases SII and part of SS	Localization (buzzer)	From 3·7° to 10·0°	From 7·5° to 15·6°	Loss in 5 of 6 animals (from 3° to 11·9°)	Amount of deficit not clearly related to extent of ablation or retrograde degeneration
Strominger and Neff (1961)	(1)	Man	Right hemispherectomy	Localization (broad-band noise)	Not tested	*Auditory field* *Left* *Centre* *Right* *Average* Test 1: 8·56 2·58 3·63 4·92 Test 2: 4·41 2·91 2·67 3·33		Average scores were just above upper end of range for 9 "normal" control subjects (1·69 to 3·18)

* Unpublished results.

TABLE VIII

Experimenters	*Animal*		*Lesion*	*Task*	PERFORMANCE			*Comment*
	Number tested	*Identification*			*Pre-operative*	*Post-operative*	*Change*	
Masterton and Diamond (1964)	(4)	Cat (M-133) (M-160) (M-200) (M-227)	Bilateral ablation of AI, AII, Ep, I-T, SII	L vs. R; LR vs. RL (lateralization)	L vs. R(+); LR vs. RL (transfer without additional training)	L vs. R(+); LR vs. RL (no transfer)	No transfer from L-R to LR-RL. Difficulty in LR-RL when intensity of L and R was equal	L vs. R not retained post-operatively but relearned; LR vs. RL relearned only when intensity as well as time differences used in retraining
Masterton and Diamond (1964)	(2)	Cat (M-297) (M-299)	Bilateral ablation of AI (two stages)	Same as above	Same as above	L vs R(+); LR vs. RL (+; poor transfer)	Deficit in transfer L-R to LR-RL	

taken as the score on any given trial. It may be seen from Table VII that this subject localized with great accuracy although his performance was not as good as that of normal subjects. Largest errors occurred in the auditory field contralateral to the side of the hemispherectomy (Strominger and Neff, 1967).*

REFLEX ORIENTATION TO SOUND IN SPACE

In a number of studies of decorticated animals, statements have been made indicating that the animals showed reflex orienting responses towards a source of sound (Bard and Rioch, 1937; Dusser de Barenne, 1924). In these studies, the observations on reflex orientation to sound were made somewhat casually; careful tests were not made.

We have not studied reflex orientation to sound in decorticated animals, but we have observed a complete or almost complete absence of the orienting response in cats with large bilateral ablations of the auditory cortex or bilateral transections of the brachium of the inferior colliculus. Smaller ablations of cortex had less effect upon the orienting response. After the largest ablations, startle or alerting responses occurred without turning of the head or body toward the sound; from visual observation only, it sometimes appeared that these responses were enhanced after operation and that habituation was slow or absent (Table IX).

The above results are in fair agreement with the earlier reports of Riss (1959) and Thompson and Welker (1963). The latter investigators noted some orienting responses after large bilateral ablations (Table IX).

GENERALIZATIONS

The generalizations listed below are not all based only upon the experiments that have been described. Some, for example (1), have long been known from the results of psychophysical experiments. Furthermore, careful scrutiny of the results in Tables I–IX and reading of the summary of results given above will reveal the difference between those generalizations that are founded upon reasonably solid experimental evidence and those that must, for the present, remain more speculative.

(1) Localization with two ears is much more accurate than with one ear even for complex sounds and when head movements are permitted. That different cues (changes in intensity and quality only and not time or phase)

* We wish to thank Dr. John F. Mullan, Director of the Division of Neurosurgery, Department of Surgery, University of Chicago Medical School for making this and other patients available for testing localizing ability.

TABLE IX

Experimenters	Animal Number tested	Animal Identification	Lesion	Task	PERFORMANCE Pre-operative	Post-operative	Change	Comment
Riss (1959)	(1)	Cat (1)	Bilateral ablation AI, part of Ep	Reflex orientation to sound	Orientation rapid and accurate	Orientation good	Response changed slightly	Animal apparently oriented primarily through use of visual cues
Riss (1959)	(4)	Cat (3, 4) (5, 6)	Bilateral ablation of AI, AII, Ep	Reflex orientation to sound	Orientation rapid and accurate	Orientation poor or absent	Almost complete loss of reflex orienting response	
Thompson and Welker (1963)	(4)	Cat	Bilateral ablation of AI, AII, Ep, I-T, part of SS	Reflex orientation to white noise (0·1 sec. or 2 sec. duration)	Not tested; operated animals compared with normals; normals show prompt accurate orientation	Responses less frequent and of greater latency	A fairly severe deficit in accuracy of localizing and in consistency of response	
Neff *et al.* (1966)*	(7)	Cat	Bilateral AI, AII, Ep, I-T, and, in some cases parts of SII and SS	Reflex orientation (buzzer)	Responses typically prompt and accurate in direction	No clear orienting responses in any case	Complete or nearly complete loss of reflex orientation to sound	Startle or alerting responses were observed on most trials with all animals. In some cases startle responses appeared to be greater after operation

Neff *et al.* (1966)	(5)	Cat	Bilateral AI, AII, Ep	Reflex orientation (buzzer)	Responses typically prompt and accurate in direction	Little or no orienting responses	Complete or nearly complete loss of orienting responses	Startle responses on most trials
Neff *et al.*	(8)	Cat	Bilateral I-T or I-T plus cortex immediately anterior or posterior to I-T	Reflex orientation (buzzer)	Responses typically prompt and accurate in direction	Good orienting responses	Little or none	Startle responses on most trials
Neff *et al.* (1966)	(3)	Cat	Bilateral BIC or BIC + IC	Reflex orientation (buzzer)	Responses typically prompt and accurate in direction	No orienting responses noted in any case	Complete or nearly complete loss of reflex orientation to sound	Startle responses present
Neff *et al.* (1966)	(2)	Cat	Unilateral AI, AII, Ep	Reflex orientation (buzzer)	Responses typically prompt and accurate in direction	Responses less prompt and less consistent	Partial deficit	
Neff *et al.* 1966)	(2)	Cat	Unilateral I-T	Reflex orientation (buzzer)	Responses typically prompt and accurate in direction	Good responses	None	

* Unpublished observations of W. D. Neff, F. Colavita, W. Chen, and P. L. Neff (1966).

are being used is apparent from the behaviour in tests of both man and lower animals. The human subject with deafness in one ear, or one ear covered so as to prevent sound reaching it, shows improvement with practice; an animal with one ear destroyed must learn a new problem.

(2) For accurate localization of sound in space or for lateralization with small differences in time of arrival at the two ears, it is necessary that the nerve impulses from the two ears interact at one of the first centres in the auditory nervous system. Transection of the trapezoid body produces deficits in both localizing and lateralizing behaviour. These findings support the inference drawn from results of anatomical and electro-physiological experiments, that the superior olivary complex is an important centre for binaural interaction.

(3) After binaural interaction has taken place at the level of the superior olivary complex, interference with the ipsilateral transmission to higher centres of the neural message from the region of one superior olive produces a deficit in both lateralizing and localizing behaviour. In some animals, after unilateral lesions of the lateral lemniscus, many trials are required to relearn to localize sound in space. After similar lesions, animals that have learned to discriminate sound presented to the left ear from sound presented to the right (L vs. R) fail to transfer to LR vs. RL where the time differences are such that in the normal animal LR sounds like L (sound lateralized to left) and RL sounds like R. From these findings of localization and lateralization experiments, it may be inferred that after the lateral lemniscus transection, the animals must learn a new discrimination. The results of the lateralization experiments indicate that LR does not sound like L and RL does not sound like R.

Ipsilateral transections of auditory pathways at the tectal level may produce similar results to those described above for lesions of the lateral lemniscus.

If the animals described above are learning a new discrimination post-operatively, what is the nature of the new discrimination? There are several possibilities: assuming that the original discrimination was made on the basis of time cues, (*a*) the new discrimination may have to be made on the basis of intensity or quality cues; or (*b*) the new discrimination may have to be made on the basis of time-difference between stimulus to left ear and stimulus to right ear (L vs. R) where there is no longer fusion so that two signals are heard as two successive sounds instead of as one localized in space or lateralized to left or right.

(4) Behavioural evidence is meagre, but from that available so far it does

not appear that the inferior colliculi play any important role in learned localization of sound in space. The binaural interaction seen in electrical recording from single units in the inferior colliculus may represent only a reflection of interaction that has already occurred at a lower level. Further experiments may reveal that the inferior colliculi are more than relay centres for information about binaural interaction.

(5) The cortical projection areas of the primary auditory afferent system are essential for localization of sound in space and for lateralization. From the results, not only of cortical ablation studies but also of other studies reviewed above, the following inferences may be made:

(i) The cortical projection areas of the auditory system (auditory cortex) are of primary importance for the most accurate localization of sound in space. It is a good first hypothesis that the importance of the cortex may be because neural mechanisms at this level are, in comparison with lower neural centres, better designed to receive and process small time-differences between afferent nerve impulses.

(ii) It is only at the level of the auditory cortex that single cells or systems of cells receive the binaural information that has been processed by both superior olivary complexes, assuming the latter to be the first centres where binaural interaction occurs in the auditory nervous system. On the basis of information now available, it is suggested that the auditory cortex is essential for receiving and handling binaural information in the nervous system that is the result of differences in time, phase, intensity or quality arriving at the two ears. Between the lowest centres (superior olivary complexes) that receive and process neural impulses from the two ears, and the auditory cortex, there may be a centre that receives the bilateral information and perhaps recodes it before transmission to the cortex. At present, we have no evidence for such a centre.

SUMMARY

Experiments are described in which the ability of animals to localize or lateralize sound was measured before and after transection of pathways or ablation of centres in the auditory nervous system. Among the principal findings are: (1) For accurate localization or lateralization of sound, it is essential that nerve impulses from the two ears interact in one of the first centres of the auditory system. Considerable evidence points towards the superior olivary complex as the centre for this initial interaction; (2) Evidence from unilateral transection of the lateral lemniscus or of the brachium of the inferior colliculus suggests that normal lateralization and localization may require that the binaural neural message, however it may be recoded at the medulla level, be projected bilaterally to the cortex; (3) After large bilateral ablations of cerebral cortex that include all or nearly all of the projection areas of the primary auditory system, the ability to localize or lateralize sound may be completely lost. Severe deficits occur with more restricted ablations; (4) Large unilateral ablations of auditory cortex lead to impairment of localization. A case is cited in which localization of sound was studied after right hemispherectomy in man. In comparison to normal human subjects, a small impairment was found; the impairment was greatest for sounds in the auditory field opposite to the side of hemispherectomy.

Acknowledgement

Research on neural mechanisms of hearing in the Center for Neural Sciences is supported by the National Science Foundation and the National Aeronautics and Space Administration.

REFERENCES

Aase, S. W. (1962). Monaural sound localization in human subjects. Master's dissertation, University of Chicago.

Arnott, G. P. (1953). Impairment following ablation of the primary and secondary areas of the auditory cortex. Doctoral dissertation, University of Chicago.

Arnott, G. P., and Neff, W. D. (1950). *Am. Psychol.*, **5,** 270.

Bard, P., and Rioch, D. M. (1937). *Johns Hopkins Hosp. Bull.*, **60,** 73–147.

Dusser de Barenne, J. G. (1924). *Proc. R. Soc. B*, **96,** 272–291.

Galambos, R., Schwartzkopff, J., and Rupert, A. (1959). *Am. J. Physiol.*, **197,** 527–536.

Hall, J. L. (1964). *Tech. Rep. Res. Lab. Electron. M.I.T.*, p. 416.

Hilali, S., and Whitfield, I. C. (1952). *J. Physiol., Lond.*, **122,** 158–171.

Masterton, R. B., and Diamond, I. T. (1964). *J. Neurophysiol.*, **27,** 15–36.

Masterton, R. B., Jane, J. A., and Diamond, I. T. (1967). *J. Neurophysiol.*, **30,** 341–359.

Masterton, R. B., Jane, J. A., and Diamond, I. T. (1968). *J. Neurophysiol.*, in press.

Moushegian, G., Rupert, A., and Whitcomb, M. A. (1964*a*). *J. Neurophysiol.*, **27,** 1174–1191.

Moushegian, G., Rupert, A., and Whitcomb, M. A. (1964*b*). *J. acoust. Soc. Am.*, **36,** 196–202.

Nauman, G. C. (1958). Sound localization: the role of the commissural pathways of the auditory system of the cat. Doctoral dissertation, University of Chicago.

Neff, W. D. (1960). In *Neural Mechanisms of the Auditory and Vestibular Systems*, pp. 211–216, ed. Rasmussen, G. L., and Windle, W. F. Springfield: Thomas.

Neff, W. D. (1961*a*). In *Sensory Communication*, pp. 259–278, ed. Rosenblith, W. A. New York: Wiley.

Neff, W. D. (1961*b*). In *Brain and Behavior*, vol. I, pp. 205–262, ed. Brazier, M. A. Washington, D.C.: American Institute of Biological Sciences.

Neff, W. D. (1962). *Psychol. Beit.*, **6,** 492–500.

Neff, W. D. (1964). *Int. Audiol.*, **3,** 170–173.

Neff, W. D. (1967). In *Hearing Processes and Disorders*, pp. 201–210, ed. Graham, A. B. Boston: Little, Brown.

Neff, W. D. (1968). *Int. Audiol.*, in press.

NEFF, W. D., and DIAMOND, I. T. (1958). In *Biological and Biochemical Bases of Behavior*, ed. Harlow, H., and Woolsey, C. N. Madison: University of Wisconsin Press.
NEFF, W. D., FISHER, J. F., DIAMOND, I. T., and YELA, M. (1956). *J. Neurophysiol.*, **19,** 500–512.
NELSON, P. G., and ERULKAR, S. D. (1963). *J. Neurophysiol.*, **26,** 908–923.
RASMUSSEN, G. L. (1946). *J. comp. Neurol.*, **84,** 141–219.
RISS, W. (1959). *J. Neurophysiol.*, **22,** 374–384.
ROSENZWEIG, M. R., and AMON, A. H. (1955). *Experientia*, **11,** 498–500.
ROSENZWEIG, M. R., and SUTTON, D. (1958). *J. Neurophysiol.*, **21,** 17–23.
RUPERT, A., MOUSHEGIAN, G., and WHITCOMB, M. A. (1966). *J. acoust. Soc. Am.*, **39,** 1069–1076.
SNIDER, R. S., and STOWELL, A. (1944). *J. Neurophysiol.*, **7,** 331–357.
STOTLER, W. A. (1953). *J. comp. Neurol.*, **98,** 401–432.
STROMINGER, N. L., and NEFF, W. D. (1967). *Anat. Rec.* (abstract).
THOMPSON, R. F., and WELKER, W. I. (1963). *J. comp. physiol. Psychol.*, **56,** 996–1002.
WEGENER, J. G. (1964). *J. aud. Res.*, **4,** 227–254.

DISCUSSION

Rose: I am somewhat confused by the interpretations of Masterton, Jane and Diamond—whose work was considered by Professor Neff—concerning the role of the inferior colliculus. J. M. Goldberg and R. Y. Moore ([1967]. *J. comp. Neurol.*, **129**, 143–155) suggest strongly that the inferior colliculus is an obligatory synaptic region for the ascending auditory fibres in both the cat and the squirrel monkey. Even assuming that there may exist in fact a small acoustic component which reaches the medial geniculate directly, there is still little doubt that the vast majority of the fibres in the lateral lemniscus terminate in the inferior colliculus. It is thus difficult to see how it can be suggested that the inferior colliculus is an effector system.

Neff: If I interpret correctly what Diamond and his colleagues have written in their discussions of results of experiments in which they studied auditory discrimination after ablations of the inferior colliculus or of the brachium of the inferior colliculus, they accept the anatomical findings that most fibres of the auditory system synapse in the inferior colliculus (Jane, J. A., Masterton, R. B., and Diamond, I. T. [1965]. *J. comp. Neurol.*, **125**, 165–191; Diamond, I. T. [1967]. In *Contributions to Sensory Physiology*, vol. 2, ed. Neff, W. D. New York and London: Academic Press). Basing their opinion on some electrophysiological and behavioural studies (e.g. Goldberg, J. M., and Neff, W. D. [1961]. *J. comp. Neurol.*, **116**, 265–290) and on consideration of the phylogenetic development of the system (Bishop, G. H. [1959]. *J. nerv. ment. Dis.*, **128**, 89–114), Diamond and his co-workers leave the question open of whether or not there may be some fibres of the auditory system that may not synapse in the inferior colliculus on their way from medulla to cortex. Concerning the role of the inferior colliculus as a centre involved in behaviour elicited or controlled by auditory signals, Diamond and his co-workers emphasize that it is more than a relay centre projecting via the brachium of the inferior colliculus and the medial geniculate to the neocortex. The other (probably older) connexions of the inferior colliculus to midbrain centres and to thalamic centres other than the medial geniculate appear to be important in such behaviour as the selection of sound rather than light to guide behaviour when both are equally available as guiding signals.

Davis: Do cats prick up their ears and show any signs of spontaneous scanning in your experiments?

Neff: When one ear is destroyed, the cats behave almost like any normal cat that is restrained in a cage; they pace back and forth behind the restricting wire screen of the starting box. One sometimes has the impression that when the sound comes from one side, the restrained animal may tend to pace more than usual on that side of the starting box, but they don't stand still and orientate towards the sound. After the sound is turned off and the starting box is opened the animals turn and go towards the source of sound. After one ear has been destroyed it is not obvious that the animal starts to move its head and body in a different fashion from when both ears were intact. From his behaviour (successful approaches to the sound source) it is obvious that he is learning to make use of new sound cues.

Evans: As I will point out in my paper (p. 272), at least three laboratories (those of Professor Rose, Dr. Goldstein and ourselves) have now shown that over half of the units in the primary auditory cortex are very sensitive to the location of the sound source, and the majority of these units respond to contralateral sound sources. For many of these units the responses were selective—that is, only correctly orientated stimuli were effective; for the remainder the responses were preferential—that is, they gave strongest responses to preferred "orientations" of the stimulus.

Could you say more about how important is ablation of the primary area, compared with the surrounding areas, from the point of view of discrimination of the position of the sound source?

Neff: Our data are limited to results from very few animals, but the AI appears to be very important for localizing behaviour. The AI is the only cortical area that causes small losses in localization when ablated bilaterally. It may be most important only because it gets denser projection from the medial geniculate than any of the other sub-areas; or it may be important because it contains more of the units that you mention, units that show binaural interaction; I do not know. In our experience bilateral removal of other sub-areas has no effect upon localizing behaviour, even though the amount of cortex removed was as great as or greater than the AI areas.

Evans: Did not bilateral ablation of the insular-temporal areas have an effect on the ability to localize the position of the sound source?

Neff: It had no great effect on localization; however, on pattern discrimination it has a very severe effect (Neff, W. D. [1961]. In *Sensory Communication*, pp. 259–278, ed. Rosenblith, W. A. New York: Wiley; [1961]. In *Brain and Behaviour* vol. I, pp. 205–262, ed. Brazier, M. A. Washington, D.C.: American Institute of Biological Sciences).

Whitfield: You mentioned the assessment of time-difference by the cortex, and this is an attractive hypothesis. However, if we think in terms of the hypothesis of Hall and of van Bergeijk ([1962]. *J. acoust. Soc. Am.*, **34**, 1431–1437) in which time-difference is converted at the olivary level into difference of neural activity of the two sides, then we have another problem. Although the *ratio* of activity on the two sides is independent of intensity, the absolute level of activity is not—at least in the brainstem. However, the units in the cortex to which Dr. Evans referred are very intensity-independent, in that if you present a sound on the "correct" side of the animal's head the unit will respond even though the sound is of quite low intensity; the same sound presented on the "wrong" side produces no response, even when the sound is very loud. Could the cortex perhaps be extracting this ratio information?

Another point is that from investigations into AII, we have been surprised to find fewer of these units than in AI, which might link up with what you were saying about localization, but I do not yet know that the difference is significant.

Neff: I agree, Dr. Whitfield, that it is not correct to imply that the neural cues for localizing or lateralizing sound are coded as time-differences at the cortex. They probably have been recoded in some fashion by lower centres. The evidence I have summarized concerning the effects of lesions of the lateral lemniscus suggests that a first recoding has taken place somewhere in the superior olivary complex.

Rose: Judy Hirsch in our laboratory recently completed a study (not yet published) in which she shows that the amplitude of the classical evoked potential in the auditory cortex of the cat can be a reliable function of the interaural time-differences when pure tones of low frequencies are employed.

Schwartzkopff: W. D. Keidel, V. O. Keidel and M. E. Wigand ([1961]. *Pflügers Arch. ges. Physiol.*, **270**, 347–369) have studied evoked potentials in the cortex and found time-dependent intensity differences. They supposed that at the level of the cortex, where time-differences have been translated into differences of intensity, there may be place differences which are the "translation" of time or intensity differences. Am I right in this?

Neff: There have been reports of changes in evoked response patterns at the cortex when changes are made in the time of arrival of signals at the two ears. For example, M. R. Rosenzweig ([1951]. *Am. J. Physiol.*, **167**, 147–158) has described amplitude changes in the response recorded by gross electrodes, and P. D. Coleman ([1959]. *Science*, **130**, 39–40) has suggested changes in place.

Whitfield: S. Jungert ([1958]. *Acta oto-lar.*, Suppl. 138) described crossings-over between the lemniscal pathways at different levels. Does it make any difference at what level the lemniscal pathway is cut? Experimentally, I imagine this is rather difficult.

Neff: We have not done this systematically. We have some information—for example, the effects of transecting the lemniscal pathways just rostral to the superior olive and at the level of the brachium of the inferior colliculus (see Tables III and IV, pp. 215, 217).

ROLE OF THE PINNA IN LOCALIZATION: THEORETICAL AND PHYSIOLOGICAL CONSEQUENCES

DWIGHT W. BATTEAU*

Department of Mechanical Engineering, Tufts University, Medford, Massachusetts

WHEN we began the study of the role of the pinna in human hearing, we had one reliable datum, namely, that distortion of the pinna produces a distortion of the subjective locale of a sound source. In particular, large shifts of apparent position in the altitude of a sound are easily produced by bending down the tops of the pinnae.

An experiment to demonstrate this was an example of true elegance. W. McLean jingled keys in front of me with my eyes closed and I pointed accurately to them. Then he had me bend the tops of my pinnae down, and jingled the keys again; I missed by 90°. Since anyone can perform this experiment, I recommend it to you.

The immediately obvious statement can be made that the pinnae have a role in localization of sound, for anything that alters a perception has a role in it. This is an instance of the general theorem that if a change in the condition of *a* affects the condition of *b*, then the condition of *b* is a function of *a*.

In a more mathematical approach, we could say that a transformation is performed by the pinna on the incoming sound front, and that this transformation is significant in localization. Our problem then was to determine the nature of the transformation and to delineate its function.

We were inclined to look for changes in spectral characteristics at first, but a little thought indicates that this is a poor supposition because of the extreme variety of sounds which one can localize, many of them heard for the first time. Then one day we were watching the oscilloscope pattern of an acoustic pulse played into an artificial ear mounted around a microphone, and as the position of the sound source was shifted we saw a feature move on the oscilloscope pattern. At that point the inevitable occurred and we said "Ah, a transformation of time relationships!".

We then set about to measure the factors of the transformation, which

* Deceased.

must clearly involve multiple time delays and attenuations. To make the measurements, which involved short times, we modelled the human pinna five times normal size and made measurements which indicated that the principal azimuth delay had a range of zero to 80 microseconds from back to front, and the principal altitude delay had a range of 100–300 microseconds. Through an arithmetic error, and poor judgment, since the delay times are (now) obvious, we said azimuth was up to 160 microseconds in the moving picture that the United States Navy made of our work. If you see that film, please forgive us and make the correction (McLean, 1959).

We then set about formulating the functions relevant to the transformation and described a sequence of summed delays by the following general equation (applicable to pinna and other multi-path transformations):

$$H(t) = \sum_{n=0}^{\infty} a_n f(t - \tau_n)$$

$H(t)$ = the sound heard

a_n = attenuation in the nth path

τ_n = the time delay in the nth path

t = time

This initial form of expression was awkward, but we wrung out the existence of the inverse to the transformation indicated and found some interesting consequences (Batteau, 1967). The mathematical form of the inverse transform was constructed in two ways, one quite brutally straightforward, and the other by a sequence of powers of two. The power-of-two sequences is, of course, an "octave" sequence in music, and may be related to the construction of the musical scale involving octaves. We then made the conjecture that musically pleasing tonal structure is related to simplicity in the computation of recognition.

Today, instead of the initial form, we use the more convenient Laplace transform notation and write the following equation:

$$H(s) = P(s) \sum_{n=0}^{\infty} a_n e^{-s\tau_n}$$

$H(s)$ = Laplace transform of $H(t)$

$P(s)$ = Laplace transform of the acoustical source

$e^{-s\tau_n}$ = delay of τ_n in the nth path

s = Laplace operator

in which the transformation $T(s)$ is clearly applied to the stimulus $P(s)$.

$$T(s) = \sum_{n=0}^{\infty} a_n e^{-s\tau_n}$$

Localization is then performed by finding the transformation applicable to the sound heard, which is appropriate to the given locale.

It is provocative to notice that our descriptions indicate that the way a signal transforms is important to recognition, and that construction of the inverse, or implication of it, provides recognition. This is not new in physics, where it is clear that physical knowledge derives from perception of how particles or gross systems transform, and the existence of such things as electrons is postulated from observations of influence. It does, however, seem to be new in general thinking to state explicitly that "in order to know, one must be able to construct the inverse to the transformation of perception". The proof is simple; if the inverse cannot be constructed, then the correspondence between received information and that which gave rise to it cannot be made.

At this point, other interesting events develop, for we may now explain the "cocktail party effect". By applying a second appropriate transformation, amounting to a matched filter, the sound from a particular locale may be attended to, or selected by increase of signal-to-noise ratio over other locales to which the transformation in use does not apply.

A typical attention function by which this can be done may be written as follows:

$$a(s) = e^{-s\tau_N} \sum_{n=0}^{N} a_n e^{+s\tau_n}$$

which indicates that a delay of τ_N is necessary to pay attention, and that the infinite sequence indicated must be truncated.

When we had derived these expressions, we realised that they were valid for room reverberation and the "precedence" effect, and for speech or reverberation in the vocal tract. We thus had a single, relatively simple expression which accounted for observable acoustic perception and function.

At this point, since we had exact mathematical forms, it became possible to inquire into hearing physiology beyond the pinnae, or the internal ear. Resonant perception of frequency is inadequate to the functions described, and since they are verifiable (for example in localization synthesis) we must discard any supposition that mechanical resonances give tone perception. The alternative means of identifying a tone is a distributed sensor—which the basilar membrane suggests—with a non-reverberant termination—which the cochlear spiral suggests. By using the distributed signals in computation, tones may be identified. Thus because of function and physiology, we are inclined towards considering a distributed sensor from which the required functions are computed in the nervous system.

Here again, we encounter a provocative situation. The electrical signals observed in the nerves of the hearing system are totally inadequate to perform the computational functions described. These signals have been thoroughly studied in cats and the conclusion has been that they are only "statistically related" to the sounds providing the stimulus. Thus we also need another model for nerve function.

In examining the possibilities for transferring information in a nervous system, we were struck by the possibility that transitions between energy levels in organic molecules could provide this function. The observed electrical activity could then be attributed to a metabolic process restoring to an excited state the population undergoing transition due to sensation or transmission of information.

If we assume a population of charges initially in a ground or low-energy state, we can imagine a metabolic process occurring as a pulse to elevate it to an energized state. The charges will decay thermally from that state if undisturbed, or decay at a rate dependent on stimulus. Fortunately, these rates are sensitive to many things: temperature, chemicals, radiation, and, to a lesser extent, pressure (or negative pressure of tension).

In examining these sensitivities, we found that *all* of the senses could thus be constructed, but that sensing mechanical quantities would best be done by *shearing* to produce large negative pressures with little work. When we looked at the anatomy of such receptors, we found that they were indeed constructed so that pressures would produce shearing. The Pacinian corpuscle is an example of such a device.

With this beginning, we undertook to design some simple experiments related to the hypothesis. Using an equation from Schrödinger, which is the inverse of the chemical reaction rate equation of Arrhenius, we could calculate the probability of transition as a function of temperature. The equation is as follows:

$$t = \tau \exp -\left(\frac{w}{kT}\right)$$

t = expected excited lifetime

τ = a natural constant (said to be about 10^{-13}sec.)

w = depth of energy well (ergs)

k = Boltzmann constant ($1 \cdot 38 \times 10^{-16}$ergs/K°)

T = absolute temperature

Fortunately, we could design an experiment so that τ need not be known in order to calculate w. By varying the temperature of the system we could obtain the ratio of expectations as follows:

$$\frac{t_1}{t_2} = \exp -\frac{w}{k}\left(\frac{1}{T_1} - \frac{1}{T_2}\right)$$

Since both T_1 and T_2 have been determined and t_1 and t_2 are measured, we can calculate w without involving τ.

We wished to make a simple experiment to see if our ideas were at least plausible. We assumed that the electrical action potential of a nerve is an artifact of the metabolic energy pumping process. We then observed that, if this were true, it should occur at about the expected lifetime of the excited state in a large population. Thus the interspike interval in a nerve unstimulated (except thermally) should give us t, the expected lifetime. We proceeded to design an experiment based on these ideas.

We used the B-cell nerve of the ear of the Catocala moth for the experiment and calculated w as about 0·6 ev or stimulatible by infrared of about 1·1 micron wavelength. Unfortunately, we never finished the test by using infrared. The results were published by Batteau and Hemmes (1966).

Continuing the consequences of the hypothesis, we wondered if a muscle could be used as a sensor, and experimented with the gastrocnemius of *Rana pipiens*, and the associated sciatic nerve. We found that spontaneous twitches were maximized at a temperature of around 83°F, and that electrical activity in the nerve could occur after the twitch, before the twitch, be absent in the vicinity of the twitch, or present for no twitch. The results were briefly reported by Batteau and Eyrick (1967).

In conclusion, I would like to remark that the mathematical studies related to hearing led from consideration of the physiology and nerve function of the hearing system, by extension to other biological systems, to a consideration of the general construction of nervous systems. The design of experiments follows quite simply in all cases, and we would say that synthesis of locales has been done, and can be computed, that the physiology of the inner ear now makes sense in view of the necessary functioning, and that the nerve character seems to be reasonable, and subject to systematic experiment. Finally, on the subject of sensing, the nerve model provides for mono-cellular senses (which it is necessary to provide somehow in biology, since they exist) and for direct read-out of genetic codes, where the seat of instinctive behaviour must lie.

SUMMARY

An inquiry into the role of the pinna in human localization provided a mathematical form which can be used to show how localization takes place and how attention can be directed. It throws light on the mechanism of attention, and provides both a means of questioning the physiology of the inner ear, and a reasonable answer.

The immediate theoretical consequences relating to basic means of "knowing" are then also opened to question, and the means of doing this in a nervous system provides questions regarding its biomolecular functioning. In exploring these questions, reasons have been found to conclude that the electrical activity of normal nerves is metabolic and that information is sensed and transmitted by energy-level shifts (as in the maser or laser).

REFERENCES

BATTEAU, D. (1967). *Proc. R. Soc. B*, **158**, 158–180.

BATTEAU, D. W., and EYRICK, T. B. (1967). *The Significance of Energy Level Transitions in Nerve Function.* Interim Technical Report for U.S. Navy Office of Naval Research, Contract 4863(00).

BATTEAU, D. W., and HEMMES, W. M. (1966). *Molecular Sensation.* First Semi-Annual Report for U.S. Navy Office of Naval Research, Contract 4863(00).

McLEAN, W. (1959). *Ears in Action.* A motion picture made by the U.S. Naval Ordnance Test Station, China Lake, California.

DISCUSSION

Lowenstein: Have you studied the effect of strapping the pinna to the head in man?

Batteau: We have, and localization is affected and altered. We have also filled the pinna with wax, and the sound appears to come from inside one's head if both pinnae are filled. If you fill one pinna and strap the other down, the localization sense is lost; if you leave both pinnae unfilled, it is retained. It is interesting that monaural localization has been known since 1901 (Angell, J. R., and Fite, W. [1901]. *Psychol. Rev.*, **8**, 225–246).

I had the interesting experience of seeing a young man whose ear had been surgically remodelled, and it had not been done perfectly. He denied that he had difficulty in localizing, but when he was checked, his accuracy was about 20° in error on that side and about the normal 3° or 4° on the other.

Lowenstein: Is a greater amount of sculpture of the pinna an advantage, then?

Batteau: I do not think so. It is just that you must have sufficient.

Lowenstein: I ask because I am impressed by the poor sculpture in the pinna of the anthropoid apes. One would have thought that there should be a premium on localization among the apes.

Batteau: The thick localizing structure of the pinna is at the bottom, very convoluted, all located in a small area which uses high frequencies. The larger upper part apparently is then directed, on localizing, to improve low-frequency perception. The pinna functions in the way I have described only at high frequencies—above 6 kilocycles.

Schwartzkopff: What would happen if you introduced movements of the pinna into your model?

Batteau: You would change the encoding, and of course this increases the confidence in the signal, if you know all the encoding transformations. It provides redundancy, in other words.

J. D. Pye: As far as I can see the pinna mechanism described by Professor Batteau does not work at all for a pure tone, because any number of sine waves of the same frequency combine together to make a single sine wave and the ear does not know the difference. So if you have to have a pure tone, and to try to locate it, as Rhinolophids and Hipposiderids do among the bats, would it help to move the ear, as they do?

Batteau: Yes, it would help, because this would rotate the co-ordinate system. The minimum number of tones for a single ear to localize is four tones. For two ears, if you use both phase and amplitude, the number is two, and you have one tone left over.

Whitfield: To describe a speech wave completely requires a great number of bits, yet it seems that the ear transmits very little of this information, perhaps something less than 20 bits per second. What the ear appears to do at any given moment is to utilize quite a small amount of information and relate it to what it has already stored from its past experience, which may vary in this context from a few milliseconds to some years. It then constructs the most probable situation compatible with the data available. The information flow needed to do this is quite small and in fact behavioural experiments, treating the system as a whole, seem to indicate that information flow is somewhere between 10 and 20 bits per second.

Batteau: I approached the bit requirement by calculating the number of bits necessary to pay attention continually, because I was interested in the continuous computation of the attention function. This gave a bit requirement of 70,000 bits per second.

Whitfield: But this does not seem to happen.

Johnstone: Professor Batteau suggested that the basilar membrane can be regarded as a "distributed sensor" or phase-delay line. The further characteristic of such a line, if it is to have good discrimination, is that the advancing front must be very sharp. Now the advancing front of a travelling wave is around 80 db per octave in slope, which is about as sharp as the neural tuning curves found by N. Y.-S. Kiang ([1965]. *Discharge Patterns of Single Fibers in the Cat's Auditory Nerve.* Cambridge, Mass.: M.I.T. Press). Perhaps the difficulty of matching our primary auditory tuning curves to the Békésy wave comes from our persisting in thinking of the Békésy wave in terms of its overall amplitude rather than just the travelling part, with its very steep front, which is working down the phase-delay line. I would suggest that this, coupled perhaps with the distribution of hair cells as advanced by Dr. Spoendlin, plus some sort of an additive circuit in the dendrites, may be adequate to explain the tuning curve recorded by Kiang, provided one can work out in some way the delay characteristics of these dendrites and hair cells.

Batteau: This is essentially what I was saying, yes.

Hallpike: If the cochlea is to act as a delay line, the longer the line and the more of its length is used, the more efficient it would be, presumably? And if all this is related only to frequencies above 6 kilocycles, the wave will not travel very far along the delay line; it will all stop at the base.

Batteau: It is not necessarily true that efficiency is related to length; I am thinking of the owl. The lower frequency component of the wave moves the basilar membrane as a whole, and is relieved by the round window. For high frequency I am not prepared to say. There is a cross-over, perhaps at 300 cycles.

Hallpike: I would have thought that the disturbance to the cochlea produced for example by the jingling of keys would be all more or less lumped together at the base.

Batteau: I can say that the delay per unit length is at least as great as what it would be in water. The delay is 1,500 metres per second in water, and for 45 millimetres, the length of the cochlea, it is about 3 milliseconds.

Hallpike: A second point: most people think that the pinna helps to amplify sound, but you are mainly ascribing a localization function to it or rather, to its various excrescences. What do you feel about the pinna as an amplifier; would it not need to be bigger than it is?

Batteau: The calculated efficiency of the pinna in the domain in which we are looking is 3 db, and it is steerable. It collects all sounds over its surface. The attention function which you apply to the particular point of orientation will give 3 db for any point.

Hallpike: That is just about perceptible. What about the elephant's pinna; would it be better to have one that size?

Batteau: As far as I am concerned, no! All the functions I have studied require a stationary system, at least during the computation; something that is relatively small, stationary, with quick localization.

Lowenstein: The elephant's pinna at least has another known function—it acts as a heat regulator.

Davis: May I mention one more quantitative relation, namely the form of the Békésy envelope and the phase of the travelling wave. The phase at which the cut-off of amplitude occurs is approximately 2π of the frequency concerned. In other words, about one cycle of the sound represents the delay at which the amplitude has just passed its maximum. If all this has to happen above 6 kcyc./sec. the available delay for a second signal can be estimated; it will be something like 0·15 msec. This is less than the refractory period of the nerve fibre after it has responded. Within that time, a second fibre will have to respond, not the same fibre. This is one limitation, from the electrophysiological point of view, in this kind of thinking.

Batteau: We have observed 2–3 μsec. discrimination in location monaurally, and also in a synthesis by wiggling a pulse by 2–3 μsec. So 0·15 msec. is easily believable as a useable interval.

Davis: During this period, it must be a second fibre which responds.

Batteau: Can you tell the difference?

Davis: You are talking about the behavioural recognition of very small time-differences between the two ears, not about repetitive discharge in a single nerve fibre.

Johnstone: If two pulses are presented to the basilar membrane very close together, the odds are that they will interfere with each other mechanically. Therefore when the subject in psycho-acoustic experiments is asked whether he can distinguish between two closely spaced clicks, and the answer is "yes", what difference is he actually talking about? I would submit that the basilar membrane cannot be regarded as a series of lumped, individual elements, but something in which each segment is intimately coupled to the next. The question must be approached not by considering one pulse going in and then another; whether or not the nerve has recovered when the second pulse arrives, the second pulse *must* be different from the first, and therefore you can tell them apart. Perhaps then, for these very short sharp pulses, and very small time-differences, it is not a question of the same nerve responding; it does not have to respond, because since the next pulse is different, another nerve will inevitably respond. When we get to widely spaced pulses, we may switch over to a different mechanism, but with pulses that are only short times apart, there will be mechanical interaction at the basilar membrane.

J. D. Pye: Professor Batteau said that four tones were necessary for localization by a single ear; must they all be above 6 kHz?

Batteau: If you are localizing by this particular function, you have to receive at least a quarter of a wave across the ear. The size is about 3 inches, so for this to be one-quarter wavelength, the wavelength must be 12 inches, which is about 1 kilocycle; so if the lowest tone were at 1 kilocycle, possibly you could localize successfully.

J. D. Pye: With a smaller ear, such as a bat has, you would have to go higher still?

Batteau: Yes.

Schwartzkopff: Professor Batteau mentioned owls, and the traditional view has been that the fact that owls have a tremendously long basilar membrane is an indication of good frequency analysis. Yet nobody has shown that owls do analyse frequency. If I understand Professor Batteau correctly, we might now work on the hypothesis that the basilar membrane is an organ to analyse *time*; if so, in the asymmetric ears of owls the coding, which in man is equal for both ears, should be different for the two ears. This must give an additional piece of information; I do not yet see how it is used, but certainly one ear will co-operate with the other; this code remains the same as for frequency but now frequency is transferred in another form, as a sequence of pulses which is resolved along the length of the basilar membrane.

Batteau: Since our own ears are symmetrical on the midplane and you cannot tell "where" from "what", for high-pitched sounds, you would expect ambiguities to exist maximally on the midplane, which experimentally is found to be true. The owl avoids this by having asymmetrical pinnae.

Whitfield: Surely, Professor Schwartzkopff, you cannot assume that the length of basilar membrane is related to time discrimination? It would depend on all the propagation constants of the system.

Schwartzkopff: Certainly, but it might be an indication; one could do experiments on this basis.

Johnstone: Perhaps length could be related to both frequency discrimination and temporal patterning.

SECTION V

CENTRAL MECHANISMS

CENTRIFUGAL CONTROL MECHANISMS OF THE AUDITORY PATHWAY

I. C. WHITFIELD

Neurocommunications Research Unit, University of Birmingham

THE existence of fibres of central origin terminating in the cochlear nucleus has been known for some seventy-five years (Held, 1893). However, neither Held nor Lorente de Nó (1933), who described a further such bundle, determined the source of these fibres. The first complete description of a centrifugal auditory pathway seems to have been that of the olivo-cochlear bundle (Rasmussen, 1942, 1946). Subsequent work (Rasmusssen, 1955, 1958; Desmedt and Mechelse, 1958, 1959) has established the existence of centrifugal pathways at all levels from the cortex to the periphery (Fig. 1).

The early view of these pathways was that they exercised an inhibitory function on the peripheral structures. Galambos (1956) showed that stimulation of the olivo-cochlear bundle in the floor of the fourth ventricle suppressed the N_1 auditory nerve response to clicks measured at the round window, and Desmedt (1960) demonstrated an inhibitory effect of lemniscal stimulation on the gross click response in the contralateral nucleus which was independent of any effect at the cochlea itself.

The first behavioural experiments (Hernández-Péon and Scherrer, 1955; Sheatz, Vernier and Galambos, 1955) seemed to support this view, but later experiments (e.g. Marsh *et al.*, 1961; Worden and Marsh, 1963; Dunlop, Webster and Day, 1964) suggested that the effects were more complicated than a simple suppression of activity.

The possibility that the centrifugal pathways might bring about facilitation as well as inhibition of neurons in the cochlear nucleus was suggested by Comis and Whitfield (1966). These workers found that acetylcholine, locally applied to neurons of the cat cochlear nucleus, was capable of lowering the threshold of such neurons to sound stimuli (Fig. 2). Application of cholinergic blocking agents such as dihydro-β-erythroidine had the opposite effect, the sound threshold being raised. The cochlear nucleus stains strongly for the presence of true cholinesterase (Whitfield, 1968) and there is reason to think that this is associated with the centrifugal rather than the centripetal fibres.

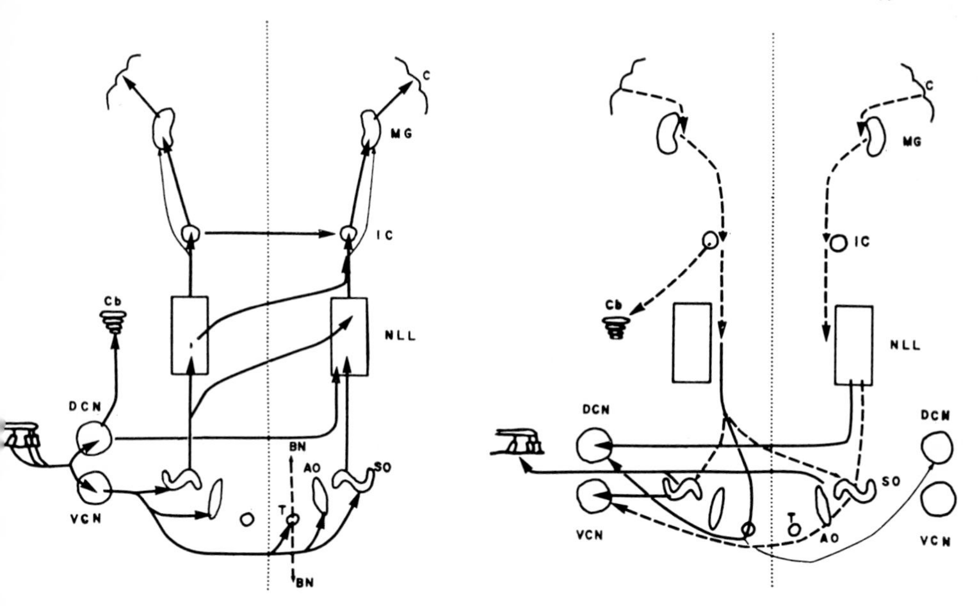

FIG. 1. Centripetal (left) and centrifugal (right) connexions of the auditory pathway. Connexions not irmly established anatomically are shown by dashed lines.

DCN, dorsal cochlear nucleus; VCN, ventral cochlear nucleus; Cb, Cerebellum; AO, accessory olive; O, lateral olivary nucleus; T, nucleus of the trapezoid body; BN, brainstem motor nuclei; NLL, nuclei of he lateral lemniscus; IC, inferior colliculus; MG, medial geniculate body; C, cortex. (From Whitfield, 967, by permission of the publisher.)

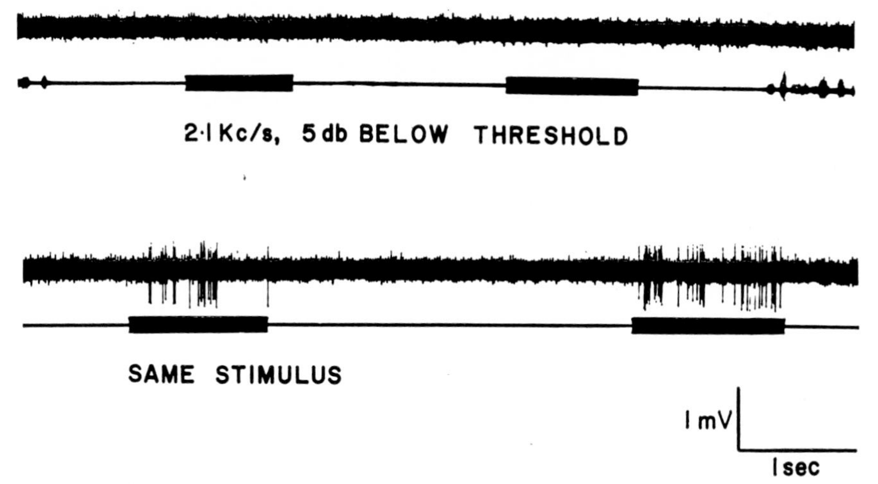

FIG. 2. The effect of acetylcholine applied to a neuron in the antero-ventral cochlear nucleus in lowering its threshold to sound stimulation.

Above: normal record: Below: unit responding, in the presence of acetylcholine (5×10^{-12} moles/sec.), to the previously subthreshold sound stimulus. Thickening of lower trace indicates "sound on". (From Whitfield, 1967, by permission of the publisher.)

Four major centrifugal pathways have been described by Rasmussen (1960) as terminating in the cochlear nucleus of the cat. These are (*i*) collaterals of the olivo-cochlear bundle to the ventral cochlear nucleus, (*ii*) fibres from the lateral accessory olivary nucleus (S-segment) to the ventral cochlear nucleus of the same side, (*iii*) fibres from the ventral nucleus of the lateral lemniscus to the dorsal cochlear nucleus of the opposite side, (*iv*) fibres from the inferior colliculus and dorsal nucleus of the lateral lemniscus to the dorsal cochlear nuclei of both sides (mainly ipsilateral). These by no means exhaust the centrifugal connexions of the cochlear nuclei, but the precise sources of the rest remain to be determined.

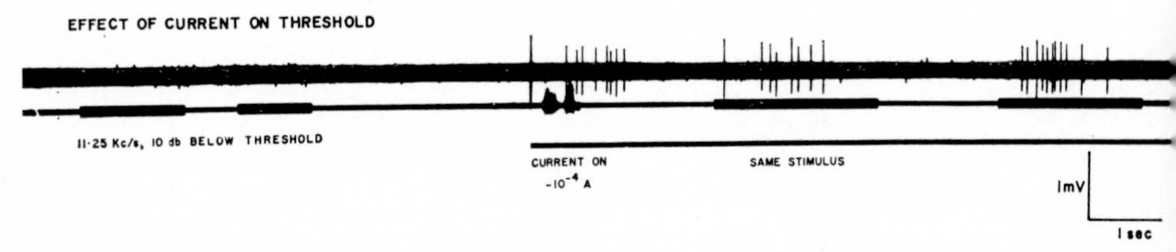

FIG. 3. The effect of stimulation of the ipsilateral S-segment of the olive on the sound threshold of a neur in the ventral cochlear nucleus.

The sound stimulus, 10 db below normal threshold, produces no response until the stimulating current applied. Note that the d.c. current stimulus itself produces no response in the absence of the sound stimul although the onset of this current does so. (From Whitfield and Comis, 1968, by permission of the publishe

It is not practicable to stimulate the centrifugal fibres as they enter the cochlear nuclei, since they are closely mixed with the outcoming centripetal fibres; impulsive stimuli would inevitably produce antidromic activation of the latter and make the results uninterpretable. We have therefore adopted direct current stimulation by means of a monopolar electrode inserted into the region of the cells of origin of the fibres, in the anaesthetized cat. Stimulation in this manner of the S-segment of the superior olive has an effect on cells of the ipsilateral cochlear nucleus which is always facilitatory. This facilitatory effect is, as far as we can determine, confined to the ventral division of the nucleus, a finding in conformity with the known distribution of the fibres (Rasmussen, 1960). Strong stimulation of the olive can cause a discharge of the target cell in the cochlear nucleus in the absence of any overt peripheral stimulus. Less strong stimulation facilitates the discharge produced by a sound stimulus, and can reduce the threshold of discharge to the latter (Fig. 3), just as does locally applied acetylcholine. The effects of olivary stimulation are, indeed, quite comparable to that of acetylcholine. In view of this similarity, the effect of cholinergic blocking agents is of obvious importance.

Gallamine, dihydro-β-erythroidine and atropine each block the effects of olivary stimulation when applied locally in the cochlear nucleus. None of these substances is entirely free from side effects; both gallamine and dihydro-β-erythroidine tend to reduce the spike height when given in large doses, so that usually only a partial block can be certainly observed (Fig. 4).

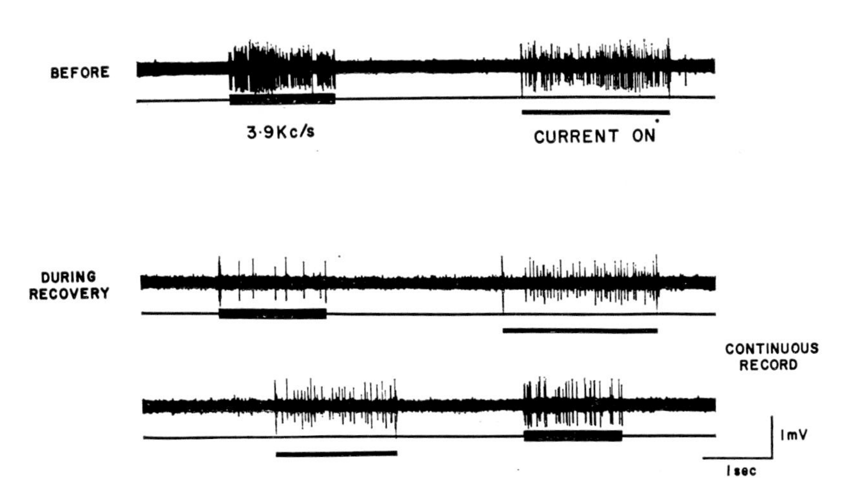

FIG. 4. The effect of a cholinergic blocking agent on the olivary–cochlear nucleus pathway.

The top trace shows the response of a neuron in the ventral cochlear nucleus to a sound stimulus (left) and to current stimulation of the olivary S-segment (right). Dihydro-β-erythroidine was then applied locally to the cochlear nucleus neuron by a microtap. After closure of the tap the gradual recovery from the block is seen in the two lower traces which form a continuous record. Stimulus conventions as in top trace. (From Comis and Whitfield, 1968, by permission of the American Physiological Society.)

Atropine is reputed to have a general depressant effect on neurons. Nevertheless, in doses which block the olivary stimulation, the excitability of the cochlear nucleus as tested by direct stimulation (Comis and Whitfield, 1968) remains normal. Rasmussen (1964) has demonstrated the presence of cholinesterase in the fibres running between the S-segment and the cochlear nucleus which we are presumably stimulating. There is thus considerable evidence for the implication of acetylcholine in the transmitter process.

Stimulation of the olivary S-segment always produces facilitation in the ipsilateral ventral cochlear nucleus. By contrast, stimulation in the contralateral dorsal nucleus of the lateral lemniscus seems to produce only inhibition, and in the *dorsal* cochlear nucleus. The same result can be obtained by the local application of noradrenaline to the cochlear nucleus neuron. We have not so far been able to interrupt this pathway by local application of blocking agents, but systemic treatment of the whole animal with reserpine has resulted in the abolition of a previously present inhibitory response.

Exploration of other parts of the contralateral lemniscal cell column leads to various results according to the position of the stimulating electrode. In that part of the column designated as the ventral nucleus of the lateral lemniscus, both inhibitory effects in the dorsal cochlear nucleus and facilitatory effects in the ventral cochlear nucleus have been obtained (Fig. 5), though not from the same point. In the most caudal part of the cell mass, close to the olive, inhibitory responses are obtained in the ventral cochlear nucleus.

The ipsilateral lemniscal column, which sends known connexions to the cochlear nucleus, has not yet been explored, but even without these it is already clear that there is a complex interaction of facilitatory and inhibitory pathways which can influence the sensitivity of individual neurons to incoming signals. What is the function of these elaborate systems? It is not particularly useful to characterize them as "feedback" loops, since at present we do not know what they feed back, or from where; indeed, they may not form loops at all. Although it seems likely that there would be connexions between the centripetal and centrifugal systems at all levels from the olive upwards, this has not been established, and it is conceivable (though I am not suggesting that it is so) that the descending pathway could remain separate though parallel throughout its length.

That the fibres play any part in the elaboration of fine detail in the centripetal pulse pattern seems very unlikely, since the latency of, for example, the S-segment–cochlear nucleus pathway is at least 30 milliseconds. It seems highly improbable, in any case, that any such significant pattern exists (Allanson and Whitfield, 1956; Viernstein and Grossman, 1961).

The now familiar idea of the afferent "throughput" of the auditory nuclei being controlled by the centrifugal system has tended to be thought of in terms of peripheral stimuli being allowed, or not allowed, to reach the higher centres of the auditory pathway. However, there is another possibility. It seems fairly clear that (in the cat) simple discriminations of frequency and intensity can be carried out in the absence of the classical auditory cortex (Diamond, Goldberg and Neff, 1962), and even after damage to the inferior colliculus as well (Raab and Ades, 1946; Neff, 1961). On the other hand, interference with the inferior collicular brachium may

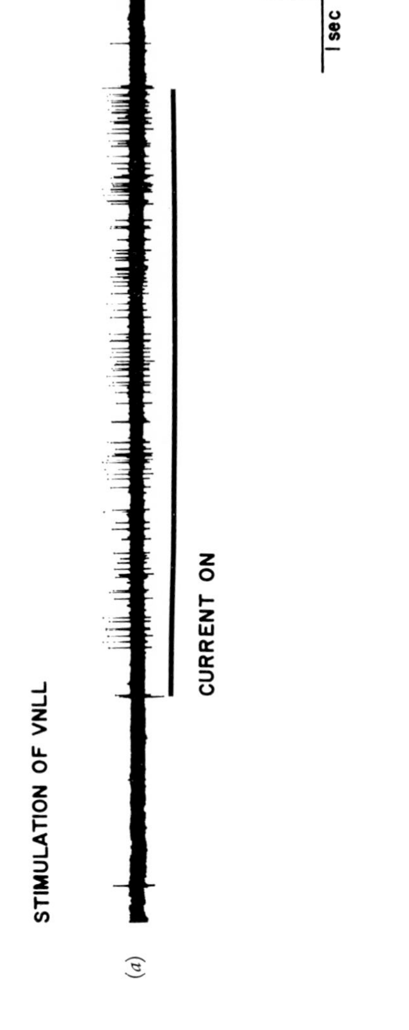

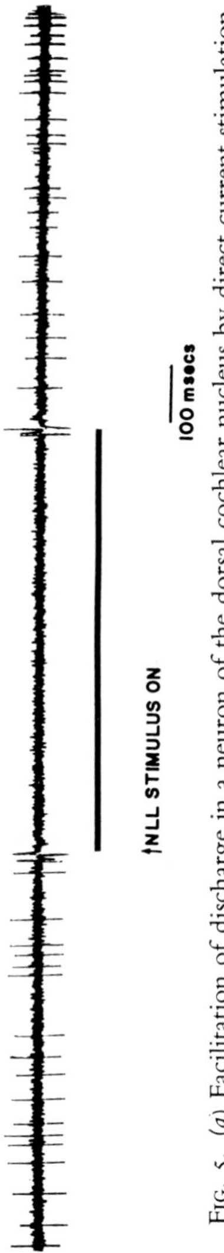

FIG. 5. (*a*) Facilitation of discharge in a neuron of the dorsal cochlear nucleus by direct current stimulation (black bar) of a point in the contralateral ventral nucleus of the lateral lemniscus.
(*b*) Inhibition of activity in a different neuron of the dorsal cochlear nucleus by direct current stimulation of a different lemniscal position.

destroy the ability to make these discriminations. This apparent anomaly could be explained if it is the descending rather than the ascending fibres which are crucial.

Neff (1961) has argued very plausibly that the types of discrimination which can be made at the brainstem level are those which involve some additional feature in the positive stimulus. If we suppose them capable of differential suppression of neural activity, the control properties of the centrifugal pathways are obvious candidates for distinguishing between the activity common to both stimuli and that arising from the additional property of the positive stimulus. It has usually been assumed that the centrifugal pathways would regulate the activity pattern permitted to ascend the main sensory pathway to the higher centres. However, the system seems somewhat over-elaborate, if its function is merely to ensure that not all the channels are used all the time. There seems no logical reason why all the information should not be allowed to ascend to the region at which the appropriate routing (discrimination) takes place.

Herein lies a clue to a possible answer. If, in a decerebrate cat, we apply a strong cutaneous stimulus to a hind foot, we can elicit the crossed extensor reflex; in a spinal animal we shall probably not succeed. However, it is not then argued that this sensory–motor reflex is at the midbrain level, but rather that tonic facilitation from above enables a spinal reflex to be elicited. Now a sensory discrimination is, looked at behaviourally, a quite analogous motor response to a sensory stimulus, but coupled with a failure of that response to a similar but different sensory stimulus. Unlike the spinal reflex, however, it is usually assumed that the sensory–motor interaction takes place at a high neural level. In view of our present findings there seems no reason to assume this. Connexions undoubtedly exist between auditory and motor nuclei in the brainstem. If we assume that it is the throughput of these connexions which is controlled by the descending pathways, then their complexity becomes reasonable, and some of the anomalies produced by lesions of the ascending pathways become less inexplicable. What I am suggesting, therefore, is that ablation studies on sensory discrimination are in need of a thorough re-examination in terms of their effect on the centrifugal, rather than the centripetal pathways. It will be of particular importance, too, to have some quantitative studies on the descending systems. The olivo-cochlear bundle has few fibres compared with the number of afferents in the auditory nerve. There are at present no such data either for the ratio of actual numbers of centrifugal to centripetal fibres at higher levels, or of the relative distribution of their fields.

SUMMARY

A number of centrifugal pathways are known to terminate on cells of the cochlear nucleus. Stimulation of the cells of origin of some of these pathways in the superior olive and in the lateral lemniscus has shown that both facilitatory and inhibitory connexions are present and that interaction between these is capable of raising or lowering the threshold of cochlear nucleus neurons to peripheral sound stimuli. There is considerable support for the possibility that acetylcholine may be the transmitter of the facilitatory olivary–cochlear nucleus pathway. Noradrenaline has a uniformly inhibitory action on cells of the cochlear nucleus, but there is as yet no corresponding weight of evidence for its being implicated in a transmission process.

It is suggested that rather than serving to shut off information from reaching the higher levels of the auditory system, the centrifugal pathways control the interaction between sensory and effector cell groups at the brainstem level. Many of the anomalies in discriminative ability uncovered by ablation studies may be explicable on this basis.

REFERENCES

Allanson, J. T., and Whitfield, I. C. (1956). In *Third London Symposium on Information Theory*, pp. 269–286, ed. Cherry, C. London: Butterworth.
Comis, S. D., and Whitfield, I. C. (1966). *J. Physiol., Lond.*, **183**, 22–23*P*.
Comis, S. D., and Whitfield, I. C. (1968). *J. Neurophysiol.*, in press.
Desmedt, J. E. (1960). In *Neural Mechanisms of the Auditory and Vestibular Systems*, chap. 11, ed. Rasmussen, G. L., and Windle, W. F. Springfield: Thomas.
Desmedt, J. E., and Mechelse, K. (1958). *Proc. Soc. exp. Biol. Med.*, **99**, 772–775.
Desmedt, J. E., and Mechelse, K. (1959). *J. Physiol., Lond.*, **147**, 17–18*P*.
Diamond, I. T., Goldberg, J. M., and Neff, W. D. (1962). *J. Neurophysiol.*, **25**, 223–235.
Dunlop, C. W., Webster, W. R., and Day, R. H. (1964). *J. aud. Res.*, **4**, 159–169.
Galambos, R. (1956). *J. Neurophysiol.*, **6**, 39–57.
Held, H. (1893). *Arch. anat. Physiol., anat. Abt.*, 201–248.
Hernández-Péon, R., and Scherrer, H. (1955). *Fedn Proc. Fedn Am. Socs exp. Biol.*, **14**, 71.
Lorente de Nó, R. (1933). *Laryngoscope, St Louis*, **43**, 327–350.
Marsh, J. T., McCarthy, D. A., Sheatz, G., and Galambos, R. (1961). *Electroenceph. clin. Neurophysiol.*, **13**, 224–234.
Neff, W. D. (1961). In *Neural Mechanisms of Auditory Discrimination*, ed. Rosenblith, W. A. New York: M. I. T. Press and Wiley.
Raab, D. H., and Ades, H. W. (1946). *Am. J. Psychol.*, **59**, 59–83.
Rasmussen, G. L. (1942). *Anat. Rec.*, **82**, 441.
Rasmussen, G. L. (1946). *J. comp Neurol.*, **84**, 141–219.
Rasmussen, G. L. (1955). *Am. J. Physiol.*, **183**, 653.
Rasmussen, G. L. (1958). *Laryngoscope, St Louis*, **68**, 404–406.

RASMUSSEN, G. L. (1960). In *Neural Mechanisms of the Auditory and Vestibular Systems*, chap. 8, ed. Rasmussen, G. L., and Windle, W. F. Springfield: Thomas.
RASMUSSEN, G. L. (1964). In *Neurological Aspects of Auditory and Vestibular Disorders*, ed. Fields, W. S., and Alford, R. R. Springfield: Thomas.
SHEATZ, G. C., VERNIER, V. G., and GALAMBOS, R. (1955). *Am. J. Physiol.*, **183**, 660–661.
VIERNSTEIN, L. J., and GROSSMAN, R. G. (1961). In *Fourth London Symposium on Information Theory*, pp. 252–269, ed. Cherry, C. London: Butterworth.
WHITFIELD, I. C. (1967). *The Auditory Pathway*. Monographs of the Physiological Society, No. 17. London: Arnold.
WHITFIELD, I. C. (1968). In *Drugs and Sensory Functions*, ed. Herxheimer, A. London: Churchill.
WHITFIELD, I. C., and COMIS, S. D. (1968). In *Cybernetic Problems in Bionics*, ed. Oestreicher, H. L., and Moore, D. R. New York: Gordon and Breach.
WORDEN, F. G., and MARSH, J. T. (1963). *Electroenceph. clin. Neurophysiol.*, **15**, 866–881.

DISCUSSION

Fex: Historically, K.-E. Hagbarth and D. I. B. Kerr ([1964]. *J. Neurophysiol.*, **17**, 295–307) started the work on the actions of centrifugal pathways on ascending sensory inflow; K.-E. Hagbarth and I ([1959]. *J. Neurophysiol.*, **22**, 321–338) asked exactly this question of whether all centrifugal activity is inhibitory or not, and we did a series of experiments recording from units in the spinal cord. We found the same thing as you, that actually what looked, with gross techniques, like pure inhibition, was a very composite effect.

Did you try to vary your preparation relative to the depth of anaesthesia or to different kinds of anaesthetics?

Whitfield: We have done some rather loose experiments with different depths of anaesthesia, but nothing startling has emerged, except that I think that there is in the normal animal what you might call tonic activity, which you can damp down. When we added a blocking agent it produced an effect opposite to that of stimulation, suggesting that there is activity in the pathway which is still present even under light anaesthesia.

Fex: What was the size of the tip of your recording electrode?

Whitfield: It was about 2 microns in diameter.

Fex: How do you know that you actually are recording from the very cell which is impinged upon by a neuron staining for cholinesterase? Surely it could equally well be an interneuron in the pathway. Are you planning further experiments along this interesting line?

Whitfield: It certainly could be an interneuron—we do not know. We have only got as far as exploring the existence of these two effects (inhibition and facilitation) and obtaining some idea of what regions of the brainstem we have to stimulate in order to obtain them. Clearly, the way in which centrifugal fibres are distributed is important, but we know nothing about this—whether there is a point-to-point distribution or whether activity from one stimulating point is spread over the whole of the terminal region.

Fex: Was it as a rule easy to find the descending tract from the S-segment to the cochlear nucleus? It should be a rather solid tract.

Whitfield: We are not actually stimulating the tract; the electrode goes into the S-segment; of course it stimulates a fair area, but as far as we can see this effect is confined to the S-segment. These effects are reasonably easy to repeat.

Salomon: In our experiments with middle-ear muscles in unrestrained cats, we had the experience that a small dose of barbiturate could completely change the behaviour of the middle-ear muscles in response to sounds, sometimes even altering inhibition of these muscles to facilitation, which in turn resulted in opposite effects on the input transmission to the cochlea. We have seen such changes occurring in the middle-ear muscle function, causing a faint sound to produce a relatively large amplitude of cochlear microphonics (inhibition of spontaneous middle-ear muscle level) and causing a strong sound to deliver relatively small cochlear microphonics (facilitatory effect). This may suggest that the middle-ear muscles are partly concerned with adjusting the cochlear microphonics to the best level for discrimination.

Whitfield: This could well be; I do not know. This brings us back to the whole question of the effect of anaesthetics on the nervous system, if we want to know what it is doing normally. One can under anaesthesia explore only the kind of things it might do. However, the problems of extending these unit studies to experiments on unanaesthetized, free-moving animals are rather formidable at the moment. But one feels that this is what should be done.

Evans: I would agree with the statement about barbiturate anaesthesia; cells in the dorsal cochlear nucleus of the cat can change their response from inhibitory to excitatory, when barbiturate is given to an unanaesthetized preparation (E. F. Evans and P. G. Nelson; to be published).

Salomon: In unrestrained cats the effect of incoming sounds on the middle-ear muscle function changed quite randomly during the day. This means that a fixed sound intensity is heard with changing loudness throughout the day.

Whitfield: We have seen repeatedly in recording from single units in the auditory cortex that during the course of half an hour the behaviour of a unit changes, and it may change from giving an inhibitory response to giving an excitatory response. Clearly, if there are pathways coming down from the auditory cortex to the lower levels, the same sort of thing will be found there too.

Erulkar: Have you seen any responses which were blocked by atropine but not by dihydro-β-erythroidine? As you know, work on the Renshaw cells and the sympathetic ganglion suggests that there may be two different types of receptor on the same cell—the "muscarinic and nicotinic" types (Curtis, D. R. [1965]. *Br. med. Bull.*, **21**, 5–9). I think that in the Renshaw cell one can record from cells which are blocked by just one of these inhibitory drugs and not the other.

Whitfield: One of the drawbacks of the microtap is that you can use only one drug at a time; we have not been able, therefore, to try different substances on the same cell, so that I cannot answer this question. We have used a number of different blocking agents in order to be satisfied that what we are observing is not due to some other effect of an agent, but we have not been able to use more than one on any given cell.

Evans: Could you comment on the latencies of the effects of stimulating the descending pathways?

Whitfield: The latency of stimulation of the S-segment is very long, of the order of what you obtain in the olivo-cochlear bundle; with the strongest stimulation used, the minimum latency is of the order of 30 msec. With weaker stimulation it may be 50 or 100 msec. The latencies of inhibition seem to be rather shorter than this but it is difficult to say how short because of the low rate of discharge. The olivo-cochlear nucleus distance is quite short; I do not know much about the actual transmission time, but I do not think the latency could all be accounted for in this way. The long latency is one fact that suggests that this pathway is not a loop with some such function as elaborating a detailed pulse pattern.

Bosher: Surprisingly, some evidence of the general function of the efferent system at the brainstem level can be derived from psycho-acoustical experiments. With others, Dr. D. E. Broadbent of Cambridge has been interested in man's ability to exclude, at least partially, from consciousness part of what he hears and his conclusion was that selective filtering—in other words, complete blocking of portions of the auditory input—does not occur at a low level of the nervous system. This ability, it would seem, was attributable to the activity of an attention mechanism. Appropriately designed experiments showed that attention could be diverted without voluntary effort, or indeed conscious awareness that this was occurring, to some event in that portion of the auditory input previously excluded from consciousness (Broadbent, D. E., and Gregory, M. [1963]. *Proc. R. Soc. B*, **158**, 222–231). There appears to be a monitoring, as it were, of the whole auditory input at a high level of the brain which produces changes in attention, when these are deemed necessary.

In this respect some recent work in another field is interesting (Oswald, I., Taylor, A. M., and Treisman, M. [1960]. *Brain*, **83**, 440–453), as it has revealed that different auditory stimuli produce evoked EEG arousal responses of varying magnitudes, some sounds and words having much greater arousal properties than others in an individual subject. This suggests that even in sleep the auditory input is monitored at a high level and when it contains something significant an attempt is made to direct the attention of the mind to it.

Investigations such as these, I feel, support Dr. Whitfield's view that the efferent cochlear system in the brainstem does not produce complete blocking of large portions of the ascending auditory input but has a more refined and more complex function.

Whitfield: Yes, this is really what I was suggesting, that the centrifugal pathways do not switch on and off what goes to higher levels, but merely control the routing of various aspects of the sensory input to particular effectors.

Fex: A. Starr and J. Wernick of Stanford University, California (personal communication) will publish a study of the effects on the cochlear nucleus of stimulation of the crossed olivo-cochlear bundle in decerebrate animals. They have seen a variety of effects on sound-evoked activity in the cochlear nucleus. Such activity has been initially depressed, with or without a rebound when the efferent stimulation was stopped. Sound-evoked activity has been facilitated and then gone back to, or below, its original level immediately after stimulation was stopped. There has also been initial facilitation going over into depression during stimulation. Facilitation of spontaneous activity in the cochlear nucleus has been seen, also when the ipsilateral cochlea had been destroyed. (K. C. Koerber, R. R. Pfeiffer, W. R. Warr and N. Y.-S. Kiang [1966. *Expl Neurol.*, **16**, 119–130] have found that at least the dorsal part of the cochlear nucleus shows little change of spontaneous activity with destruction of the ipsilateral cochlea, in barbiturate-treated cats.) So these results again tie in with your findings of mixtures of effects. I would stress that we know from Grant L. Rasmussen's studies ([1964]. In *Neurological Aspects of Auditory and Vestibular Disorders*, pp. 1–19, ed. Fields, W. S., and Alford, B. R. Springfield: Thomas; [1967]. In *Sensorineural Hearing Processes and Disorders*, pp. 61–75, ed. Graham, A. B. Boston: Little, Brown) that not only the olivo-cochlear fibres stain for cholinesterase, but that also quite a number of descending fibres at different auditory levels stain for cholinesterase. It might be that your system also is convenient for studying whether or not such fibres are true cholinergic fibres.

Whitfield: We think that the inhibitory fibres may be adrenergic, because we can produce inhibition in the cochlear nucleus by local application of noradrenaline. We have tried the fluorescent technique (of A. Dahlström and K. Fuxe [1964]. *Acta physiol. scand.*, **62**, Suppl. 232), to see if we can identify any endings, but have found nothing very satisfactory. We have also tried treating cats with reserpine, and this abolishes the inhibition produced by stimulating a point in the nucleus of the lateral lemniscus. However, I would not regard this as particularly conclusive evidence.

Davis: We might mention in this context not only the effects of sound on the auditory system but also the effects of sound on ongoing motor activity. Bickford and his colleagues have observed, and others have confirmed, that a succession of auditory stimuli (clicks) can "pull into step" the action potentials of certain muscles, particularly neck muscles or the postauricular muscle at the back of the ear: the motor units tend to discharge in a time-locked fashion with respect to these clicks (Bickford, R. G., Jacobson, J. L., and Cody, D. T. R. [1964]. *Ann. N.Y. Acad. Sci.*, **112**, 204–223; Mast, T. E. [1965]. *J. appl. Physiol.*, **20**, 725–730). This is an overflow of some sort, a triggering of neighbouring units in the motor pathway. The particular interest of this is the extraordinarily short latency of some of these effects. The shortest that I have seen are from the postauricular muscle, about 7 msec., which allows little time for much reverberating in complicated loops! We see widespread effects on ongoing motor activity, with the implication that there is a multiple spreading of activity in the nervous system at this very simple, early reflex level.

Johnstone: We have worked on this postauricular muscle reflex (Lazlo, J. I., and Johnstone, B. M. [1967]. *Aust. J. exp. Biol. med. Sci.*, **45**, P7), because I was interested in its short latency; we have even recorded a take-off of 6 msec. The interesting thing is that it is immensely arousal-dependent. It is so quick that it cannot go through many synapses; but to keep the reflex at a constant level we have to set the subject an extremely well-controlled task, otherwise the reflex goes down. This is not habituation, although we suspected it from the beautiful time decay curves. If you startle the subject the response comes up again. The reflex is tremendously arousal-dependent, which can only come from some depressing centres. The point I want to make, which Dr. Davis has brought up, is that this reflex must be a long way from the cortex because of the short time delay, but it is tightly controlled by influences coming down from much higher up—for what purpose, I do not know.

Davis: It is a question of what is controlling what. This is an effect of sound, controlling the timing of ongoing activity; but there must be ongoing activity in order to bring it out. One of my colleagues makes a very good subject for studying this reflex, because he is able to move his ears, and turns the "sonomotor reflex" on or off, depending on whether he is holding his ears back or whether he is allowing them to relax.

Lowenstein: The reticular formation has not been specifically mentioned.

Whitfield: I have the feeling that sooner or later we shall sort the reticular formation into things which we can identify functionally. Fibres probably come from a lot of other places that we have not yet located. These pathways on the medial side of the lemniscus would probably have been called reticular, earlier on.

AUDITORY RESPONSES EVOKED IN THE HUMAN CORTEX

HALLOWELL DAVIS

Central Institute for the Deaf, St. Louis, Missouri

MY purpose, in this final section of our symposium, is to outline the properties of a slow, diffuse electrical response of the human brain to auditory stimuli. Actually the response is evoked equally well by any sudden stimulus, whether auditory, visual, tactile, somatosensory or electrical, and therein lies a large part of its interest.

I assume that we are all familiar, in principle at least, with the electrical responses of the primary projection areas in the cortex. These responses, while slower than axon spikes and classed as "slow wave" responses, occur within the first twenty to fifty milliseconds after the stimulus. They are,

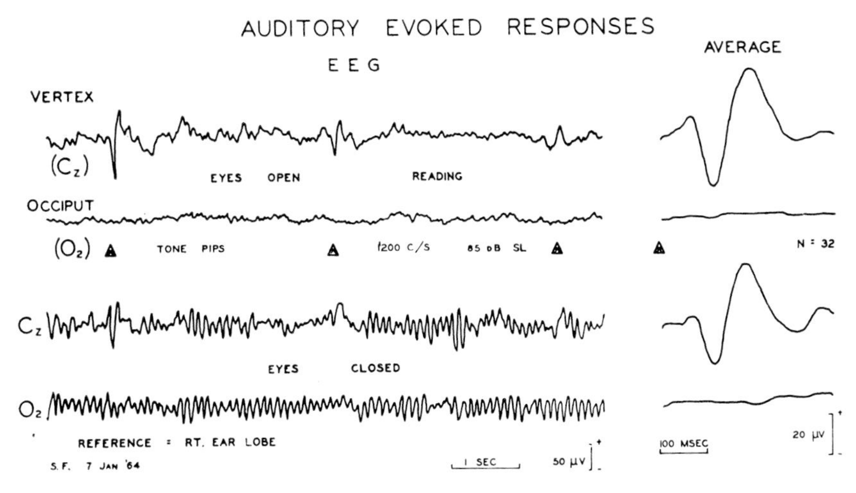

FIG. 1. The EEG was recorded from vertex and from occiput with eyes open and with eyes closed while tone pips were delivered by loudspeaker at intervals of 3·2 sec. (as indicated by the triangles).

Upward deflection indicates vertex more positive, both in the raw EEG and in the average response. Average responses were collected simultaneously from vertex and occiput. Stimuli were tone pips at 1,200 Hz at a level 85 db above the subject's threshold. (From Zerlin and Davis, 1967, by permission of the Editor of *Electroencephalography and Clinical Neurophysiology* and the Elsevier Publishing Company.)

in general, confined to the direct primary projection areas and the adjacent secondary areas and they have been widely used to map out those areas. Such responses have been recorded from the human cortex at operation or by implanted electrodes, and in the case of visual, cutaneous or somatic stimulation they can be detected through the intact skull by the now-familiar technique of averaged or summed responses. Unfortunately the

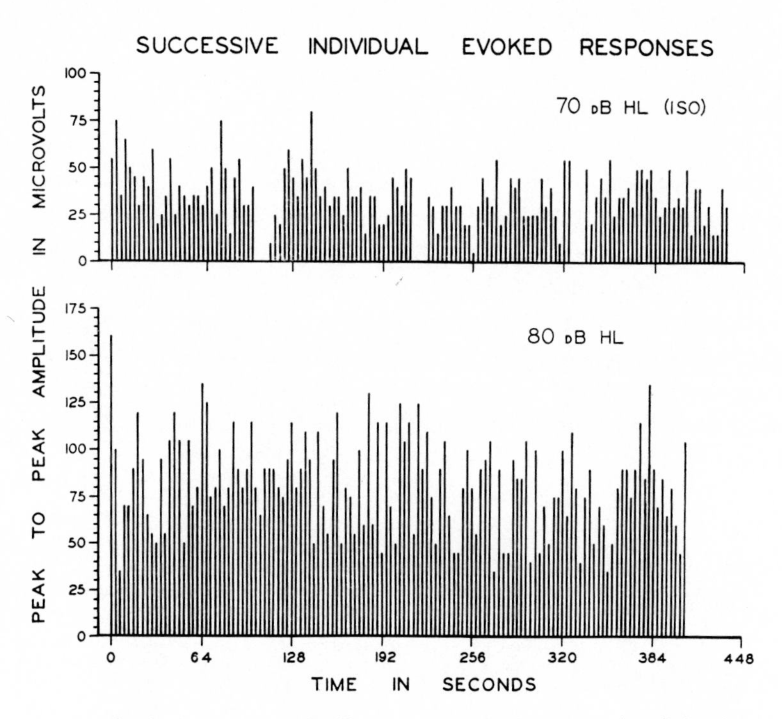

FIG. 2. Individual responses evoked by 1,200 Hz tone pips at 70 and 80 db hearing level (ISO), measured (N_1-P_2) by hand, are shown here in serial order. The interval was 3·2 sec. (From Zerlin and Davis, 1967, by permission of the Editor of *Acta Oto-Laryngologica*.)

auditory cortex in the human being is buried in the Sylvian fissure and is overlaid externally by the temporalis muscle. Several investigators believe that they have detected primary auditory evoked responses, but it is very difficult to investigate them because of their extremely low voltage and the masking, in many situations, by muscular action potentials of the so-called "sonomotor reflex" (Bickford *et al.*, 1964). I shall therefore speak only of a quite different slower later response that can be recorded widely over the calvarium, best over the frontal lobe and best of all at the vertex. There, with reference electrode on mastoid or ear lobe, the voltage is relatively high and interference by muscle potentials or eye movements is low. I call these slow diffuse non-specific responses the V (or vertex) potentials, following the suggestion of Bancaud, Bloch and Paillard (1953). (Cf. also Gastaut, 1953.)

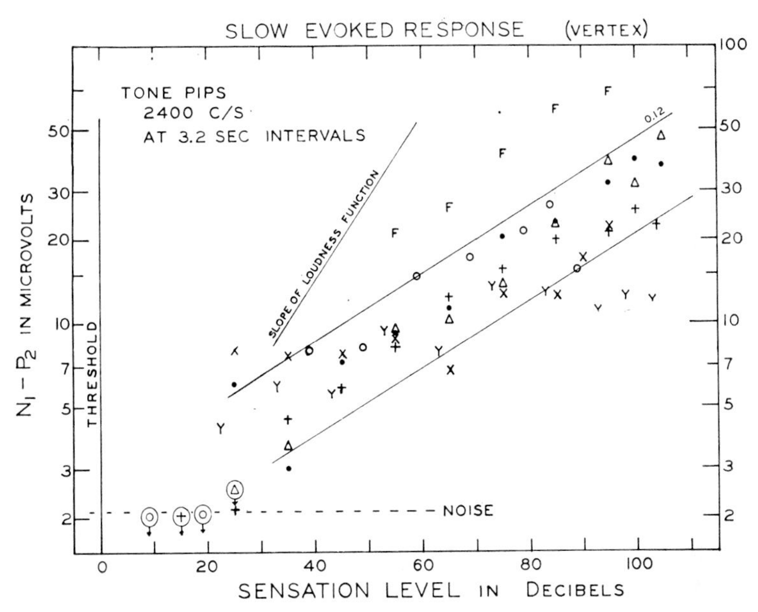

FIG. 3. Input–output relations of the V potential, shown in logarithmic coordinates.

Each point represents the average of 2 blocks of 32 responses each. The straight lines are fitted by eye. Their slope is much less than that of the loudness function (Stevens, 1957), also shown in the figure as if the ordinate scale were the logarithm of magnitude of sensation. Note the exceptionally high voltages of subject "F", who is the same subject as in Figs. 1 and 2. (From Davis and Zerlin, 1966, by permission of the Acoustical Society of America.)

The V potentials were first recognized and clearly described by the late P. A. Davis in 1939, but they were practically inaccessible for systematic study until the advent of the average response computer. Fig. 1 shows how the V potential stands out clearly from the background EEG in an individual with exceptionally large responses (Zerlin and Davis, 1967). The responses are clear at the vertex but practically absent, even in the averaged record, at the occiput. The slow tri-phasic waveform, with peaks or troughs at about 100, 180 and 300 msec., is shown here and also in Figs. 5, 6, 7, and 8. In this subject the variability of single responses could be studied (Fig. 2).

The distribution of amplitudes was random, with a large standard deviation. Variability of amplitude from one response to the next, from one subject to another, and as a function of numerous other conditions is an outstanding characteristic of the V potentials. There is also some variation

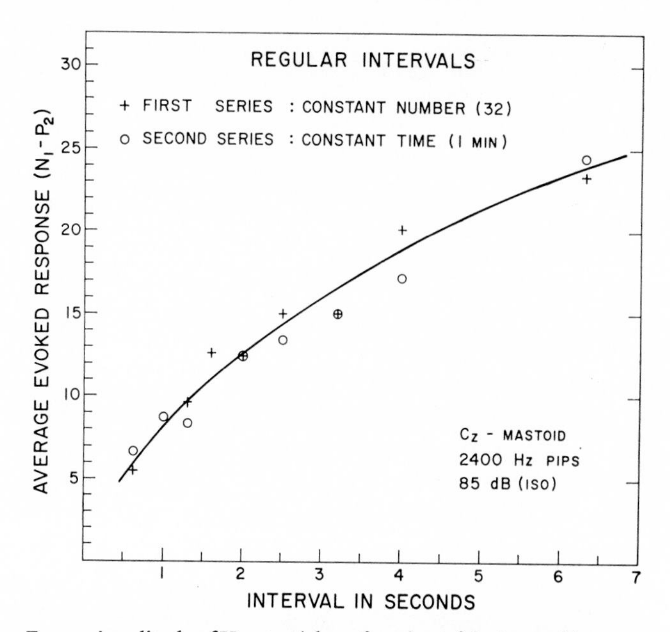

FIG. 4. Amplitude of V potential as a function of the interval between tone pips. The scale is approximately in microvolts. Plot points are averages across five subjects, three of whom are common to the two series. (From Davis *et al.*, 1966, by permission of the Editor of *Electroencephalography and Clinical Neurophysiology* and the Elsevier Publishing Company.)

in waveform and latencies from subject to subject (cf. Fig. 7), but the pattern and latencies for a given subject are quite stable unless there is a major change of state, such as going to sleep. It is the amplitude that is variable.

In spite of its variability, the amplitude, which is best measured as peak-to-peak voltage from "N_1" (vertex negative at about 90 to 100 msec.) to "P_2" (positive at about 180 msec.), depends on the intensity of the stimulus and also the interval between stimuli. Fig. 3 shows the relation to intensity. It illustrates the variability, both across and within subjects, and at the same time a rising trend with intensity. The scales are both logarithmic. The parallel lines represent a power law, like the psychophysical "law" of

Stevens (1957), with a very small exponent. The slope is much less steep than the slope of the loudness function, which is also shown in the figure. In a more recent study we have found even smaller exponents, and actually it is not easy to be sure that a power law is a better description of the data

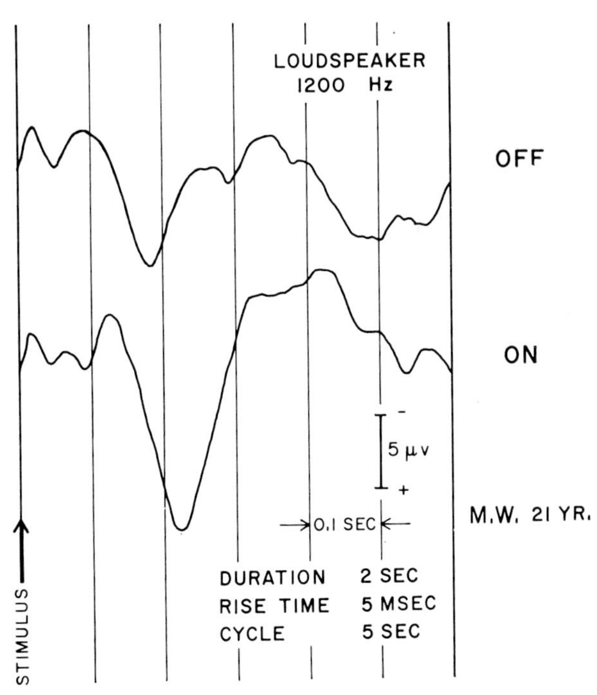

FIG. 5. These tracings show the similarity of the off-response to the on-response. The lower amplitude of the off-response is due in some part to its shorter recovery time (2 sec.) as compared with 3 sec. silent interval before the on-response. The latency of this on-effect is unusually long. Intensity of tone-burst 80 db SPL. (From Davis and Zerlin, 1966, by permission of the Acoustical Society of America.)

than a linear relation of voltage to decibels. We can be quite sure, however, that the voltage increases much less rapidly than subjective loudness.

The V potential requires a long interval between stimuli, perhaps 10 seconds, if the responses are to be of maximal amplitude. Fig. 4 shows the relation to interval. The plot points are the averages across several subjects so that the scatter of the individual observations is not apparent. This slow recovery curve is quite well established. We have not found any other physiological or psychological process with a similar time-course.

The V potential is an on-response and also, if a steady stimulus has lasted long enough, an off-response. The off-response is about one-third of the amplitude of the on-response, but is similar in form, as shown in Fig. 5. The response is initiated very early during the (linear) onset of a tone of moderate intensity, when the tone has reached only one-sixth of its final

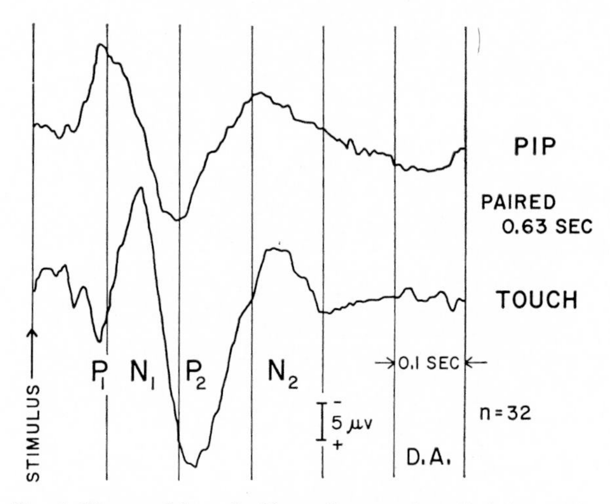

FIG. 6. V potentials evoked by auditory and tactile (vibratory) stimuli. As usual, the latency is longer and P_1 is more prominent for the tactile stimuli. The stimuli were presented in pairs, tone pip then touch, every five seconds. The touch response was unusually large in this case for a stimulus interval of only 0·63 sec. Vertex negative is upward. (From Davis *et al.*, 1966, by permission of the Editor of *Electroencephalography and Clinical Neurophysiology* and the Elsevier Publishing Company.)

amplitude. The duration of the plateau of a tone is unimportant beyond 30 msec. following the *onset* of the tone, although a really fast rise followed by a 30 msec. plateau is a slightly more effective stimulus than a linear 30 msec. rise (Onishi and Davis, 1967). We should recall that the loudness of a brief tone-burst continues to increase with duration up to about 100 msec.

Stimuli of different modalities produce similar V potentials, as shown in Figs. 6, 7 and 8. The latencies may differ because of nerve conduction time or a slow peripheral process as in vision. When auditory stimuli of different frequency or spectral composition are paired, they interact like similar stimuli with respect to their necessary recovery period. Likewise a vibratory tactile stimulus to the fingers depresses the response to a tone, and *vice versa*, almost as effectively as two stimuli in the same modality or two sounds to the opposite ears. Often the interaction is complete. On

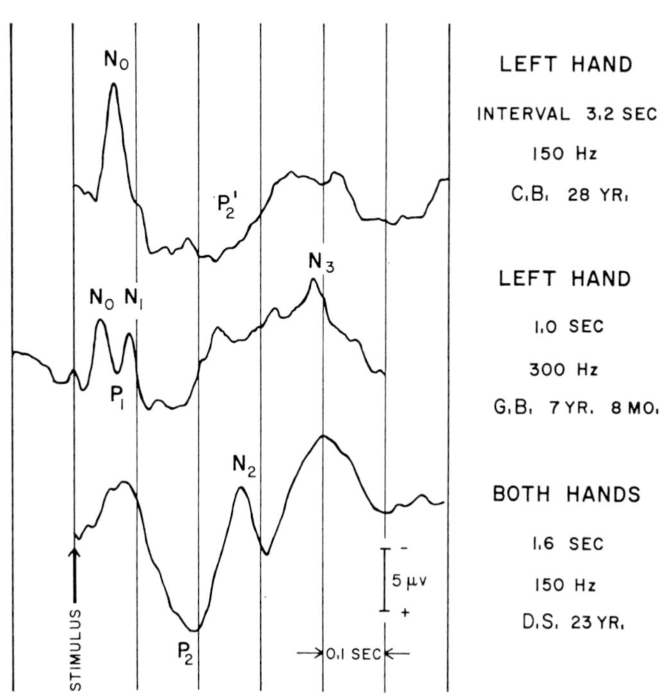

FIG. 7. In the upper trace N_2 is absent; in the middle trace it is very small. The lower trace shows a strong N_1-P_2-N_2 complex and also N_3. The middle trace shows a strong early N_0-P_1 complex. Auditory V potentials show similar variations. These different patterns are quite characteristic for each subject.

the other hand, electrical shocks to the median nerve did not interact at all with tone pips (filtered clicks), as shown in Fig. 8. The outline of these observations is in press (Davis *et al.*, 1968), and we are continuing our study of cross-modality interactions.

It is difficult to think of a physiological process in the cortex which, even though it is taking place outside the well-known primary and second-

INTERACTIONS OF PIPS AND SHOCKS

PIPS : 1200 CPS, 74 DB SL
SHOCK TO WRIST
N = 32

VERTEX TO EAR LOBE

SHOCK - SHOCK

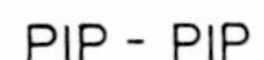

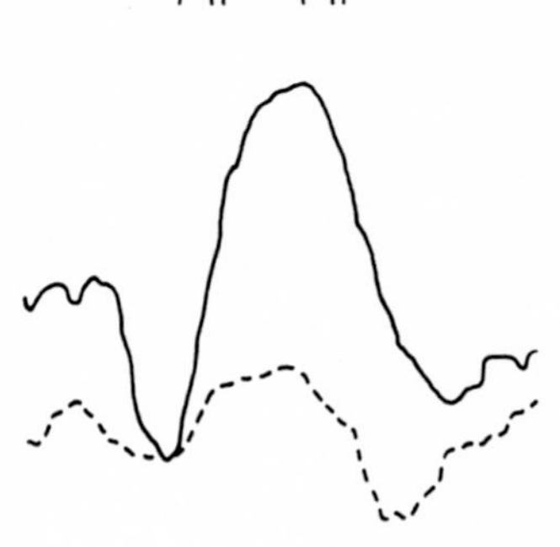

First
Stimulus

Second
Stimulus

100 MSEC

INTERVAL = 0.55 SEC

5 μV

SHOCK- PIP

PIP - SHOCK

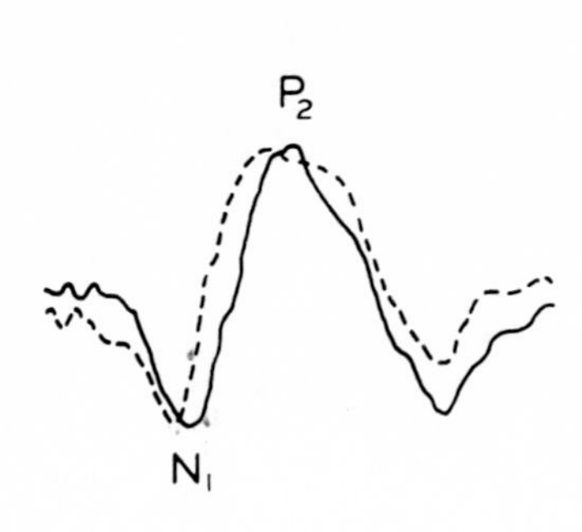

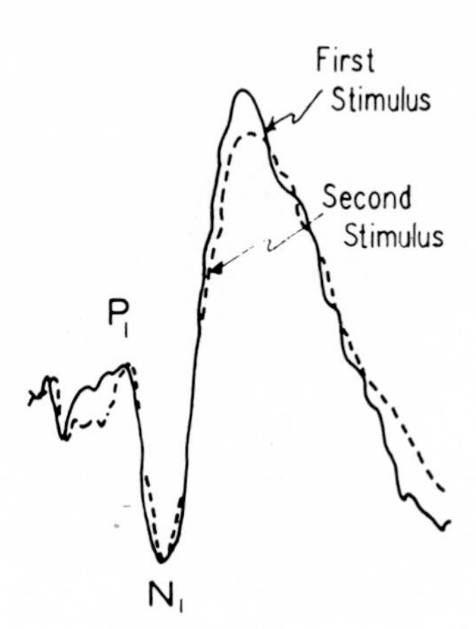

FIG. 8. The stimuli were tone pips at 1,200 Hz from a loudspeaker at 74 db above threshold and electric shocks, applied to the left median nerve at the wrist, which were adjusted in strength to give V potentials of approximately the same amplitude.

The stimuli were presented in pairs, every 5 seconds. The intra-pair interval was 0·55 sec. Upward excursion indicates vertex more positive relative to ear lobe. The responses have been retraced with superposition of the N_1-P_2 complexes. Actually the latency of the response to shock is longer than to tone pip, chiefly because of the longer peripheral nerve conduction time. The strong depression of a second response to a stimulus of the same modality is clear but there was no interaction across these two modalities in any of the eight subjects in this experiment.

ary sensory areas, is not related in its amplitude to the subjective intensity of the sensation. The amplitude of the V potential is highly variable and seems to carry little information. If information is carried, it would seem to be primarily temporal information concerning changes in sensory stimulation, with some emphasis on novelty. I shall not discuss certain very late waves, after 300 msec., or d.c. shifts in the interval between stimuli, which seem to relate very clearly to expectation (Walter *et al.*, 1964)

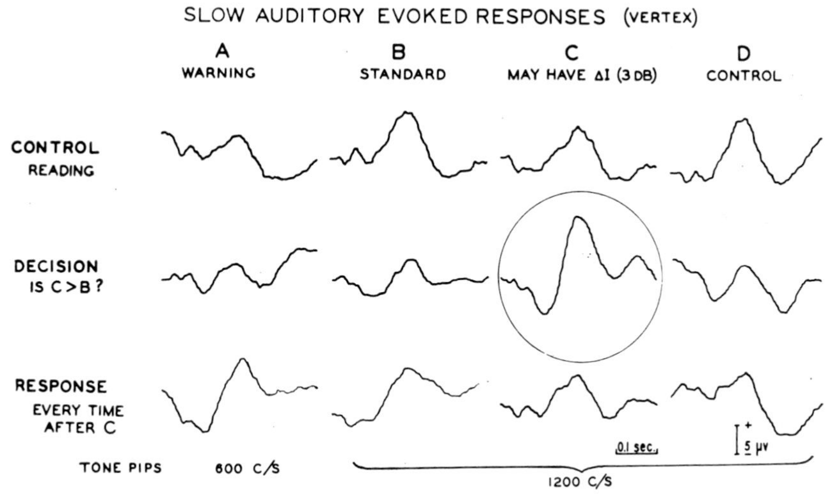

FIG. 9. Slow responses evoked by auditory stimuli and recorded from the vertex.
Tone pips at about 70 db hearing level were delivered at the start of each trace. The intervals, A–B, B–C, C–D, D–A, were all 2·5 seconds. Responses to 32 such cycles were averaged. ΔI (3db) was added 10 times and also subtracted 10 times in the "decision" series. Upward deflection indicates that the vertex is becoming more positive relative to the right mastoid. This enhancement appeared in the records of five of our six subjects. (From Davis, 1964, by permission of the publishers. Copyright 1964 by the American Association for the Advancement of Science.)

or to novelty and information content (Sutton *et al.*, 1967). I shall merely mention in closing that most subjects can influence the amplitude of electrically evoked V potentials by concentrating their attention on one wrist or the other (Satterfield, 1965). The stimuli to which attention is given tend to be larger—but a strong effort of attention is needed. A variant of such an experiment is illustrated in Fig. 9. Here a rather difficult discrimination was required. It is worth noting that the increased response is evoked by the test stimulus and not by the standard of reference that preceded it, although obviously attention to the standard as well as to the test stimulus was required.

These V potentials present a very puzzling set of properties. They illustrate, at the very least, that the cortex can react to a given stimulus in different ways in its different sub-systems.

SUMMARY

A slow diffuse response can, by average-response techniques, be recorded widely over the calvarium and best at the vertex. This "vertex potential" is complex, with peaks or troughs at about 100, 180 and 300 msec. It is an on-response, or an off-response following a long steady tone. Several seconds must elapse before another full-sized response can be evoked. Its latency is quite precise. Its amplitude is only a few microvolts and varies widely and apparently at random from one single response to the next. The averaged response also varies across trials and across subjects. The amplitude increases very slowly with sound intensity. Tones of different frequency spectra depress one another's responses like those of the same frequency. Auditory, tactile, visual and electrical stimuli evoke very similar vertex potentials, and there are varying degrees of interaction across the various sensory modalities. The vertex potential can be modified by the subject directing his attention to or away from the stimuli. The vertex potentials relate much more clearly to time than to either intensity or quality. They seem to signal a new event.

Acknowledgement

Preparation of this manuscript was supported by Public Health Research Grant No. B-3856 from the National Institute of Neurological Diseases and Blindness.

REFERENCES

BANCAUD, J., BLOCH, V., and PAILLARD, J. (1953). *Revue neurol.*, **89**, 399–418.

BICKFORD, R. G., JACOBSON, J. L., CODY, D. T., and LAMBERT, E. H. (1964). *Trans. Am. neurol. Ass.*, **89**, 56–58.

DAVIS, H. (1964). *Science*, **145**, 182–183.

DAVIS, H., MAST, T., YOSHIE, N., and ZERLIN, S. (1966). *Electroenceph. clin. Neurophysiol.*, **21**, 105–113.

DAVIS, H., and ZERLIN, S. (1966). *J. acoust. Soc. Am.*, **39**, 109–116.

DAVIS, H., ZERLIN, S., BOWERS, C., and SPOOR, A. (1968). *Electroenceph. clin. Neurophysiol.* Society proceedings, in press.

DAVIS, P. A. (1939). *J. Neurophysiol.*, **2**, 494–499.

GASTAUT, Y. (1953). *Revue neurol.*, **89**, 382–399.

ONISHI, S., and DAVIS, H. (1967). *J. acoust. Soc. Am.*, **42**, 1189–1190.

SATTERFIELD, J. H. (1965). *Electroenceph. clin. Neurophysiol.*, **19**, 470–475.

STEVENS, S. S. (1957). *Psychol. Rev.*, **64**, 153–181.

SUTTON, S., TUETING, P., ZUBIN, J., and JOHN, E. R. (1967). *Science*, **155**, 1436–1439.

WALTER, W. G., COOPER, R., ALDRIDGE, V. J., MCCALLUM, W. C., and WINTER, A. L. (1964). *Nature, Lond.*, **203**, 380–384.

ZERLIN, S., and DAVIS, H. (1967). *Electroenceph. clin. Neurophysiol.*, **23**, 468–472.

DISCUSSION

Lowenstein: Dr. Davis, you say there is no subjective correlate of the vertex phenomenon; have you not been able to elicit any description of what is experienced after the time-lag? I gather that the time-lag between giving the signal and obtaining the evoked response is quite long.

Davis: This delay is 100 msec. to the peak of N_1, about 200 msec. to P_2; there is also a later wave which occurs after the N_2 and apparently relates to whether the subject receives information or not. This is Sutton's work (Sutton, S., Tueting, P., Zubin, J., and John, E. R. [1967]. *Science*, **155**, 1436–1439), and apparently there may be a response to the absence of an expected signal, which is even more curious.

Lowenstein: It may be rather primitive to mention this, but I have the impression that my awareness of the stopping of a clock is delayed quite considerably. We do not notice it immediately it stops and I suppose it is possible there is a constant time-lag between this and our becoming aware of the ensuing silence.

Davis: With these later waves there is time for a "double-take" ("yes, now I know the answer"). Another effect is the contingent negative variation which is a shift in the baseline that can be brought about by conditioning. It begins after a warning signal, and is resolved by the imperative signal. Grey Walter has been investigating that (Walter, W. G., Cooper, R., Aldridge, V. J., McCallum, W. C., and Winter, A. L. [1964]. *Nature, Lond.*, **203**, 380–384). That is still another electrical variation, a still later effect, but I have been talking about the earlier part of the vertex potential for which we have not been able to find any subjective correlate.

Salomon: Could one get a clue about when the decision was made?

Davis: I think not; I do not know how you would determine this.

Whitfield: You spoke of the vertex response being diffuse. There are two ways in which it can be diffuse; either it occurs all over the surface simultaneously, or it is fairly localized but deep. Is there any information about this? The fact that you obtain it right on the vertex might indicate that it is deep.

Davis: I have no information of my own here. Grey Walter has found that with electrodes implanted in human subjects for other purposes, it can be identified as appearing locally but not quite simultaneously at different points. The slow potential wave seems to sweep from forward to back, over the frontal lobe.

Neff: To what extent have evoked responses at this latency been investigated with implanted electrodes in animals?

Davis: I hope that people will do experiments on this, particularly in the monkey. I have the impression that it may be a particularly human sort of response, because it seems to relate so strongly to the frontal lobe. I am not sure that we have a complete analogue in animals. We did some experiments in chinchillas (Rothenberg, S., and Davis, H. [1967]. *Percept. Psychophys.*, **2**, 443–447) to find a short-cut for determining thresholds, but did not get very good late responses, although we could see earlier components clearly. We then found that the inferior colliculus was apparently providing most of the output; and with this technique and the thin skull of the chinchilla we could pick up cochlear microphonics, action potentials in the auditory nerve, and the cochlear nucleus and the inferior colliculus firing off in sequence!

Kleerekoper: When you asked subjects to report on the difference in loudness of subsequent pips, was there any correlation between the size of the potential change and the scoring performance?

Davis: We looked for this and it did not appear. However, the scores were uniformly fairly good.

Erulkar: Are vertex potentials present or modified in response to, say, tactile stimulation in patients who are deaf? Or in a partially deaf subject, are they still present as well as in response to other sensory stimuli?

Davis: We use the vertex phenomenon as an audiometric tool, particularly in deaf children; and we find the tactile response useful to show the pattern that we should seek in the output from a particular child. In children not only are responses more variable, but in the infant they are considerably slower. Also they vary with waking and sleeping and also with the maturation of the nervous system. I had hoped that we would be able to distinguish tactile from aural responses through some of these interactions, in order to determine whether, when the child responds in an audiometric test, he is really hearing or is experiencing a tactile sensation in the ear. The problem of explaining to a person who is congenitally deaf what the sound is that he should be reporting is a rather difficult task! Unfortunately the aural and tactile responses are so similar that we have not been able to make that discrimination.

Fex: I realize that most stimulations seem to give the vertex response, so this is hard to achieve, but have you been able to get patients to stimulate themselves? One might anaesthetize the surface of the forefinger so that tactile sensation there would not interfere; they could press a button to give themselves a pip. This would increase the precision with which they could time the on-coming signal.

Davis: We have not tried this, but it could be done. If one were engaged in correlating subjective experience with the signal, this might be a useful method. The kinesthetic system might elicit a response without any cutaneous sensory stimulation. Dr. Freeman Mast in our laboratory looked for something happening just before a movement—that is, at the initiation of voluntary movement. In 1964 he observed such a potential wave but he was not satisfied with the evidence and never published it. A wave appeared in the motor area corresponding to the hand being used during the 100 msec. before a voluntary movement began.

Fex: E. V. Evarts ([1966]. *J. Neurophysiol.*, **29**, 1011–1027) has seen change of activity just before a movement, with microelectrodes recording from pyramidal tract neurons in the motor cortex in conscious rhesus monkeys.

Rose: Would you elaborate on the off-response and its relation to the duration of the tone? Is the off-response always present?

Davis: The tone must be long enough so that its termination is not in the refractory period of the on-effect. We have not explored the necessary duration of the tone more than to establish that it seems to follow the same rule as for the interval between two tone pips. Although we do not often look for it, we have every reason to think that the off-response always occurs, although at about one-third of the amplitude of the corresponding on-effect.

Whitfield: If it is fundamentally smaller, is this not due to the fact that there has been an on-response immediately before it?

Davis: Other things being equal, it seems to be smaller.

Hood: How dependent is the on-effect on the rise-time of the stimulus? I suppose that if the stimulus were to rise very slowly, there would be no on-effect, only an off-effect.

Davis: We have done a set of experiments on rise-times from 3 to 300 msec. (Of course, at 300 msec. the tone is still rising while the response appears, because the peak of the N_1 response is at 100 msec.) Certainly within 30 msec. the rate of rise makes only a slight difference. It is a little more effective if it is 3 msec. than if it is 30 msec. (For the full description of these experiments, see Onishi, S., and Davis, H. [1967]. *J. acoust. Soc. Am.*, **42,** 1189–1190.)

Lowenstein: Have you any clinical characteristics of subject "F" (Fig. 3, p. 261) that might explain her exceptional performance?

Davis: She was one of those from whom we expect to get good responses, girls in the student teacher-training class who are new to laboratory situations, and are all eager and maybe a little tense about it. The one subject who failed to show an increase in the "decision" experiment was this kind of person. It was her first visit to a laboratory of any sort; she was brought in by a student friend. She did not speak English very well, being a Latin American. She was very tense and she gave large responses which showed no increase on the critical stimuli. She was probably already doing all she was capable of, so that we failed to record any further increase related to the decision.

Johnstone: You said that electrical stimulation of the median nerve gave the same sort of response. Desmedt and co-workers (Desmedt, J. E., Debecker, J., and Manil, H. [1965]. *Bull. Acad. r. Méd. Belg.*, **5**, 887–936) have stimulated the finger electrically and get a short sharp response, which would precede your N_1; their figure was about 20–30 msec. Why can responses to electrical stimuli be picked up but not tactile or auditory responses of this sort?

Davis: They had the electrode over the somatic area and were picking up the primary response preceding the vertex potential. You can pick up tactile responses the same way. With auditory stimuli the responses are purely non-specific because the primary auditory cortex is down in the Sylvian fissure, orientated unfavourably and also overlaid by the temporalis muscle.

CORTICAL REPRESENTATION

E. F. Evans

Medical Research Council Group, Department of Communication, University of Keele

Classically, cortical representation has been considered in terms of the representation of the cochlea, and as a consequence, of frequency. Woolsey has proposed as a result of a long series of detailed electrical and anatomical studies (summarized: 1960, 1961) that there are at least five representations of the cochlea in the auditory cortex of the cat: the primary area (AI), the secondary area (AII); the posterior ectosylvian gyrus; the suprasylvian fringe area; and the insular-temporal cortex. Although the recent investigations of the auditory cortex by Neff and his colleagues (reviews: 1960, 1961) using ablation techniques suggest that the primary receiving area might not be the most important of these areas, the majority of the neurophysiological data is confined to this area of the cat cortex.

The present paper will therefore be limited to the main cortical representation considered in relation to the primary auditory cortex of the cat (Fig. 1). It will attempt to discuss the evidence for the representation of the cochlea in the cortex and its implications, and to suggest that the classical interpretation should be replaced by a notion of cortical representation analogous to that recently proposed for the visual system.

Much of the data discussed here is derived from studies of the responses of single units in the primary cortex of the unanaesthetized and unrestrained cat, in collaboration with I. C. Whitfield and H. F. Ross in the Neurocommunications Research Unit at the University of Birmingham.

REPRESENTATION OF FREQUENCY

Although there had previously been some indirect evidence for a point-to-point projection of the cochlea on to the cortex, this notion was given a firm foundation in the classical experiments of Woolsey and Walzl (1942). They found that electrical slow wave responses on the cortex could be evoked by directly stimulating small areas of the exposed cochlea. From their data they concluded that they had demonstrated a point-to-point projection of the cochlea on to the primary cortex. The projection was such that high frequencies were represented anteriorly and low frequencies posteriorly. From measurements of the threshold curves of points on the cortex of the dog, under conditions of local strychnine application, Tunturi (1950) concluded that there was an accurate representation of frequency along the primary area. Octaves were represented in equal lengths of cortex in an antero-posterior direction. This same technique was utilized in the cat by Hind (1953), but he found a much less strict relationship between frequency and position on the cortex.

From these data in particular the notion of a tonotopic arrangement of frequencies on the auditory cortex has been elaborated. This notion has often been implicit or explicit in "place" theories of frequency discrimination (e.g. Bremer, 1953; Adrian, 1947; Licklider, 1951; Allanson and Whitfield, 1956). Thus Adrian (1947) wrote: "as far as the main 'primary' area is concerned, therefore, the map of sounds on the brain is made by the peripheral apparatus which analyses the sound into its component frequencies, and the central apparatus, which indicates them by the position of activity along the cortex and indicates their intensities by the number of impulses arriving at each point". There are now a number of serious objections to this view.

In the first place, the experiments on which the theory of tonotopic organization was based, were undertaken expressly to examine the organization of the *afferent connexions* to the cortex (Woolsey and Walzl, 1942; Tunturi, 1950; Hind, 1953). Deep barbiturate anaesthesia was used for the explicit purpose of suppressing the spontaneous cortical activity which would otherwise obliterate the evoked slow waves. Experiments, again with gross electrodes, but under reduced or no anaesthesia, have indicated that, under these conditions, there are at best only blurred gradients of frequency representation in the primary cortex of the cat (Licklider, 1941; Bremer, 1953; Gross and Small, 1961) and even of the dog (Tunturi, 1960). Gross and Small (1961) concluded that the overlap of the area of maximum response for different frequencies was greater than the difference between the areas; moreover, there appeared to be no orderly distribution of points of maximum response over the cortex as a function of frequency.

Secondly, the important studies by Neff and his colleagues (summarized 1960, 1961) have undermined the validity of the functional conclusion usually drawn from a tonotopic organization of the cortex, namely that it serves for frequency analysis. Selective ablation of the known auditory areas in both hemispheres of cats does not abolish the ability of the animal to discriminate between tones of different frequency and intensity. The functions which are eliminated by these lesions are more complex, namely,

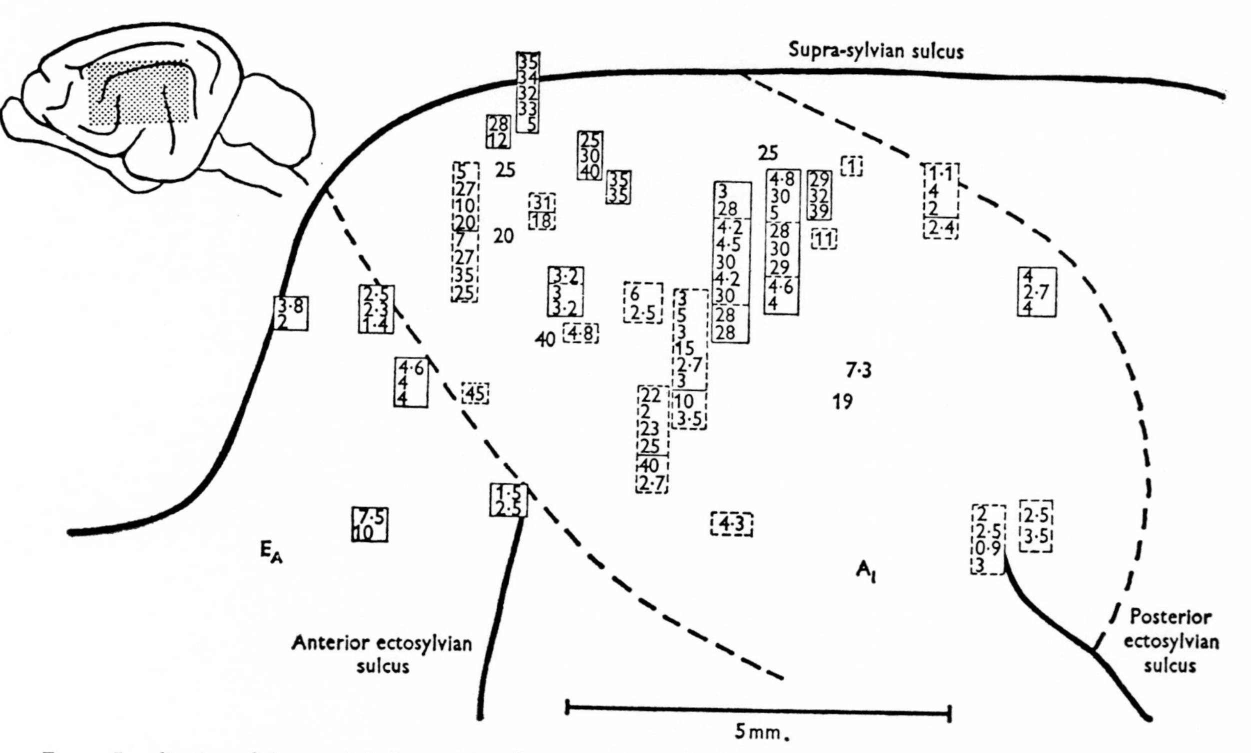

FIG. 1. Localization of characteristic frequencies of 97 units on a representative left auditory cortex of the cat.

The interrupted lines delimit the primary (AI) and the secondary (E_A) cortex (after Woolsey, 1960). Figures represent characteristic frequencies (in kcyc./sec.) of units obtained at the positions shown. Those enclosed in rectangles indicate units obtained at a site in the centre of each rectangle; subdivisions indicate more than one puncture at that site, units being arranged in order of depth. Figures within interrupted outlines or rectangles are of units obtained in the right auditory cortex. (From Evans and Whitfield, 1964, by permission of the Editors of the *Journal of Physiology*.)

the capabilities of discrimination between temporal patterns of tones, between tones of different duration, and between sound sources of differing locations in space.

More recently, the question of a tonotopic organization of the auditory cortex has been re-examined at the single neuron level by means of micro-electrode analysis (Erulkar, Rose and Davies, 1956; Hind *et al.*, 1960; Evans and Whitfield, 1964; Evans, Ross and Whitfield, 1956). These experiments indicate that only a most general trend can be seen in the localization of unit characteristic frequencies* over the primary cortex. In the posterior and anterior halves of the cortex, units of low (1–10 kcyc./sec.) and high (10–70 kcyc./sec.) characteristic frequencies respectively were predominantly found. On the other hand, low frequency units could be found throughout the entire extent of the primary cortex, apparently in continuity with those found in the anterior ectosylvian gyrus, E_A, Fig. 1. This can be clearly seen in Fig. 2, which shows a plot of 105 unit characteristic frequencies obtained from five cats as a function of their position in an antero-posterior direction in AI (Evans, Ross and Whitfield, 1965). From a regression line fitted to the points, 90 per cent limits for frequency were calculated (interrupted lines in Fig. 1). These limits were 3·8 octaves apart, compared with the total range of characteristic frequencies observed of 6 octaves (1–70 kcyc./sec.). An analysis of these data revealed that the considerable variance of the points about the regression line was not significantly different from that of unit characteristic frequencies within individual penetrations. This indicated that the variance observed received an insignificant contribution from errors incurred in the establishment and superimposition of the penetration sites. In keeping with this finding, a very considerable spread of characteristic frequencies may be seen in a single cortex (open circles and triangles, Fig. 2), and in a single penetration (e.g.: 4 octaves; open triangles).

In its simplest form, a "tonotopic organization" of the cortex would mean that there existed a strict relationship between position in the cortex and the frequency to which the elements constituting it responded. It would also imply that the individual elements had bandwidths of response which were narrow compared with the total frequency range. The strict relationship would be a consequence of the organized layout of a cortical data-processing system, consisting of a "continuum" of basically similar subsystems working in parallel and spread out over the frequency range of the system. In this strict model, a plot of frequency against position would

* The characteristic frequency of an auditory neuron is that frequency at which the threshold for response is lowest, in terms of the tone intensity.

have to show a monotonic relationship, such as linear or octave. The arrangement of neurons in the cochlear nucleus of the cat approaches closely this definition of a tonotopic system. A microelectrode encounters neurons of progressively decreasing characteristic frequency, as it penetrates

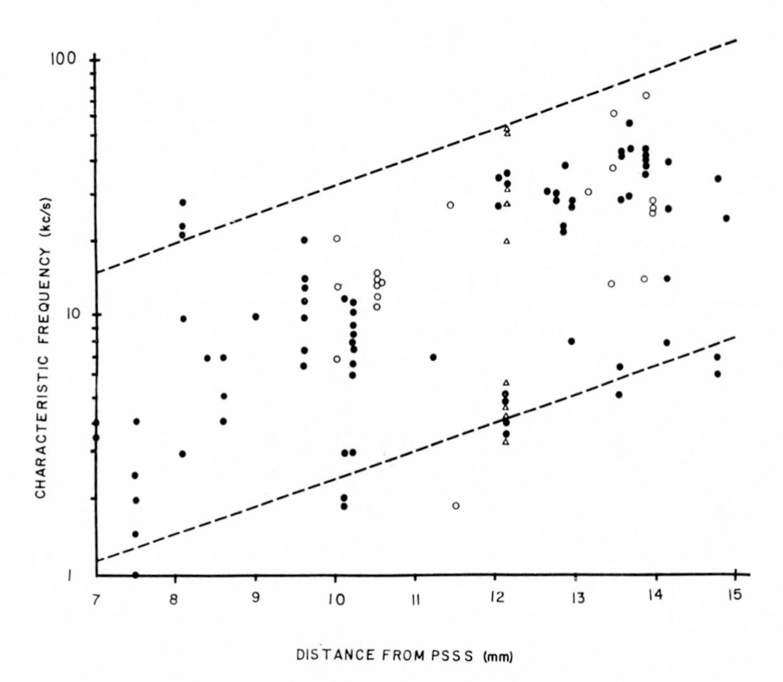

FIG. 2. Characteristic frequencies of 105 units, plotted against their position in the primary auditory cortex.

Points were obtained from nine cortical hemispheres of five cats, and superimposed upon a representative cortex. The abscissa represents the position of the electrode penetration in which the unit was found, measured in a postero-anterior direction from the posterior suprasylvian sulcus (PSSS) of the representative cortex. The points are distributed over most of the antero-posterior extent of the primary cortex. The 7 mm. position on the abscissa represents the mean location of the posterior ectosylvian sulcus, i.e. 2 mm. anterior to the posterior limit of AI. The 15 mm. position represents the anterior limit of AI, approximately 2 mm. anterior to the anterior ectosylvian sulcus. Open circles and triangles mark points obtained in a single cortex. The open triangles (at 12·1 mm.) represent 9 units encountered in a single penetration.

each subdivision of the cochlear nucleus from the dorsal aspect (Rose, Galambos and Hughes, 1959). Fig. 3 shows a typical plot of characteristic frequency (on a logarithmic scale) against the position of the neurons along the microelectrode track (Evans and Nelson, unpublished data).

Such a strict relationship as the above obviously does not hold in the primary cortex (Fig. 2). An attempt has however been made (e.g. Woolsey, 1960; Hind *et al.*, 1960) to retain the idea of tonotopicity, but in the more limited form of a progressive increase or overall trend in the mean characteristic frequency of elements from the posterior to the anterior limits of the primary cortex. In this way, the idea is preserved of a "continuum" of data-processing subsystems spread over the cortex, but with a degree of overlap. This modified model, however, becomes meaningless in terms of frequency analysis if the degree of overlap is comparable with the overall length of the system. The system is then not frequency-specific in the

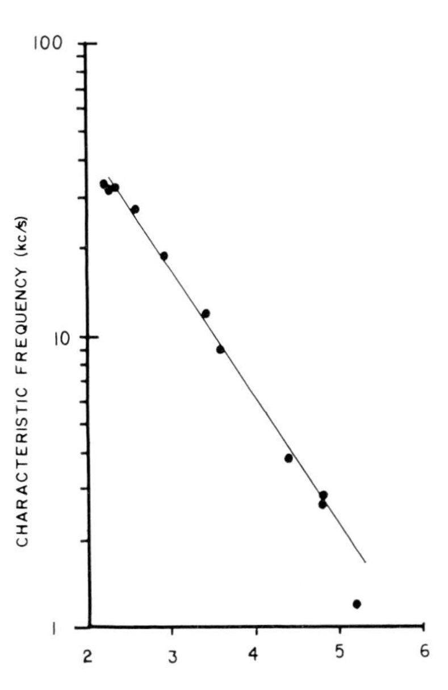

FIG. 3. Tonotopic organization of neurons in the ventral cochlear nucleus of the cat.
Ordinate: characteristic frequencies of consecutive units encountered in a single microelectrode penetration through the ventral nucleus in a dorso-ventral direction. *Abscissa:* location, along the electrode penetration, of the units with the characteristic frequencies indicated. Measurements from the dorsal surface of the cochlear nucleus.

sense that the concept of tonotopicity wishes to retain. The results of Evans, Ross and Whitfield (1965) offer quantitative evidence that the primary auditory cortex of the cat has a system variance which is prohibitively large. The 90 per cent prediction limits of frequency for any one position (Fig. 2) were nearly four octaves, compared with the system total of six octaves. Furthermore, the 90 per cent prediction limits of position of any one frequency were $\pm 2 \cdot 9$ mm. This means that if one wishes to encompass even 9/10 of the units of a particular characteristic frequency, one must delineate a length of cortex equal to 60 per cent of the total length of the primary area. Finally, cortical neurons are not the frequency selective elements that a "frequency analyser" model would require. Although some units do have response areas as narrow as those found at lower levels of the auditory system, the majority have relatively broad bandwidths of response (Katsuki, Murata *et al.*, 1959; Hind *et al.*, 1960; Evans and Whitfield, 1964).

These findings are clearly not consistent enough with the requirements of even the modified tonotopic model for it to remain meaningful in its original sense. It is concluded, therefore, that the frequency trend which is encountered in the primary cortex probably represents a residuum of anatomical arrangement from sub-cortical levels, rather than a property which has a functional significance for frequency analysis.

As mentioned above, the behavioural data compiled by Neff and his colleagues indicate that aspects of auditory function which are dependent upon an intact auditory cortex are as follows: the discrimination of temporal attributes of the stimulus, such as the temporal patterning and the duration of tones, and the discrimination of the spatial location of sound sources. Recent investigations of single unit activity in the primary auditory cortex of the cat (Katsuki, Watanabe and Suga, 1959; Bogdanski and Galambos, 1960; Evans and Whitfield, 1964; Whitfield and Evans, 1965; Goldstein, Hall and Butterfield, 1968) have begun to examine the responses of neurons to stimuli more complex than steady tones. These studies indicate that many cortical units have functional attributes which would fit them well for the roles which the conclusions of Neff and his colleagues would require—namely, preferential sensitivity to the temporal and spatial features of the stimulus.

REPRESENTATION OF TEMPORAL FEATURES OF THE STIMULUS

One characteristic of the behaviour of neurons in the primary cortex is their remarkable sensitivity to complex sounds such as clicks and voice sounds (Katsuki *et al.*, 1960; Bogdanski and Galambos, 1960; Evans and Whitfield, 1964). In the unanaesthetized cortex, as many as 20 per cent of the neurons encountered would respond only to these stimuli: for these neurons, simple tonal stimuli were ineffective. Evans and Whitfield (1964) found that 10 per cent of the neurons which would respond to tonal stimuli would respond only if the tone frequency was changing (Fig. 4; Whitfield and Evans, 1965: category d). Frequency modulated tonal stimuli were in fact very effective stimuli for the majority of neurons responding to tones; vigorous and consistent responses could be obtained from neurons stimulated inconsistently by unmodulated tones (Whitfield and Evans, 1965: categories b and c). Furthermore, many of these neurons, and all of the neurons stimulated only by frequency modulated tones, exhibited responses which were preferential to certain directions of frequency change. Some cells therefore responded only or best when the tone frequency swept toward higher values, others to downward frequency sweeps (e.g. Fig. 5). Some cells exhibited responses oriented to one direction of frequency change at certain frequencies and to the opposite direction at other frequencies.

These units could signal quite small changes in frequency, and excursions of a semitone on the musical scale ($\pm 2 \cdot 5$ per cent) were usually adequate for response (Fig. 4). A number of units exhibited responses restricted to

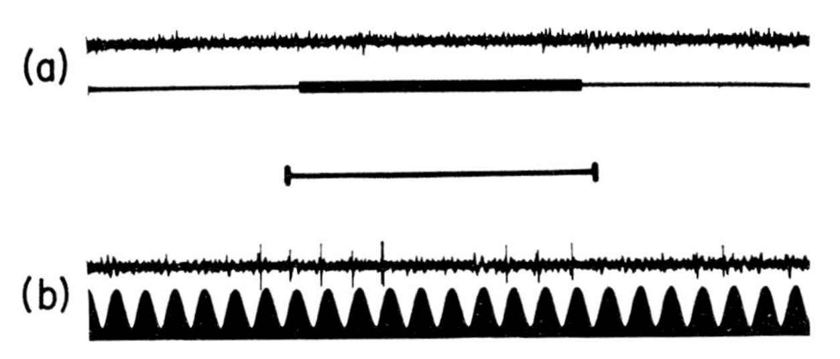

FIG. 4. Unit in primary auditory area responding to tones only when the frequency is changing.

(*a*) No response to steady tone of 2·5 kcyc./sec.

(*b*) Intermittent response to modulation of the 2·5 kcyc./sec. tone at a rate of 10 cyc./sec. over a depth of $\pm$ 2·5 per cent. Envelope of lower trace represents the excursions of frequency.

Time bar: 1 sec.

a very narrow band of frequencies indeed under these conditions (Whitfield and Evans, 1965: category a). They responded, as would be expected, to tones whenever the frequency was inside their response area*. However, with modulations of frequency across the boundary of the response area, at rates higher than a few cycles per second, the responses became limited to the point at which the tone frequency actually crossed the boundary.

The average range of effective sinusoidal modulation rates was limited to between about 1 cyc./sec. and 20 cyc./sec., although some units had much more restricted ranges. The attenuation of the response at higher rates of modulation (Fig. 5) apparently resulted from the repetitive nature of continuous sinusoidal modulation; intermittent "ramp" or sawtooth modulations representing similar or higher rates of change of frequency were always effective (Fig. 5, lower pair of traces). These units therefore

* The "response area" is defined as the frequency-intensity domain within which a stimulus is effective in eliciting a response from the neuron. It is analogous to the "receptive field" of a neuron in the visual system.

showed a greater or lesser selectivity of response for rate of frequency change, direction of change, and repetition rate.

For many other units, the repetition rate and duration of *steady* tonal stimuli were critical parameters. The responses of many units habituated rapidly to repeated presentations of the same stimulus (Fig. 6*a*). A few exhibited the converse behaviour, that is, they required a repetitive stimulus for a consistent response (Fig. 6*b*). In those units responding to the cessation of the tone, a longer stimulus usually evoked a more vigorous response

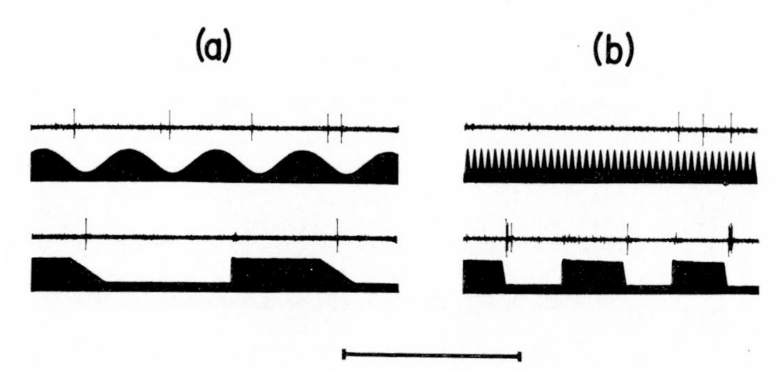

FIG. 5. Orientation of response of unit in primary auditory cortex to downward frequency changes. Lower trace of each pair indicates the excursions of frequency. Modulation depth: ± 5 per cent. Intensity of tone: 15 db above threshold.

(*a*) Upper pair of traces: consistent responses to downward phases of sinusoidal modulation at 2 cyc./sec. Lower pair: similar responses to linear downward frequency changes of 200 msec. duration ("ramp" modulations).

(*b*) Upper pair of traces: infrequent responses to sinusoidal modulation at 25 cyc./sec. Lower pair: vigorous responses to intermittent ramp modulations at a rate of change of frequency equivalent to sinusoidal modulation of 25 cyc./sec. (20 msec. downward ramps).

Time bar: 1 sec.

(Fig. 6*c*,*d*). However, a small number of these units showed the reverse preference, that is, for short tone duration (Fig. 6*e*).

Of the 474 units examined in the complete series of experiments of Evans and Whitfield on the primary cortex, about 60 per cent responded in some manner to tonal stimuli. Of these, 45 per cent gave a sustained excitatory response and 14 per cent a sustained inhibitory response to tones of longer than 100 msec. duration. The excitatory response in this case frequently consisted of a vigorous initial transient excitation followed by a sustained discharge at a slower rate. 17 per cent of the total units signalled only the onset of a tonal stimulus; 10 per cent signalled only the termination, and about 2 per cent signalled both the onset and termination of the stimulus. These responses were by no means invariant for any given unit. A different type of response could occur as the result of a change in a stimulus parameter, such as intensity or frequency (Fig. 7).

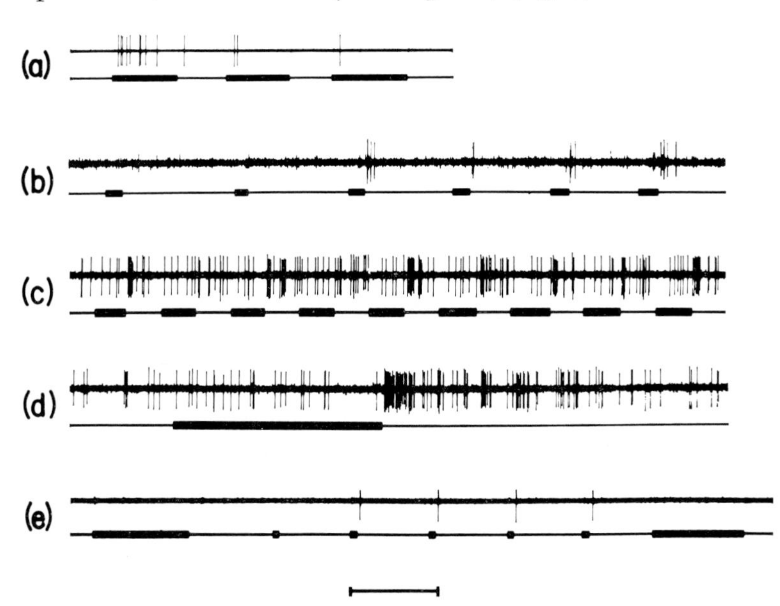

FIG. 6. Influence of temporal pattern of auditory stimuli on responses of different units in primary cortex.

Thickening of the lower trace of each pair indicates the presence of a tone.

(*a*) Habituation to repeated stimuli. (*b*) Responses only to repeated presentations of the same tone. (*c*) and (*d*) Brief and prolonged "off" responses to short and long tones, respectively, in the same unit. (*e*) "Off" responses limited to tones of short duration.

Time bar: 1 sec.

REPRESENTATION OF THE SPATIAL LOCATION OF THE SOUND STIMULUS

From studies of the evoked slow wave responses given by the primary auditory cortex of the cat to different locations of the sound source in space, Rosenzweig (1954) concluded that a differential response between the two cortical hemispheres signalled the direction from which the sound originated. On the basis of a similar study, Coleman (1959) predicted that each cortex would contain elements in which ipsilateral and contralateral information would be represented. Subsequent single unit experiments in the anaesthetized cat (Brugge, Dubrovsky and Rose, 1964) and in the unanaesthetized cat (Evans and Whitfield, unpublished observations; Hall and Goldstein, 1968) have confirmed Coleman's prediction.

Table I indicates the results of a study by Evans and Whitfield of 45 consecutive units obtained from the primary cortex of two cats, where the responsiveness of each unit was crudely determined as a function of the location of the sound source. The stimuli were generally transient sound complexes such as finger flicks; in many units it was confirmed that the

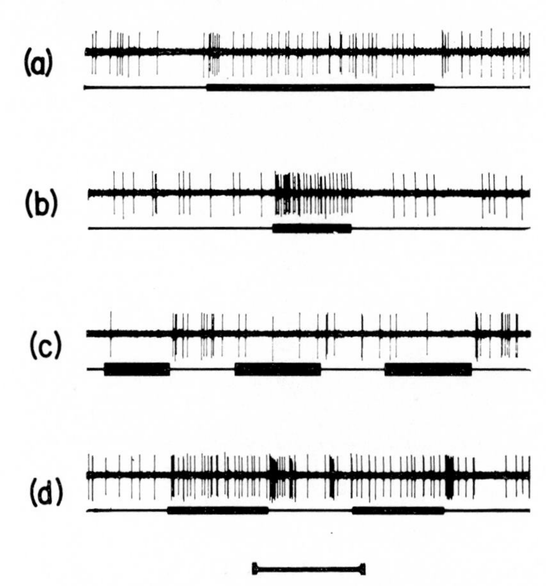

FIG. 7. Dependence of nature of responses of a unit in primary auditory cortex on the frequency of the stimulus tone. Tone intensity: 50 db above threshold.
(*a*) 1·5 kcyc./sec.: transient response to onset of tone. (*b*) 2·5 kcyc./sec.: sustained excitatory response. (*c*) 9 kcyc./sec.: inhibitory and "off" responses. (*d*) 13 kcyc./sec.: sustained excitatory and "off" responses.
Time bar: 1 sec.

results were valid for tonal stimulation generated by a moveable loudspeaker. Forty-four per cent of the units could be stimulated equally well by sound sources located anywhere about the cat's head, although positions anterior to the ears—that is, out of the "shadow" thrown by the pinna—were generally the most effective, particularly when stimuli of high frequency were used. Fifty-six per cent of the units gave preferential responses for particular locations of the stimulus. Preferences for contralateral locations of the stimulus were most common (29 per cent); conversely, preferences for ipsilateral locations were uncommon (9 per cent).

Thirty-one per cent of the units could be stimulated *only* if the sound sources were in particular locations, and again contralateral locations predominated. For these units responding preferentially or selectively to contralateral or ipsilateral stimuli, the response attenuated sharply in the region of the median sagittal plane of the cat's head: the response could disappear completely, even to high intensity stimulation, on moving the source a few

TABLE I

CLASSIFICATION OF 45 CONSECUTIVE UNITS FROM AI OF TWO CATS, AS A FUNCTION OF THEIR PREFERENCE FOR CERTAIN LOCATIONS OF SOUND

(i)	Units with preference for location:	20	(44%)
(ii)	Units with some preference for a certain location:	25	(56%)
	Contralateral	13	(29%)
	Ipsilateral	4	(9%)
	Midline	3	(7%)
	Visual field	5	(11%)
(iii)	Units requiring a certain location for any response:	14	(31%)
	Contralateral	5	(11%)
	Ipsilateral	3	(7%)
	Midline	2	(4%)
	Visual field	4	(9%)

degrees beyond the midline. A small number of units responded preferentially, or responded only (4 per cent) when the sound source was situated in the median sagittal plane anterior to the cat's head. Locations more than a few degrees on either side of this plane were ineffective. About 10 per cent of the units were surprisingly responsive to stimuli originating within the visual fields, and most of these responded only when the animal was observing the source of the sound (e.g. the movement of the fingers producing the click stimuli). Blowing on the animal's head, thus causing it to close its eyes, would eliminate this response, as would near-total darkness; finger movements producing an inaudible stimulus also had no effect. These units therefore appear to be bimodally sensitive and are activated whenever a cat turns its head to observe a sound stimulus. It is very probable that they represent the "attention units" described by Hubel and his co-workers (Hubel *et al.*, 1959) as accounting for 10 per cent of the units in their initial study of single units in the auditory cortex of the unanaesthetized, unrestrained cat. These were units which were difficult to drive by orthodox or complex stimuli until the animal's "attention" was drawn to the stimuli. The "case histories" of these units indicate, however, that the criterion of "attention" was that the animal looked at the stimulus.

Three units were encountered which were sensitive to the location of the sound source in the contralateral field, but only when it was above (in two cases) and below (in one case) the Horsley-Clarke horizontal plane passing through the external auditory meatus.

Hall and Goldstein (1968) have obtained similar data from unanaesthetized, immobilized cats under more controlled conditions. They were able to investigate the nature of the neural interactions produced by stimuli presented monaurally or simultaneously to the two ears. They found proportions of units responding preferentially to stimuli in the contralateral, ipsilateral or midline positions, similar to those reported above. The great majority of the units received input from both ears, and the binaural interactions were in the main those of summation. A few units exhibited inhibitory interactions which were often dependent upon the frequency of the tones.

CONCLUSION

The organization of neurons in the primary auditory cortex of the cat does not appear to be tonotopic in any sense that can have functional significance for frequency analysis.

On the other hand, cortical neurons are particularly sensitive to certain parameters of an auditory stimulus. The majority exhibit marked responses to the onset or termination of a tonal stimulus, or to both. Many units respond vigorously to changes in the frequency of a tone. One-third of the cortical neurons can be stimulated only by tones of changing frequency or by more complex sounds. The responses of many cortical units are dependent upon particular durations and repetition rates of tonal stimuli. Over half of the neurons are sensitive to the location of the sound stimulus.

A high proportion of cortical cells therefore emphasize (and one might presume have the function of abstracting) temporal and spatial features of the sound stimulus. They could well serve to provide specific items of information such as: the stimulus is on; it is off; the frequency is changing; it is changing in a certain direction; at a certain rate; there is a certain repetition rate; the sound originates from a certain location; and so on.

An analogy between this situation and the findings of Hubel and Wiesel (e.g. 1962) in the visual cortex is very tempting. There must be a hierarchy of information-processing in or below the auditory cortex, for the responses characteristic of cortical cells in the steady and modulated stimulus situations are rarely found in the cochlear nucleus (Rose, Galambos and Hughes, 1959; Evans and Nelson, 1966*a*,*b*). However, attempts at finding in the auditory cortex the level of functional architecture encountered in the visual cortex have been disappointing. Vertical columnar organization of the auditory cortex appears to be much less distinct than that of the visual cortex. There appears to be no obvious organization of types of response within vertical columns (Evans and Whitfield, 1964; Abeles *et al.*, 1967). On the other hand, there is evidence that the columns may comprise cells responding to similar frequencies (Hind *et al.*, 1960; Onishi and Katsuki, 1965; Abeles *et al.*, 1967). These columns are possibly narrower than those in the visual cortex, and a penetration must be very accurately normal to the cortical surface to detect cells of similar characteristic frequencies arranged throughout its depth (Onishi and Katsuki, 1965). Oblique penetrations often appear to pass through a series of regions, each comprising cells of similar characteristic frequencies. Neighbouring regions, however, generally have very dissimilar characteristic frequencies. For example, the sequence of the characteristic frequencies of the units in two different penetrations was (in kcyc./sec.): 4·2, 4·5, 30, 4·2, 30: (Fig. 1; Evans and Whitfield, 1964); 27, 3·3, 5·6, 4·1, 31, 4·5, 52, 52, 19·5 (Evans, Ross and Whitfield, 1965).

More work along the lines of columnar reconstruction from multiple electrode penetrations is obviously required before closer analogies may be drawn between the auditory and other sensory cortices. More interesting still perhaps, will be the outcome of investigations of single unit behaviour in the non-primary auditory cortex.

SUMMARY

(1) The evidence for a tonotopic organization of the primary auditory cortex of the cat is critically reviewed. In the light of recent microelectrode investigations of the problem, it is concluded that the organization of the primary cortex is not tonotopic in any sense that could have the functional implication of frequency analysis.

(2) Single unit data indicate that many neurons in the primary cortex are specifically or preferentially sensitive to important stimulus parameters. These parameters largely determine the temporal "shape" and pattern of the stimuli. Particularly represented by these neurons is information on the onset and termination, the duration, the repetition rate, and the rate and direction of frequency change of stimuli.

(3) Over half of the neurons encountered in the primary cortex are preferentially or specifically sensitive to particular locations of the sound source with reference to the head of the animal. The majority of these cells exhibit preference for contralateral sound sources.

(4) An analogy is drawn between the apparent ability of cortical neurons to "abstract" temporal and spatial features of a stimulus, and the situation found in the visual cortex. More pertinent data are required, however,

before the analogy can be pressed very far in the direction of functional architecture.

Acknowledgements

This research was supported by U.S. Army contract number DA-9-591-EUC-2092, 2803, 3253. The Wellcome Trust awarded a grant to construct the soundproofed room in which the investigations were conducted. It is a pleasure to acknowledge the technical assistance of Mr. T. G. Williamson in this work. I am most grateful to Dr. M. H. Goldstein, Jr., for preprints of the work cited.

REFERENCES

ABELES, M., DALY, R. L., GOLDSTEIN, M. H., Jr., and McINTOSH, J. S. (1967). *J. acoust. Soc. Am.*, **42**, 1207.

ADRIAN, E. D. (1947). *The Physical Background of Perception.* Oxford: Clarendon Press.

ALLANSON, J. T., and WHITFIELD, I. C. (1956). In *Third London Symposium on Information Theory*, pp. 269–286, ed. Cherry, C. London: Butterworths.

BOGDANSKI, D. F., and GALAMBOS, R. (1960). In *Neural Mechanisms of the Auditory and Vestibular Systems*, chapter 10, ed. Rasmussen, G. L., and Windle, W. F. Springfield: Thomas.

BREMER, F. (1953). *Some Problems in Neurophysiology.* London: The Athlone Press.

BRUGGE, J. F., DUBROVSKY, N., and ROSE, J. E. (1964). *Science*, **146**, 433–434.

COLEMAN, P. D. (1959). *Science*, **130**, 39–40.

ERULKAR, S. D., ROSE, J. E., and DAVIES, P. W. (1956). *Bull. Johns Hopkins Hosp.*, **99**, 55–86.

EVANS, E. F., and NELSON, P. G. (1966*a*). *Fedn Proc. Fedn Am. Socs exp. Biol.*, **25**, 463.

EVANS, E. F., and NELSON, P. G. (1966*b*). *J. acoust. Soc. Am.*, **40**, 1275–6.

EVANS, E. F., ROSS, H. F., and WHITFIELD, I. C. (1965). *J. Physiol., Lond.*, **179**, 238–247.

EVANS, E. F., and WHITFIELD, I. C. (1964). *J. Physiol., Lond.*, **171**, 476–493.

GOLDSTEIN, M. H., Jr., HALL, J. L. II, and BUTTERFIELD, B. O. (1968). *J. acoust. Soc. Am.*, **43**, in press.

GROSS, N. B., and SMALL, A. M., Jr. (1961). *Expl Neurol.*, **3**, 375–387.

HALL, J. L. II, and GOLDSTEIN, M. H., Jr. (1968). *J. acoust. Soc. Am.*, **43**, in press.

HIND, J. E. (1953). *J. Neurophysiol.*, **16**, 475–489.

HIND, J. E., DAVIES, P. W., WOOLSEY, C. N., BENJAMIN, R. M., WELKES, W. S., and THOMPSON, R. F. (1960). In *Neural Mechanisms of the Auditory and Vestibular Systems*, chapter 10, ed. Rasmussen, G. L., and Windle, W. F. Springfield: Thomas.

HUBEL, D. H., HENSON, C. O., RUPERT, A., and GALAMBOS, R. (1959). *Science*, **129**, 1279–1280.

HUBEL, D. H., and WIESEL, T. N. (1962). *J. Physiol., Lond.*, **160**, 106–154.

KATSUKI, Y., MURATA, K., SUGA, N., and TAKENAKA, T. (1959). *Proc. imp. Acad. Japan*, **35**, 571–574.

KATSUKI, Y., MURATA, K., SUGA, N., and TAKENAKA, T. (1960). *Proc. imp. Acad. Japan*, **36**, 435–438.

KATSUKI, Y., WATANABE, T., and SUGA, N. (1959). *J. Neurophysiol.*, **22**, 603–623.

LICKLIDER, J. C. R. (1941). Ph.D. Dissertation, University of Rochester, N.Y.

LICKLIDER, J. C. R. (1951). *Experientia*, **7**, 128–134.

NEFF, W. D. (1960). In *Neural Mechanisms of the Auditory and Vestibular Systems*, chapter 12, ed. Rasmussen, G. L., and Windle, W. F. Springfield: Thomas.

NEFF, W. D. (1961). In *Sensory Communication*, chapter 15, ed. Rosenblith, W. A. New York: Wiley.

ONISHI, S., and KATSUKI, Y. (1965). *Jap. J. Physiol.*, **15**, 342–365.

Rose, J. E., Galambos, R., and Hughes, J. R. (1959). *Bull. Johns Hopkins Hosp.*, **104,** 211–251.
Rosenzweig, M. R. (1954). *J. comp. physiol. Psychol.*, **47,** 269–276.
Tunturi, A. R. (1950). *Am. J. Physiol.*, **168,** 712–727.
Tunturi, A. R. (1960). In *Neural Mechanisms of the Auditory and Vestibular Systems*, chapter 13, ed. Rasmussen, G. L., and Windle, W. F. Springfield: Thomas.
Whitfield, I. C., and Evans, E. F. (1965). *J. Neurophysiol.*, **28,** 655–672.
Woolsey, C. N. (1960). In *Neural Mechanisms of the Auditory and Vestibular Systems*, chapter 12, ed. Rasmussen, G. L., and Windle, W. F. Springfield: Thomas.
Woolsey, C. N. (1961). In *Sensory Communication*, chapter 14, ed. Rosenblith, W. A. New York: Wiley.
Woolsey, C. N., and Walzl, E. M. (1942). *Bull. Johns Hopkins Hosp.*, **71,** 315–344.

DISCUSSION

Lowenstein: In your experiments with frequency modulation you presumably did control experiments with amplitude modulation on the same units at the same time?

Evans: We did in a few cases, but only with amplitudes up to about 75 per cent modulation. However, the fact that one could get these responses to frequency-modulated stimuli when there were no responses at all to the switching on or off of any steady tone, is a control for intensity.

Erulkar: One can see clearly the ascending hierarchy in the central auditory pathway for the analysis of complex sounds. I have been doing frequency modulation studies of units in the cochlear nucleus with Dr. G. L. Gerstein and Dr. R. A. Butler ([1968]. *J. Neurophysiol.*, in press), and Drs. P. G. Nelson, J. S. Bryan and I did studies of frequency and amplitude modulation for units in the inferior colliculus (Nelson, P. G., Erulkar, S. D., and Bryan, J. S. [1966]. *J. Neurophysiol.*, **29**, 834–860). The interesting point is that at the cochlear nucleus it is relatively rare to find a unit which will respond to the direction of the modulating frequency. In the majority of cases, a unit will respond to a particular frequency in a similar fashion whether that frequency is presented during a descending or ascending scale. Indeed, in response to slow frequency modulation of the stimulus the firing patterns of most of the recorded neurons may be predicted from their responses to tone bursts of fixed frequency. This type of prediction is rarely possible at higher levels of the auditory system. When trapezoidal modulation of frequency is used for stimulation some units in the cochlear nucleus respond perhaps to the sign but not to the magnitude—not to the slope of the change. In the inferior colliculus, in contrast, there are many units which respond to the rate of change as well as to the direction. At both levels almost every unit that is recorded can be activated by some form of sound stimulus. At the cortical level, Professor Rose, Dr. Davies and I found that only 14 per cent of the units from which we recorded could be activated by our acoustic stimuli—by pure tones (Erulkar, S. D., Rose, J. E., and Davies, P. W. [1956]. *Bull. Johns Hopkins Hosp.*, **99**, 55–86). We ascribed this to anaesthesia, but as Dr. Evans pointed out, we now know that there are other factors too. I think that we have not yet discovered what the pertinent stimulus for the cortex is, but whatever it is, it is obviously highly complex.

I cannot agree that there is no functional significance in terms of tonal discrimination of the pathway at higher levels; there is order there, as Dr. Evans admits; it may not be to the same extent as at lower levels, but the information has to get up to the cortex in some form, perhaps not to convey that the stimulus is one tone compared to another tone, but that the particular tone is involved in some other more complex process. I feel that there is a substratum of tonotopic localization at the cortex, but it is overlain by a structural organization responsible for the more complex effects which you and Dr. Whitfield have shown. I would agree that it is at the lower levels that the majority of the analysis of tonal discrimination takes place, as Professor Neff has shown, but it is necessary for the information to get up to the cortex in some organized way.

Whitfield: When we analysed the distribution of frequency in the cortex under various conditions it seemed fairly evident that the observed trend was to some extent a residual effect of the incoming fibres, but the non-zero slope of the regression line is due in part to the fact that very high frequencies are confined to a smaller region of the cortical surface than low frequencies (low frequencies for this purpose being something up to the order of 30 kilocycles). Below 30 kilocycles there is a fairly even distribution, but the fact that these very high frequencies are in a small group tends to skew the data.

If I may comment on the question of auditory cortical areas other than the primary areas, we have been studying (Rigby, D. C., Ross, H. F., and Whitfield, I. C. [1968]. *J. Physiol., Lond.*, **194**, 67P) what you might call "new AII"; this is not the original second area that Woolsey and Walzl proposed, but the truncated area taking out the anterior and posterior ectosylvian regions; in other words, it is the region lying between the ectosylvian sulci. We have not many units but we find a picture substantially similar to that Dr. Evans showed in AI; when we plot position along the cortex against characteristic frequency, the regression line slopes in the direction of having more high frequency units anteriorly, as it does in AI, and not in the opposite direction, as has previously been inferred from stimulation experiments of auditory nerve fibres. The distributions of the points, too, are very similar. There are as many short latency responses (15 msec.) in this region as in AI and we have found no significant differences between AI and AII in the behaviour of the units. There is a slight suggestion that there are less position units; that is the only difference so far.

Schwartzkopff: Dr. Evans, have you repeated the experiments of Tunturi and others—for instance, Hind—to see whether, applying your methods, you get the same results, or have we to say that the earlier results were wrong?

Secondly, have you any information on different *horizontal* layers in the cortex? At the Brain Research Institute in Moscow Dr. Rabinovich was inserting combined electrodes into about five different layers; not in primary areas, but in associate areas. He could show very interesting differences in learning experiments. It would be interesting to know whether an organization in horizontal layers occurs. This would be complementary to the columnar organization, of course.

Evans: I do not think that the early experiments were wrong; the conclusions drawn from them *at the time* seem appropriate to the data. However, it is clear that subsequently the wrong conclusions have been drawn from these experiments because it has been forgotten that the expressed purpose of the early experiments was to look at the organization of frequency representation in the *afferent pathway* to the cortex and not in the cortical arrangement itself. There can be little doubt that there is such an arrangement in the afferent pathway, although it may not be as well-defined as Tunturi concluded. As I have pointed out in my paper, similar experiments, but under reduced or no anaesthesia, have not been able to find a well-defined frequency representation, at least in the cat. As for the dog, A. R. Tunturi ([1960]. In *Neural Mechanisms of the Auditory and Vestibular Systems*, chapter 13, ed. Rasmussen, G. L., and Windle, W. F. Springfield: Thomas) published a picture taken from an experiment on the unanaesthetized dog which showed areas of response which were as diffuse as those found in the experiments on the cat.

On your second point, I know of only one source of evidence for a horizontal stratification, from the experiments of S. Onishi and Y. Katsuki ([1965]. *Jap. J. Physiol.*, **15**, 342–365) who have described functional differences occurring by layers, the deepest units having narrow response areas, units nearer the surface having multiple peaks in their response areas, and superficial units having very broad response areas. However, it is not clear from their data what percentage of units at any given level are represented by these responses; furthermore, other workers have not observed this kind of stratification.

Rose: It is high time that we take a new look at the organization of the auditory cortex. It would seem clear that tonotopical frequency arrangements do not provide by themselves a satisfactory key for understanding this organization simply because there are several such arrangements within the region. I do not think, though, that we should legislate as to which findings are and which are not important. Any consistent finding is presumably very important indeed and must eventually be understood. Much of the present knowledge about cortical units pertains to their response areas; there are other stimulus parameters which need to be explored and the findings may hopefully provide new clues. I see little reason to dismiss some older findings as insignificant simply because we search for a better understanding of a complex problem. The classical findings on tonotopical organization as obtained in deep anaesthesia are, I think, quite valid. They are based on the results of electrical stimulation of nerve fibres in the cochlear coils, on the distributions of evoked potentials in response to pure tones and finally on response areas of cortical units. In my experience, if an electrode is normally orientated in respect to the cortical surface, the best frequencies of units in a given column tend to cluster around a frequency which may be called a dominant frequency, though from time to time a unit may occur whose best frequency is far apart from the dominant one. Since it is rather rare to isolate well in a single puncture a large number of responsive units, the phenomenon is not easy to study. The frequency picture as you say is not simple but I doubt that it is hopelessly confused or confusing.

Evans: I think we basically agree. I have been trying to say not that the residuum of organization by frequency that we see in the cortex is insignificant; but that it is not significant in the sense that has been generally proposed, namely, for frequency analysis. We are not legislating on the horizontal organization of the cortex at all, but we find that the organization is so blurred that it hardly seems to have a functional use in the original sense, and we are asking: what use could it have? My view of the organization of the vertical columns is that, although I would be very happy to find similar characteristic frequencies of cells within the same column, the data at present available are not conclusive on this point.

Lowenstein: Am I right in thinking that all cortical projections—that is, in the whole range of modalities—tend to be sharp and well-focused under deep anaesthesia, whereas in the unanaesthetized animal they are blurred?

Rose: Deep anaesthesia tends to simplify the picture. It may prevent observation of important events. It is difficult to conceive, though, that the phenomena which it permits us to observe are altogether absent in the normal state. It would seem right to integrate the observations obtained with and without anaesthesia rather than to give preference to one or another approach.

Tumarkin: Is this not a question of what one means by a pattern? Why insist on an orthogonal pattern that goes one particular way or another; can we not accept that a mass of uniformly distributed units can represent many different patterns, just as a number of lights can compose alternative pictures?

Fex: Have you ever, Dr. Evans, in your preparation seen a response area from a particular unit change into quite another response area, when you recorded later from the same unit? And if so, did you see a shift of the so-called characteristic frequency of a unit? This would again indicate that one must be careful in talking about characteristic frequencies.

Evans: In the sense that part of the response area that was once excitatory could become inhibitory, yes we have, but a significant shift in the characteristic frequency did not occur in many units, if at all. It is, however, extremely difficult in practice to plot the response area of a cortical unit in the unanaesthetized animal because the responses of these cells are so labile; it is often very hard to define the characteristic frequency. There is also the problem that a significant proportion of units have very broad response areas; R. Galambos ([1960]. In *Neural Mechanisms of Auditory and Vestibular Systems*, p. 143, ed. Rasmussen, G. L., and Windle, W. F. Springfield: Thomas) called them the "broadband" units. In our series, they were usually inhibitory units—that is, they were inhibited by tones—and it was very difficult to assign any characteristic frequency at all. In fact we had to exclude them from our analysis, so the position is even worse than it looks in my Fig. 2 (p. 276).

Lowenstein: If someone took a bird's eye view of the cochlea, would he find the same thing, namely that the lower frequencies are more or less effective everywhere and the higher frequencies are bunched in the basal turn?

Evans: This is so in the cochlea in terms of frequencies as low as a few hundred cycles per second, but by "low frequency units" at the cortical level I mean frequencies up to 10 kilocycles per second.

Fex: There actually were few really low frequency units (200–1,000 cyc./sec.) at the cortex, then; how does this come about?

Evans: I do not know. This seems to be a characteristic of higher levels of the auditory system of the cat. Either they are very infrequently represented in the cortex or they are not represented at all. But I think that this situation is not an uncommon finding. Did you in fact find this, Professor Rose?

Rose: The cortical area devoted to very low frequencies is quite narrow in AI of the cat by comparison with the areas devoted to higher frequencies. It would appear that either the cat is particularly good at higher frequencies (say at 20 kilocycles) or that more cortical tissue is needed as the frequency rises.

Evans: This problem becomes acute if frequency is expressed on an octave scale, because then there is a huge "gap" below 1 kcyc./sec. that remains to be filled. However, H. F. Ross (personal communication) has been looking at what might be a better frequency "scale" for the auditory system, and finds that a square-root frequency scale is more appropriate (for example, for comparing the response bandwidths of single units of different characteristic frequencies). If one uses such a scale, the problem is not quite so acute.

Rose: The relatively very small amount of tissue devoted to very low frequencies is already apparent in the cochlear complex. Neurons with very low best frequencies are crowded in a narrow strip. If one were to probe the complex at random the probability of observing a best frequency lower than some 4 kilocycles would be small.

I would like to ask how good the cat actually is behaviourally at higher frequencies?

Neff: Those doing behavioural experiments have not often used tones in what would be considered the high frequency hearing range of the cat, yet it has been shown that the cat hears tones at least up to 60,000 cyc./sec. (Neff, W. D., and Hind, J. E. [1955]. *J. acoust. Soc. Am.*, **27**, 480–483; Miller, J. D., Watson, C. S., and Covell, W. P. [1963]. *Acta oto-lar.*, Suppl. 176; Elliot, D. N., Stein, L., and Harrison, M. J. [1960]. *J. acoust. Soc. Am.*, **32**, 380–384). We also know that frequency discrimination is reasonably good up to 8,000 cyc./sec. (Butler, R. A., Diamond, I. T., and Neff, W. D. [1957]. *J. Neurophysiol.*, **20**, 108–120). In experiments the results of which have not yet been published, we have attempted to train cats to discriminate between tonal patterns (low–high–low *vs.* high–low–high) in which the tone components of the pattern were made up of frequencies chosen from different parts of the frequency scale—for example, 200–350–200 cyc./sec. *vs.* 350–200–350 cyc./sec.; or the same kind of pattern with the two frequency components consisting of 800 and 1,200 cyc./sec., 2,000 and 3,000 cyc./sec., 3,000 and 4,500 cyc./sec., 4,000 and 7,000 cyc./sec., or 8,000 and 12,000 cyc./sec. We have had little success in training animals to learn the pattern discrimination in the frequency range above 4,000 cyc./sec. although they learn to discriminate frequency change in this range, and they learn to discriminate pattern in the lower frequency range.

We are also training cats to localize pure tone signals in space. Again, we have had some difficulty in training animals to localize an 8,000 cyc./sec. tone.

These results must be considered as tentative because only a limited number of animals have been tested and different training procedures must still be tried. Nevertheless, I think we shall find that the cat cannot learn to use high frequency sounds in the same ways that it uses lower frequencies. This raises the question of why an animal that has great sensitivity to high frequencies does not seem to utilize them fully. My belief is that they do use their sensitivity in the *detection* of high frequency sounds, and from casual observation, I would suggest that the response of the cat to high frequencies in its "natural" environment may be primarily an escape response due to fear. If you turn on a high pressure air jet in the laboratory, many cats will show a fright reaction to the noise. Such noise has been used as punishment in learning experiments. I think the cat is responding to the rather intense high frequencies in the jet noise that we do not hear or hear only poorly.

Rose: One is tempted to speculate that the large amount of tissue devoted to higher frequencies at the various synaptic levels need not reflect the capacity of the animal to discriminate these frequencies particularly well but may result from a need for more tissue to handle a higher tone. This could be true if the volley principle should apply for coding of higher frequencies. Clearly, the number of fibres required to code the frequency information would be rising steeply the higher is the tone, if the same discriminatory capacity is to be realized. One would expect this to be true already in the cochlear nerve. In this respect some figures are of interest. The cat has some 40,000 fibres in the cochlear nerve and this seems to be a conservative estimate; man has about 30,000 fibres. Man can hear over a range of about 8·5 octaves while the cat hears over a range of some 10 octaves (the upper limit for man and cat is taken here as 20 kilocycles and 60 kilocycles respectively). Now, why should the cat have more fibres than man? The cat does not discriminate frequencies any better and in fact it seems to discriminate less well. Of course, conditions for testing the cat and man are quite different and it may be fairer to assume here that there are no significant differences in discriminatory capacity. But if this be so one is inclined to assume that cat has more fibres because it can hear 1·5 octaves higher. A difference of 10,000 fibres for 1·5 octaves means some 6,000 fibres per octave. This figure, however, is obviously wrong as an overall average since this would mean that the cat should have some 60,000 fibres for its 10 octaves and man some 51,000 for his range. The difficulty vanishes if one assumes that the number of fibres needed per octave is not a constant but rises with the rise in frequency.

Evans: I am not sure that this relation you propose between frequency and the number of elements subserving it holds for the cat cortex, at least. As Dr. Whitfield mentioned earlier (p. 288), the cortical distribution of cells responding to higher and higher frequencies gets smaller and smaller. Cells having characteristic frequencies of below 10–30 kcyc./sec. are distributed evenly over the cortex, whereas cells responding to more than 30 kcyc./sec. are found only in a small area in the anterior part of the cortex.

Rose: Quite so, but you consider here only the highest octave and we know little about the discriminatory capacity at the limit of the cat's hearing. We must consider really low frequencies. We cannot say that 10 kilocycles is a low frequency.

Evans: If you mean that cells of characteristic frequencies below 1 kcyc./sec. are compressed into a small cortical space whereas those above are widely distributed, then the relationship you propose may hold in general terms; however I know of no evidence for the discrete localization of neurons of characteristic frequency below 1 kcyc./sec.

Engström: This relationship depends on the numbers of fibres in the acoustic nerve, but those that Professor Rose gave are slightly inexact, and there are figures of up to 40,000 fibres in the human acoustic nerve. Is there not an immense increase for high tones in the first acoustic nuclei? The dolphin has about 500,000 cells, while man has 125,000; the whale has an enormous increase in the first acoustic nuclei. This is what we can see, that a huge increase in number of neurons takes place in many mammals in the acoustic nuclei, not in the acoustic nerve.

Rose: For man I quoted figures given by Schuknecht ([1960]. In *Neural Mechanisms of the Auditory and Vestibular System*, pp. 76–90, ed. Rasmussen, G. L., and Windle, W. F. Springfield: Thomas) which seem in agreement with the data of Retzius. Schuknecht quotes Guild, who counted in man 29,000 spiral ganglion cells, and Rasmussen, who counted 31,000 fibres. Hence the average of 30,000 for man. For the cat Schuknecht gives a figure of 39,500 spiral ganglion cells. J. A. Vinnikov and L. K. Titova ([1964]. *The Organ of Corti: Its Histophysiology and Histochemistry*. New York: Consultants Bureau) give 52,000 as the number of cochlear fibres in the cat.

Neff: The essential point is that for its behaviour, the cat has a very large number of nerve fibres for handling high frequency information, but the only information that is extracted by the cat is in relation to the *detection* of sound. The cat has not, in fact, a large number of fibres available for a more complex kind of discrimination, because it takes a lot of fibres just to *detect* sound!

Lowenstein: Are you satisfied that the cat is suitable for your conditioning experiments? The cat is not a good animal for conditioning as such.

Neff: For the avoidance response to shock or for an approach response like localizing a source of sound, the cat is hard to beat. In the double-grill box, the cat learns the tonal patterns that we have described at the same rate as a rhesus monkey. In my opinion, the cat is an excellent animal to use in behavioural studies of auditory discrimination. They do not work well if restrained, but if the training apparatus allows them free movement and if the response required of them is appropriate, they will learn quickly. Experimenters have not had too much success teaching cats to push a lever to get food or avoid shock, but they will readily learn to approach food and avoid shock in the situations I have described.

Schwartzkopff: Professor Neff suggested that the cat uses its hearing at higher frequencies of about 10,000 cyc./sec. mainly for escape. Of course, that is what is seen in conditioning experiments, but in its natural environment the cat is a predatory animal, hunting mice and shrews, which are animals producing sounds of very high frequency (above 10,000 cyc./sec). I would be very surprised if this does not have something to do with its hearing. I agree that it may not be possible to do training experiments to show this, but in the cat's normal life evasive behaviour to the highest frequencies is surely less important than localizing auditory stimuli, since the cat hunts mainly in the dark.

Dr. Engström mentioned the higher number (500,000) of second-order neurons (acoustic nuclei fibres) in the dolphin. It is rather dangerous to compare very specially adapted animals. If we compare normal birds with those highly adapted for auditory functions, we find, for example, the auditory nerve in the owl contains about twice as many fibres as that of the crow, making allowance for the difference in size. However, in the brain centres there is an increase up to tenfold, but only in very specialized owls, which may be comparable to the dolphin. The dolphin may have four or five times as many fibres as man, because it is specialized for a certain function of hearing, which apparently needs processing of auditory information at a low level. The neuronal response to a click must not travel for a long time before being processed, or the answer would come too late—like in the bat, which has its main processing in the inferior colliculus and not in the cortex.

Engström: Has the owl twice as many cells as the crow in the secondary nuclei?

Schwartzkopff: In the secondary nuclei the specialized owls have four times as many, and in the superior olive about ten times as many. This increase is not found in all owls: the little owl (*Athene noctua*) essentially hunts by sight and shows no increase. It is closely related to the barn owl (*Tyto alba*) and the horned owl (*Asio otus*), in which there is this tremendous increase. I have no figures for the midbrain, but it does not seem that the increase affects the midbrain very much.

Neff: I agree that the use of high frequencies is important to the cat for detecting prey as well as avoiding predators, but with the data we have at the moment, I would question whether it helps the cat very much in localizing prey. Even though higher frequencies are excellent for localizing, I am not sure that the cat can use them very well, but at present we have inadequate information on this.

Lowenstein: Even though shrews and mice produce sounds at 10,000 cyc./sec., they may avoid making them when a cat is about! I thought that it was substrate noise—the rustling of leaves—which alerted the cat, and that is in a far lower frequency spectrum. This would fit in with Professor Neff's findings so far.

Whitfield: The cortical units to which Dr. Evans referred and which are directionally sensitive are commonly found in the 16–20 kilocycle range; this observation seems relevant to Professor Neff's query on whether cats actually use these frequencies for localization. If it turns out they do not, we shall obviously have to think again about the function of such units.

Neff: We do not know enough yet; we are still in the middle of these behavioural experiments on localization.

Davis: May I make some general comments here. We have heard the word "processing" used a number of times, and have discussed what seems to be the significance of the cochlear nucleus, the inferior colliculus, and so on. I have not heard the old-fashioned term "relay station", which implied that these nuclei simply passed information on, the cortex being the all-important centre for discrimination and integration. We now have a much better, and I am sure truer, perspective on the processing which takes place in several stages in the various lower structures. It seems self-evident that more than one type of processing must be envisaged—that is, processing in different ways to extract different aspects of information from the total input. Certainly, by the time we get to our perceptions, various features such as differences of pitch, intensity, quality and other aspects of sounds must require completely different types of "processing". Intensity must require summation of the input over many different nerve fibres: while pitch discrimination requires an enhancement of contrast, and this implies mutual or "lateral" inhibition.

These are two types of processing; localization is a third type. One of our problems is to identify the nature of the processing and to decide whether we have inferred the basic qualities and corresponding processes correctly from our own subjective experience. The emphasis in these discussions has been on localization —that is, the location of the "thing heard" in an external frame of reference— and on change—in other words, on events in the temporal frame of reference. I was much impressed by Dr. Evans' concluding point that the job of the cortex may be essentially to provide the temporal and spatial frames of reference for sound stimuli. I would like to underline that, because I find it fits so well with my own less complete thinking along these lines. Regarding change, a short-term memory for comparison with the immediate past certainly seems to be a requisite mechanism which the cortex provides.

Ultimately, these different components have to be synthesized somewhere into the whole which we finally perceive; we do not perceive these elements separately but perceive a particular sound that has an ensemble of qualities—including a localization in space, as a rule. Whether the final reassembly takes place in the cortex or subcortically is an open question.

GENERAL DISCUSSION

ABSOLUTE SENSITIVITY OF THE HEARING MECHANISM

Lowenstein: So far we have not dealt with the absolute sensitivity of the hearing mechanism. There are of course theories, and one ought to know whether they have a basis in fact. I am referring to the sensitivity of, say, mechanical displacement at the oval window, of the order of a fraction of the diameter of a hydrogen atom. I should like to have that value either exorcised or confirmed.

Johnstone: We have made some measurements on this and perhaps I can supply a value (Johnstone, B. M., and Boyle, A. J. Y. [1967]. *Science*, **158**, 389). Of course I have not measured it at threshold, but one can assume linearity with amplitude in the system. After all, the microphonics increase linearly with intensity up to about 95 db.

The figure I obtained by measurement for a peak-to-peak swing of the guinea pig basilar membrane, at a distance of 1·5 mm. from the stapes, for a tone of 90 db at 18 kcyc./sec., is 1,200 Å. One can extrapolate down from 90 db and obtain a figure of 0·04 Å for the basilar membrane at threshold intensity. This is the point where we have to guess a little because I have confidence in my amplitude measurements only by a factor of 2, and for the sound pressure by a factor of 5 db.

Whitfield: It seems to me that this extrapolation is where the argument fails. It has usually been assumed that the basilar membrane behaves in a linear manner up to an intensity of 90 db, because of the lack of distortion in the recorded microphonic, but as Ross and I showed (Whitfield, I. C., and Ross, H. F. [1965]. *J. acoust. Soc. Am.*, **38**, 126–131), this is mainly because the system acts as a low pass filter and even if hair cells were producing square waves one would still be able to record a fundamental sinusoid and not a distorted wave; so the fact that the microphonic is sinusoidal is not evidence for the linear movement of the basilar membrane. There is evidence in other contexts that the basilar membrane behaves non-linearly (e.g. the summating potential), and now that the microphonic evidence is no longer valid, we need some alternative direct evidence of linearity over this wide dynamic range before we necessarily accept the rather unlikely result to which the extrapolation leads. In other words, at present we just do not know.

Schwartzkopff: Students of the Wever group (Wever, E. G., Rahm, W. E., and Strother, W. F. [1959]. *Proc. natn. Acad. Sci. U.S.A.*, **45**, 1447–1449) have shown that at lower intensities the microphonics move linearly with intensity. On the other hand, there are papers on reptiles, and we have done studies on birds, which show that the amplitude function of the microphonics is not linear, but has a slope of 0·4, which is far from linearity even at lower intensities (−80 to −40 db). We have not seen a distortion but if I understand you, Dr. Whitfield, you would not make that a pre-condition? But certainly the intensity function is not linear in lower vertebrates while it has been said to be so in mammals.

Whitfield: I am proposing that even if we assume that the displacement of a region of the basilar membrane is proportional to the electric output of a hair cell on that bit, what we record as a microphonic is not this output alone, so it is difficult to adduce the amplitude of the cochlear microphonic in evidence.

Johnstone: I would emphasize that I checked linearity over a range of 20 db, because I realize that 90 db may be high. We have not done a full intensity function simply because we cannot cover a great dynamic range—my equipment is too sensitive—but we have measured between 200 Å and 20 Å, and it is linear between these values at least. This is amplitude linearity—that is, change in output with change in input—a separate consideration from wave-form linearity, which may be distorted but still have amplitude linearity.

I disagree that it is up to people to prove linearity; surely the first assumption one makes is that a system is linear.

Whitfield: I said that I would like some evidence on which to base the assumption, since it leads to the rather incredibly small value to which Professor Lowenstein has referred. It is not impossible, of course, but it does not seem the best way to achieve the result that the ear does achieve.

Johnstone: That is no problem: the condenser microphone does the same thing—it works! Admittedly, before information can be transmitted at these amplitudes, it must be integrated, because these are co-operative efforts and it is a significant signal-to-noise problem which we all know, but in fact the coherent movements of the diaphragm at 40 db in the $\frac{1}{4}$-inch condenser microphone are under 1 Å.

Lowenstein: It fits in well with values in other modalities—in the mechanical field for instance.

Schwartzkopff: Autrum has measured the displacement of the vibration receptors in the cockroach; it is also of the order of 1 Å at the threshold. This is a mechanoreceptor but of a completely different structure (see Autrum, H. [1942]. *Naturwissenschaften*, **30**, 69–85).

Davis: The great difficulty is in conceiving a model of how the transduction is effected. It is true that the diaphragm of the condenser microphone moves through distances of the same order of magnitude, and we have succeeded in measuring the pick-up of such a microphone at the human threshold level by using the average-response technique (Miller, J. D., Engebretson, A., and Weston, P. [1964]. *J. acoust. Soc. Am.*, **36**, 1591–1593). We could reproduce the sinusoidal wave forms. The response was linear right down to the threshold level. In this situation the law of large numbers is important. The overall action is a statistical affair, the movement of many molecules and the transfer of a certain amount of energy. May I point out that calculations of the energy level at the human threshold lead to exactly the same order of magnitude of amplitude of movement. I think we must accept the value; the real problem is to envisage a model, because this must include an amplifier. Energy must be supplied. The detector must be a control system of some kind, not a passive transducer, because the signal is so far down in the physical and biological noise. At these energy levels you cannot do an appreciable amount of work.

Tumarkin: Mathematically we have no difficulty in handling infinitesimal displacements. They are implicit in the concept of a continuous function. But Planck has told us that the physical world is not continuous and Heisenberg's uncertainty principle sets a limit to the accuracy with which we can envisage, let alone carry out, our measurements. We must reconcile ourselves to this atomicity and uncertainty under certain conditions, but surely this does not apply to a macroscopic structure like the tympanic membrane. Even granted that at threshold, quantum differences of displacement might exist between infinitesimally small areas, nevertheless the average displacement integrated, so to speak, over the whole area must surely obey the elementary physical laws. Indeed it would be much more difficult to imagine the drum entirely failing to respond to incoming energy until the latter reaches a level at which it can begin to move. It would be rather reminiscent of the electron's leaps from one energy level to another, and so far as I know there is no reason to believe that this occurs at macroscopic levels.

Lowenstein: We have then no overwhelming reason for exorcising the figure for displacement of the basilar membrane, but the difficulty is in visualizing a model.

POSSIBLE FUNCTIONS OF THE COCHLEAR EFFERENT FIBRES

Spoendlin: May I revert to the question of the cochlear efferents and their significance? It might perhaps be, as has been suggested, that they have no function at all, but I have difficulty in believing this when I see their enormous anatomical representation in the organ of Corti. The contact area of the efferent nerve endings with the outer hair cells is greater than the contact area of the afferent nerve endings, and it is hard to see why so much space should be wasted, when generally Nature tries to save space! The one thing we know about the function of the efferents is their inhibitory action on the acoustic input, which, however, is limited: J. E. Desmedt has calculated it to be at the maximum the equivalent of about 20 db of the overall input from the cochlea ([1962]. *J. acoust. Soc. Am.*, **34**, 1478–1496). The important anatomical representation of the efferents in the organ of Corti suggests a more complex function.

The first possibility to consider would be frequency discrimination. The efferents might be involved in a process of increasing contrast, or, as Békésy called it, a funnelling action—a general functional principle of the nervous system (Békésy, G. von [1960]. *Experiments in Hearing*, ed. Wever, E. G. New York: McGraw-Hill). If we consider the distortion of the cochlear partition by the travelling wave envelope as the proper stimulus, and the position of the travelling wave envelope's peak along the basilar membrane as the first means of pitch discrimination, a frequency discrimination threshold of only approximately 10 per cent would be expected, which is in contrast to the actual discrimination threshold of 0·2 per cent (Keidel, W. D. [1964]. In *Biochemie des Hörorganes*, ed. Rauch, S. Stuttgart: Thieme).

This gain in frequency discrimination seems to occur somewhere in the neural pathways. If, theoretically, the stimulation of a sensory unit at one place is proportional to the mechanical distortion of that particular portion, we should be able to deduce a mirror image of the travelling wave which would give an approximate picture of the neural activity in the associated nerve fibre. If we compare this estimated curve with the measured tuning curves of a given fibre of the acoustic nerve, we find that the tuning curve is much sharper than the estimated curve. A sharpening process seems therefore to happen between the peripheral receptor and the first-order neuron and it is possible that the efferents are involved in this process.

Fex: This suggestion of a sharpening process at the cochlear receptor by the cochlear efferents recalls what Dr. Johnstone mentioned in the discussion of Professor Batteau's paper (see p. 240), that the travelling wave itself has certain characteristics which seem to lend themselves very nicely to a sharpening process. I believe in that. I do not think that the efferents help towards sharpening in the sense Dr. Spoendlin suggests, and will come back to that argument.

Let me take up a question that Dr. Hallpike has asked me, of whether something could have happened to the efferents in clinical cases where one finds recruitment. One could imagine that if something has damaged the metabolism of the transmitter system of the efferent endings on the outer hair cells—whatever the transmitter is chemically—this could then functionally knock out the outer hair cells more or less completely. The cochlea would essentially be left with only the inner hair cells with which to respond to sound and this might lead to clinical findings corresponding to recruitment. Such a hypothesis is not contradicted by the findings that cutting the crossed efferents in experimental animals does not interfere with the absolute threshold for sounds; cutting the efferents may have less effect on the outer hair cells than disruption of the metabolism at the efferent synapses might cause. So one should bear this possibility in mind.

Concerning the function of the efferents, this question is discussed in my paper (p. 169). We now have good reasons to believe that the cochlear efferents are physiologically important, although we still do not know exactly how.

Johnstone: Dr. Spoendlin has surely used the wrong Békésy wave; if you are comparing the Békésy wave with a tuning curve, you must look at a single point on the cochlea; then it is necessary to use the diagram of the wave against frequency, not distance. It is certainly true, as Dr. Spoendlin points out, that the neural tuning curve is sharper than the Békésy envelope, and furthermore it becomes much sharper at high frequencies, again in apparent contrast to the travelling wave as described by Békésy. However, if instead of looking at the overall wave, one looks at the slopes of the leading edge of the travelling wave envelope, as I suggested earlier (p. 240), one sees a regular increase of slope from low frequencies to high.

From Békésy's Figs. 12–23 and 12–16 ([1960]. *Experiments in Hearing*, ed. Wever, E. G. New York: McGraw-Hill), the lowest (apical) point tested (maximum 200 cyc./sec.) shows a maximum slope of about 20 db per octave, whereas our (Johnstone, B. M., and Boyle, A. J. Y. [1957]. *Science*, **158**, 389) tuning curve has a maximum slope of 80 db/octave for a (basal) point with a maximum of 20 kcyc./sec.

This increase from 20 db at low frequencies to 80 db at high frequencies is not too different from the changes in the neural slopes. Hence perhaps one does not need to invoke the olivo-cochlear bundle for sharpening. However, it may be that the olivo-cochlear bundle is concerned with *temporal* discrimination and therefore we cannot see the effect of the olivo-cochlear bundle by the use of single-tone frequencies. We know that if we obtain tuning curves with and without the olivo-cochlear bundle, there is little or no difference, and I would suggest that this is the wrong experiment, and you should use some other indication of structuring a response of a neuron; then if you activate the olivo-cochlear bundle, you might show a result. Perhaps time-discrimination between two clicks would be a suitable test.

Fex: Concerning the sharpening in the cochlea at the level of the sensory cells, a few words about the work of N. Y.-S. Kiang ([1965]). Research Monograph No. 35. Cambridge, Mass.: M.I.T. Press) should be added. Kiang used anaesthetized animals and this might block the effect of the efferents. He also cut the efferents in some of his animals, with no change of the tuning curves of the auditory nerve fibres. This, again, indicates that the efferents are not needed for that particular job of sharpening in the sense suggested by Dr. Spoendlin.

Spoendlin: If we accept tonotopic localization as the main principle of frequency discrimination in the cochlea, the length of the basilar membrane and the number of frequency-specific spots on it is related to the frequency discrimination threshold. In other words, frequency is related to distance. If the stimulation of a sensory unit at one point of the basilar membrane is really proportional to the mechanical distortion we actually can expect the diagram of the travelling wave against frequency to be similar to the tuning curve. But what we would like to demonstrate is the difference between mechanical spatial frequency discrimination in the cochlea by the travelling wave envelope and the actual frequency discrimination in the cochlear nerve as, for instance, expressed in tuning curves. I agree that these two different curves cannot be compared in an absolute way, but their characteristics, which are considerably different, can be compared.

Engström: I find it very difficult to believe that a system that is as well provided with mitochondria as is the efferent system should be of so little importance. Fig. 1 is a section through an outer hair cell of the guinea pig cochlea. It shows the efferent endings full of mitochondria, presumably with a high level of enzymic activity, and forming the bulk of the neural matter at the base, more than ten times the volume of the afferent endings. Fig. 2 is a surface view of the organ of Corti of a guinea pig. It shows a large amount of efferent endings and fibres, forming a very large cross-correlation system. It is a most important problem to discover what part the cross-correlation system plays in hearing.

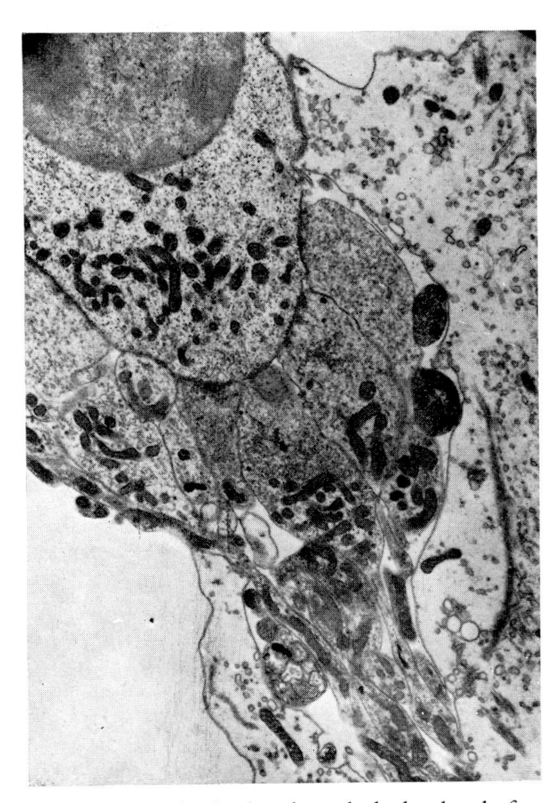

FIG. 1 (Engström). Section through the basal end of an outer hair cell in a guinea pig cochlea. The large cluster of nerve endings at the base consists mainly of richly granulated efferent nerve endings. Observe the large number of mitochondria in the endings, indicating areas of high enzymic activity.

Bosher: One of the striking points from Dr. Spoendlin's paper was his finding that the efferent nerve endings are not evenly distributed throughout the cochlea; there are far more at the base. This may be because they are concerned in some way with high tones, but also it might be related to the fact that the basilar membrane at the base is activated by every travelling wave. This continual activity at the base produced by the low-pitched sounds, it could be argued, may need to be corrected in order for the high-pitched sounds to be normally appreciated. The morphological findings must surely have some significance.

Lowenstein: Could the efferents have something to do with either sensitivity or linearity? Consider the characteristic curve of a hair cell, which is an S-shaped curve with a long linear portion. I have indirect evidence that the working point of such a cell can be shifted along the characteristic curve into the linear portion. Is it possible that in the cochlea such continuous adjustments of the working point

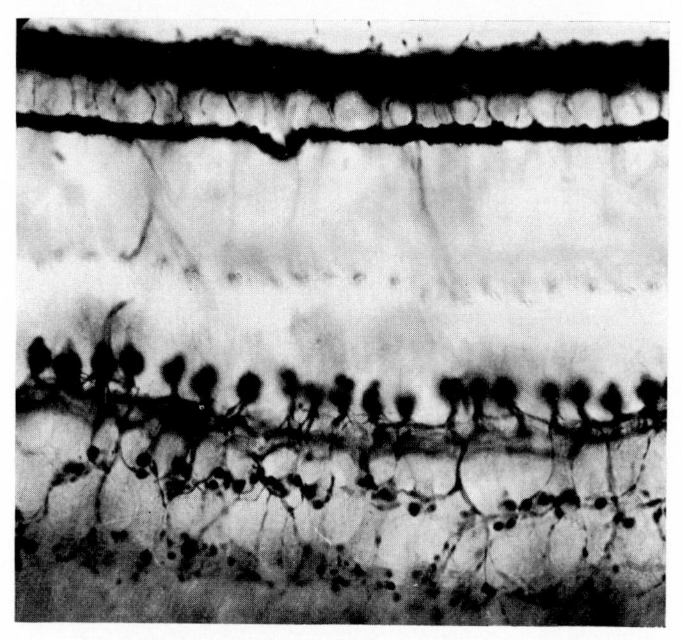

FIG. 2. (Engström). Surface view of the organ of Corti of a guinea pig, stained with a modified Maillet stain. The black areas correspond mainly to efferent fibres and endings.

of the individual hair cell are necessary (Lowenstein, O. [1955]. *J. Physiol., Lond.*, **127**, 104–117)?

Johnstone: The characteristic S-shaped curve of the hair cell in the cochlea is mechanical, not electrical, whereas in the fish, such a characteristic curve is probably due to the cell's electrical function, not to its mechanical function. In other words, the hair cell in the macula of fishes acts as a rectifier, due partly to its electrical characteristics; it has no positive "endocochlear" potential, so it is not biased centrally by a positive endocochlear potential. I think it would not act like that in the cochlea, but I suspect that it may well do so in the vestibular system or in a system where the electrical biasing point is up at one end of the characteristic curve.

Fex: This is where we disagree. If there is mechanical nonlinearity, I cannot see why it would not have an electrical sign, corresponding to the summating potential of Dr. Davis. I think that the crossed efferents shift the d.c. bias, or the polarization, of the hair cell membranes and that this would interact with such polarization to which mechanical nonlinearity could give rise. J. E. Desmedt ([1962]. *J. acoust. Soc. Am.*, **34**, 1478–1496) has shown, and I (Fex, J. [1962]. *Acta physiol. scand.*, **55**, Suppl. 189) also, that these fibres can exert an inhibition roughly corresponding to a lowering of sound pressure of, say, 20 db, which is quite a lot, not only a fine control. On the other hand I am not willing to draw conclusions about the size of current from my potentials; I would rather have intracellular records for that, although I am not sure if anything meaningful concerning the size of currents could be seen even there.

Whitfield: Stimulating the olivo-cochlear bundle certainly does influence the summating potential.

Johnstone: I agree with Dr. Fex, but this would be a very fine control. Perhaps this is right; you may find that the increase in current caused by stimulating the olivo-cochlear bundle is only about 5–6 per cent of the standing current, and hence it will correct this nonlinearity. This may be highly significant in certain circumstances. The problem is that we have here a system which is very coarse in that it has relatively few fibres and they go everywhere, but in the circumstances you have outlined it could only exert a very fine control—a very fine positioning of the characteristic curve.

From what Dr. Fex has said, and from discussion with Dr. Spoendlin and Mr. Bosher, I would like to suggest a way in which the olivo-cochlear bundle might work. Perhaps we can assume that fibres going to the internal hair cells are responsible for any sharpening and temporal discrimination, as previously discussed (see p. 300); so I shall confine myself to the olivo-cochlear bundle that goes to the external hair cells, the crossed efferents (at least in the cat). If we list six of its known properties, we find that in the cat: (1) It has a coarse innervation—Dr. Spoendlin said 500 fibres; (2) From the electron micrographs, the innervation is massive, of huge nerve endings, so one expects that it will do a lot of work or for a long time; (3) The innervation is graded from the base to apex of the cochlea, being very heavy at the base and much less at the apex.

Engström: There may be a second, longitudinal system also for the outer hair cells, in the tunnel of Corti—a cross-correlation system.

Johnstone: If we list, in addition, the electrical activity of the olivo-cochlear bundle, (4) It modulates the action potential of the afferent nerve; (5) Its effect is slow: Dr. Fex said that the effect of stimulating the olivo-cochlear bundle dies away in about 100 msec.; also, you must stimulate several times before something comes out. (6) It increases the recorded cochlear microphonics.

Fex: With intracellular recording, using an increased content of calcium in artificial perilymph (as mentioned in my paper, p. 169), you can see an evoked potential after 4 msec.; I would reserve judgment on how slow the effect is.

Johnstone: From these morphological and physiological characteristics, my explanation for this particular group of efferent fibres is that it acts to stabilize the threshold, primarily against fluctuations in the positive endocochlear potential (+EP). In guinea pigs the long-term stability of the +EP is very dependent on the condition of the animal; if the condition is poor, particularly if you tie off the arteries a little, the +EP is low, about 60 mv, and Dr. M. Lawrence has some pictures showing variations in +EP linked to respiratory changes.

If we give vasoactive substances—adrenaline, noradrenaline and acetylcholine —we can often vary the +EP from 70 mv up to 85 mv. In the normal, free animal, presumably during fear reactions or startle reactions, blood is shunted from the head to other places and there may be variation of the +EP. We know that the +EP does vary with the blood supply, possibly also locally between different sections of the cochlea, perhaps by autonomic regulation (Perlman, H. B., Kimura, R. S., and Fernández, C. [1959]. *Laryngoscope, St Louis*, **69**, 591–613). I suspect that it would also vary if the middle-ear muscles were contracted, because if the cochlear partition were distorted, the endocochlear potential would vary.

But if the +EP varies the threshold will vary, because the microphonics are proportional to the +EP and the action potential is a function of the microphonics. Thus if there is a lower +EP the threshold will be raised. My hypothesis therefore is that normally there are fluctuations of one sort or another in the positive endocochlear potential, perhaps due to changes in the stria vascularis metabolism or blood flow, and these would cause fluctuations in threshold, which would be undesirable.

The olivo-cochlear bundle to the outer hair cells does not end on the nerve fibres, as Dr. Spoendlin has shown. I suggest that its function is not to control the nerve fibres directly. It ends on the hair cells themselves, and the parameter which will keep the threshold of the action potential stable is the current through the hair cell; the alteration in the action potential or in the threshold that I have hypothesized is not due directly to the change in +EP, but to the fact that the current through the hair cell will fluctuate because the driving force—the battery —changes. So it is the current through the hair cell which must remain stable in order to stabilize the action potential. If we imagine tonic efferent impulses coming up to the hair cells and causing resistance changes there, the current through the hair cell could be increased or decreased according to these tonic discharges, and they could conceivably compensate for fluctuations in +EP, so accounting for properties (4) and (6) of the olivo-cochlear bundle.

This hypothesis has certain consequences: we can imagine the scala media as a cable, and the electrical length constant of the scala media is about 2 mm. (Johnstone, B. M., Johnstone, J. R., and Pugsley, I. D. [1966]. *J. acoust. Soc. Am.*, **40**, 1398–1404) which means that if you make a point fluctuation in voltage, 2 mm. away it has died to $1/e$ of its value. This is a large change and I would say that we need to stabilize better than this, possibly to a few per cent, which would be about

0·2 mm. Electrically then, as far as this problem of endocochlear potential is concerned, we can regard the scala media as a series of lumped elements, each about 0·2 mm. long. If you want to control +EP, all you must do is to control each of these elements; you need no finer control than that. In a cochlea of, say, 25 mm. (25,000 μm.) we have 500 fibres to subserve that function. Since there are three rows of hair cells, one efferent fibre would control about 3 × 50 μm., or 150 μm., of endocochlear potential, which is about the figure that we have obtained from the length constant. This would explain the coarse efferent innervation (property 1): no more fibres are needed; one just needs fibres going up every so often to "clamp" the current through the hair cells; there is a co-operative effort.

The enormous size of the efferent endings may have the same explanation. One would not need particularly fine control; one would possibly need a long-term control (property 2). These changes in +EP, resulting perhaps from blood vessel changes, would be long-term and of fairly slow onset, so a fast transmitter substance would not be required, which may explain why the transmitter substance has a comparably long time-constant. It would be an embarrassment if it were fast; it might be interpreted as an input signal (property 5). It has, however, a rise-time (Dr. Fex's figures suggest about 5–10 msec.) which corresponds to 100–200 cyc./sec. If the efferent bundle was anywhere near where the cochlea analysed frequencies around 100 cyc./sec. one would get into trouble, but it is not; down at the low frequency end of the cochlea one does not find the efferent endings. This might explain why they are graded in distribution (property 3).

Davis: There is the germ of an excellent idea here, although I would suggest a different emphasis, one which points to a possible experiment. One of the problems in thinking about the mechanism of the organ of Corti is the tremendous dynamic range of amplitude of movement to which it is exposed. We certainly encounter nonlinearity of movement as we go up the scale of amplitude. I fully agree with Dr. Whitfield's comments on this (p. 296). The hypothesis that the flow of current through the cell is what does the job of excitation is my own view, and I am happy to see it endorsed. It is the only working hypothesis or model that I have been able to think of. The current through the hair cells must vary over a very wide range, and it is presumably that current which controls the liberation of the chemical mediator. I suggest that it would be biologically efficient to stabilize the threshold or perhaps to prevent wastage of the chemical mediator, which presumably would occur when the cell is stimulated much above its threshold. The effect of the efferent nerve fibres in modulating the action potential seems to be just such an elevation of threshold. If that elevation is brought about, not by a change of sensitivity of the afferent nerve endings to the chemical mediator but by reducing the release of the mediator, this would give it a conserving function. The experiment which would test this hypothesis is to see whether animals in which the crossed system has been cut and has degenerated are more subject to auditory fatigue than those with an intact and functioning crossed olivo-cochlear system.

Fex: Dr. Johnstone's hypothesis is a very interesting one, and another experiment which might test it is as follows. There are the techniques worked out by A. Rupert, G. Moushegian and R. Galambos ([1963]. *J. Neurophysiol.*, **26**, 449–465) with implanted microelectrodes in the cat. I suggested a few years ago to Dr. Moushegian that he could investigate the activity of the cells of origin of the olivo-cochlear fibres under different behavioural conditions. You could thus have a cat walking around and see whether its crossed efferents vary their activity with respiration, for instance.

The other side of the issue is, how near the brink of disaster is the endocochlear potential and how dependent on the oxygen supply? When I record from the scala media I often see variation but I then feel that the technique might be deficient. By manipulating the electrode, I can sometimes get rid of what I believe is the drag on the electrode tip by the tectorial membrane; and then suddenly I get an extremely stable endocochlear potential, at my resolution of 0·1 mv. So the question is, what kind of variations of endocochlear potential would really be critical? The kind of variations that Dr. Lawrence has described and the ones you have seen would be larger than 0·1 mv, presumably?

Johnstone: Variations in the endocochlear potential have been recorded in Professor F. Kawata's laboratory (Dept. of Otolaryngology, Fukuoka Medical School, personal communication) by stimulating sympathetic fibres or by stimulating in the hippocampus in the cat. This changed blood-flow patterns, which changed the endocochlear potential, either up or down, by about 5 or 6 mv.

Bosher: May I comment on the stability of the endocochlear potential, which I have been recording by creating a small fenestra over the middle turn of the rat cochlea and inserting a microelectrode through the spiral ligament and the stria vascularis. There is no doubt that when the experimental conditions are well controlled the potential is extraordinarily stable, but as soon as there is any disturbance—say, somebody bangs a door or respiration changes because of a little mucus in the bronchi—fluctuations occur and these are often about 4–5 mv. In other words, they are frequently of the same order as the changes Dr. Fex produced when he stimulated the olivo-cochlear bundle. It might be a coincidence, but it seems a curious one.

Davis: We have probably all seen fluctuations of the order of 5 mv. The contractions of the stapedius muscles are a potent source of this kind of fluctuation. But as far as the function goes, I am not so much concerned about this, because it is a variation of only 5 mv out of a total of some 150 mv which is available as the driving force for the current that we are assuming is the stimulus. So I cannot be quite as concerned as you are, Dr. Johnstone, about the necessity for stabilizing the endocochlear potential. That is why I prefer the alternative idea that the flow of current is stabilized in order to prevent wasteful liberation of the chemical transmitter.

Johnstone: The standing current through the cell is large compared with the variation in current caused by the microphonics, a minimum ratio of 10:1; hence the percentage variation which normally occurs in the standing current through modulation by the changing resistance due to sound is not very great, and I would suggest that a change of 5 mv in the endocochlear potential may cause a change in standing current of the same order.

Davis: I would say, in view of the dynamic range of sounds that the ear deals with, "what of it?" Fluctuations of this size will surely turn out to be relatively insignificant.

Johnstone: I agree, except at threshold. There they will be significant.

Schwartzkopff: Would it be in conflict with your hypothesis, Dr. Johnstone, if I suggested that the efferents stabilize not the threshold, but the level of adaptation? That is my word for what Dr. Davis was saying. By adaptation I mean something similar to what happens in the retina; that you would have to bring the whole system to a certain level of noise, for instance, and then re-establish high sensitivity for differences on this new level.

Johnstone: When I suggest that this system stabilizes the threshold, this is just one point on the adaptation curve; it would of course stabilize the whole system. But presumably the system requires greater stabilization at threshold.

Davis: The biological purpose of this system is, I believe, to maintain optimum sensitivity to change. The organ of Corti should be set so that it is working in the middle of its operating characteristic. The output of the nerve fibre must not be saturated, nor should the stimulus be too far below threshold. It would be most effective near the centre of the operating curve because the ear would then be most sensitive to change in stimulation.

Johnstone: Yes; and to be at optimum sensitivity, it must eliminate outside changes.

Davis: I am thinking of a later step, the output of the nerve fibres in terms of nerve impulses.

Johnstone: I see what you mean: in other words, it is the nerve output which it is desired to regulate against fluctuations in endocochlear potential.

Davis: I agree completely with your analysis of the coarse innervation; and the mechanical spread of the Békésy envelope is broad, so the system should keep the whole area adjusted.

Fex: Even though the efferent system appears crude, there is a possibility for it to create subtle changes in the frequency coding, if the auditory system does work with frequency coding through the phase-locking that has been demonstrated here by Professor Rose (pp. 144–157), and first by I. Tasaki ([1954]. *J. Neurophysiol.*, **17**, 97–122). If during the auditory stimulus the output of the efferents is increasing and the sound response thus takes place during a rising phase of inhibition, the efferents might create a gradual shift in the phase-locking of the responses in the auditory nerve fibres. This could create an effect as of a shift in the frequency of the stimulating tone, as seen by the next auditory station, the cochlear nucleus.

Whitfield: Dr. Johnstone, when you talk about the endocochlear potential, where are you actually envisaging the current flow? There are at least three different potential levels involved, because the interior of the hair cell is negative to both the scala media and the scala tympani.

Johnstone: I should like to modify my earlier idea a little. Instead of stabilizing the current through the hair cell we should really talk of stabilizing the *changes* in current (caused by sound). This can be done by altering the standing current through the hair cell, as previously suggested.

Whitfield: But we were talking about fluctuations of the endocochlear potential. The reticular lamina appears to be a resistive element, so that assuming for the moment that the hair cell potential is stable, the current through the lamina will vary in direct proportion to the change in the endocochlear potential. This current will leave the hair cell through its (very extensive) sub-reticular surfaces. However, unlike the reticular surface most of the potential difference across the rest of the cell is due to an e.m.f., and comparatively little to the IR drop. In other words, the resistance is comparatively low, and most of the current can "escape" through the very long side walls. The distant base of the hair cell will be only minimally exposed to any current fluctuations. This might even be one reason why the afferent endings are down there at the very tip! It looks as though the synaptic region is already fairly well protected against such changes.

Johnstone: We can stabilize the current in two ways by altering the resistance: (1) by a series resistance; (2) by a shunt resistance. Only the current through the region of the afferent synapses need be controlled and the efferent endings act as a shunt resistance. So I suppose that the total cell current is not stabilized and in fact if the +EP falls, the olivo-cochlear bundle will decrease its activity—that is, raise the shunt resistance, thus increasing back to normal the current through the afferent region. This of course means that the total hair cell current will have fallen even further. So the direction of this action is that a reduction in +EP should lead to a reduced activity in the olivo-cochlear bundle.

Tumarkin: It is also important to keep the two ears balanced; is there any possibility that one ear communicates with the other in order to maintain a balance, using the crossed efferent system?

Fex: Yes, certainly. In decerebrate cats in which one ear is stimulated with sound, this stimulation evokes activity in the crossed efferent fibres running to the other ear. This has been shown by direct recording from single efferent fibres (Fex, J. [1962]. *Acta physiol. scand.*, **55**, Suppl. 189).

Spoendlin: Dr. Johnstone, have you any idea how the efferents would be activated, in order to stabilize fluctuations of the endocochlear potential?

Johnstone: If the steady current through the afferent region is kept constant, then the microphonic variation of that current will also be constant, even if the total cell currents, both steady and microphonic, are altered by +EP changes. Perhaps this steady current causes the spontaneous firing noted by Professor Rose. And this, when integrated over some time and place along the basilar membrane, could signal changes in +EP. Perhaps it may also have something to do with the fact that the fibres are crossed.

Bosher: Dr. Spoendlin, you have shown that there is a system of sympathetic fibres to which you were unable to assign a function. Might these be part of the system?

Spoendlin: Unfortunately they are rather far away from the scala media. They are in the osseous spiral lamina, and probably do not come up into the organ of Corti.

Davis: We should emphasize that the various parts of the surface of the hair cell are not equivalent. There is the specialization of the cuticular layer with its cilia and the thin spot; the membranes of the sides of the cell, which presumably are doing a job as batteries; there is the sodium pump which is common to practically all cells; then there is the subsynaptic membrane, probably of two different varieties at least, with their different properties. Many of our ideas about polarization, excitation and so on have been based implicitly on the idea of a uniform membrane but the hair cell emphatically is not uniform. It is specialized in different ways in different areas.

CHAIRMAN'S CLOSING REMARKS

PROFESSOR O. LOWENSTEIN

In closing this symposium, let me make the general point that it takes considerable effort to assemble such a group of people from all parts of the globe, and if an effort is taken on all sides, it should be worthwhile. What then are our criteria of worthwhileness? A symposium serves two purposes, I suggest. It is a medium through which we can obtain information—this is especially important nowadays when one hears so much about a publication explosion and the impossibility of keeping up with the current literature. An hour in such a meeting as this is worth, I should say, a term's vain efforts in reading the literature. And I think there is no doubt that on this occasion we got plenty of information.

The second function of a symposium, to my mind, is to make someone who has participated in it go back to his laboratory and do experiments following a new idea which he conceived at the symposium. Now I know that this has actually happened in this series of symposia. To quote one case: Professor Rose has said that after the symposium on "Touch, Heat and Pain", he carried out a set of experiments which were triggered off in part by the discussions of that meeting, and the results were the astonishingly beautiful facts which he has presented at this symposium. I would close these proceedings, therefore, with the hope that among us there are several who will be induced to follow new lines in their research efforts as a result of ideas generated here.

INDEX OF AUTHORS

Numbers in bold type indicate a contribution in the form of a paper; numbers in plain type refer to contributions to the discussions

Anderson, D. J. **144**

Batteau, D. W. 13, 60, 61, 119, 203, **234,** 239, 240, 241, 242, 243

Bosher, S. K. 16, 39, 87, 124, 125, 141, 256, 301, 306, 309

Bredberg, G. . **126,** 139, 140, 141

Brugge, J. F. **144**

Davis, H. 15, 16, 38, 40, 85, 87, 119, 142, 159, 161, 162, 164, 165, 166, 184, 232, 241, 242, 257, 258, **259,** 269, 270, 271, 295, 306, 307, 309

Enger, P. S. **4,** 11, 12, 14, 15, 16, 17, 38, 168, 205

Engström, H. 14, 84, 88, 122, 123, 124, 141, 204, 293, 294, 300, 303

Erulkar, S. D. 159, 162, 166, 181, 255, 270, 287

Evans, E. F. 162, 163, 232, 255, 256, **272,** 287, 289, 290, 291, 293

Fex, J. 13, 14, 88, 119, 122, 140, 142, 162, **169,** 181, 183, 184, 185, 186, 254, 255, 257, 270, 290, 291, 299, 300, 303, 306, 307, 308

Fraser, F. C. 11, 12

Hallpike, C. S. . . 14, 62, 139, 241

Hind, J. E. **144**

Hood, D. J. . . . 39, 141, 271

Johnstone, B. M. 14, 63, 86, 87, 119, 161, 182, 183, 185, 201, 240, 242, 243, 258, 271, 296, 297, 299, 302, 303, 304, 306, 307, 308

Kleerekoper, H. **188,** 201, 202, 203, 204, 205, 270

Lowenstein, O. E. **1,** 12, 13, 14, 16, 39, 40, 62, 63, 84, 85, 87, 88, 142, 161, 168, 181, 201, 203, 204, 205, 239, 241, 258, 269, 271, 287, 290, 291, 293, 294, 296, 297, 298, 302, **310**

Malar, T. **188**

Neff, W. D. 62, 162, 165, 204, **207,** 231, 232, 233, 269, 291, 293, 294, 295

Pye, Ade 88, 139

Pye, J. D. 38, 39, **66,** 84, 85, 86, 87, 88, 202, 240, 242

Rose, J. E. 141, **144,** 158, 159, 161, 162, 164, 165, 166, 167, 231, 233, 270, 289, 290, 291, 292, 293

Salomon, G. . . 84, 85, 255, 269

Schwartzkopff, J. 12, 14, 16, 37, 39, 40, **41,** 59, 61, 62, 63, 85, 86, 160, 162, 164, 166, 167, 203, 233, 240, 242, 243, 288, 294, 296, 297, 307

Spoendlin, H. 16, **89,** 119, 122, 123, 124, 125, 162, 163, 298, 300, 308, 309

Tumarkin, A. 13, **18,** 38, 39, 84, 87, 140, 142, 185, 186, 202, 203, 204, 290, 298, 308

Whitfield, I. C. 38, 40, 59, 60, 62, 141, 142, 157, 161, 164, 165, 166, 167, 184, 185, 204, 232, 233, 240, 243, **246,** 254, 255, 256, 257, 258, 269, 271, 288, 295, 296, 297, 303, 308

INDEX OF SUBJECTS

Acetylcholine, as transmitter substance of cochlear efferent fibres, 175–178, 181–184
  effect on cochlear microphonics, 175, 184
  effect on hair cells, 175, 181
Acoustic behaviour in birds, 42
  message, first coding of, 89–125
Action potential, and inner hair cells, 184–185
  role of olivo-cochlear bundle, 303
Adrenergic innervation, in acoustic message coding, 114, 117
Ageing, changes in hair cells, 132–134
Amphibians, middle-ear evolution of, 19, 22–23, 27
Anabantidae, 7
Anaesthesia, and tonotopic frequency representation, 273, 287, 289–290
Anura, 30, 36
Arch of Corti, in bats, 76
Archosaurs, middle-ear evolution of, 29–30, 37
Asio otus, 45, 294
Athene noctua, 54, 294
Auditory bulla, in bats, 77, 87
Auditory centres, in medulla, 53–54
Auditory conducting apparatus in vertebrates, evolution of, 18–40
Auditory nerve,
  activity, in birds, 51, 52–53
    in fish, 9
  connexions, in bats, 77
    in birds, 52
Auditory nerve degeneration, industrial noise causing, 136
Auditory nerve endings, on hair cells, 105–106
Auditory nerve fibres,
  afferent, distribution in organ of Corti, 107, 109, 112, 116, 122–123
    in cochlea, 105, 120
    relation to hair cells, 119
    species differences, 124
    types of, 122–123
  distribution of, in birds, 52
Auditory nerve fibres—*continued*
  efferent, and endocochlear potential, 308
    cochlear function, 298–309
    distribution of, 301
    in auditory nerve, 52
    in cochlea, 105, 107, 120
    inhibition, 116, 169–186
    in organ of Corti, 106, 107, 109, 112, 116
    size of, 305
    species differences, 124
    types of, 122–123
  electrotonus in, 124–125
  entering cochlear nuclei, 248, 255
  in cochlea, 106–107
    amounts, 292, 293
  inhibitory, 257
  interspike intervals, distribution of, 145–149
  in the organ of Corti, 104–105
  patterns of activity in squirrel monkey, 144–168
  periodic discharge, 160–161
Auditory pathways, analysis of complex sounds, 287
  and localization of sound, 212–216, 228, 231, 233
  centrifugal control, 246–258
  in birds, 52–57
  in higher brain centres, 56–57
  latency of inhibition, 256
  major, 248
  producing facilitation, 246, 255
Auditory stimuli. *See also* Sound
  cortical responses in human, 259–271
    latency, 264, 269
  EEG response, 256, 261
  effect on cochlea in bats, 70, 71
  effect on cochlear nucleus, 248
  effect on hair cells, 116–117
  effect on muscles, 257
  from ear to forebrain, in birds, 56–57
  masking, 155, 158
  monitoring of, 256
  multiple discharge, 148
  near-field and far-field, 4–6, 188

Auditory stimuli—*continued*
periodograms, 149–150
probability of cycle being effective, 151–155
reception, in birds, 44–45, 59–60
relation of period to discharge timing, 147
representation of temporal features, 278–281
response to, 119
in brain of fish, 12, 14
single unit response to in bats, 78
to inner ear of birds, 51
vertex response, 260–268, 269–271
Auditory threshold, in birds, 43
in fish, 15
sound frequency range and, 6–8
Auks, 45

Basilar membrane, 89
and pitch discrimination, 167–168
displacement of, 296, 297
in bats, 74, 88
in birds, 50
ion permeability, 186
puncture of, 142
role of, 240, 242, 296
thickening of, 75, 122
ultrastructure, 91
width of, 132
Bats, central nervous system in, 77–80
cochlea in, 73–77, 87
directional sense, 88
electrophysiological investigations, 77–80
external ear in, 67–71
hearing in, 66–87
inner ear in, 88
middle ear in, 71–73, 84
Behaviour, and hearing, 42–44, 80–81, 189
Békésy travelling wave, 164, 166, 240, 241, 299–300
Binaural sound localization, 207, 210, 225, 228
Birds, auditory pathways, 52–57
ear and auditory brain areas in, 41–63
ear function and behavioural studies, 42–44
efficiency of hearing, 43–44, 53–54
external ear of, 44–46
"extra-auditory" hearing, 60
flightless, 63
higher brain centres in, 56–57
Birds—*continued*
information transfer in, 164
inner ear in, 46–50
learning processes, 61
middle ear in, 46
neurophysiology of medulla, 54–56
pitch discrimination in, 45
sound differentiation in, 61
upper limits of hearing, 55, 59–60
Bird song, 61
Blocking agents, 248–249, 255
Body temperature, and collicular evoked potentials, 80
Bombinator, 19
Brain, auditory areas in birds, 41–63
higher centres in birds, 56–67
Brainstem, pitch discrimination and, 250–251
Breevoortia tyrannus Latrobe, 198
Bubo bubo, 43, 54

Calcium, role in efferent transmitter substance, 170, 181
Carp (*Cyprinus carpio*), effect of sound on, 193–198
locomotion of, 189
methods of study, 189–190
Cat, behavioural experiments, 291–292, 293–294
cochlea in, 106, 107
cortical representation in, 272–285
hair cells in, 123
olivo-cochlear fibres in, 124
olivo-cochlear inhibition in, 169–186
response latency in, 85–86
sound localization by, 208 *et seq.*
Cerebellum, and sound location, 216
Cerebral cortex, different horizontal layers, 288–289
distribution of frequencies, 272–278, 288
effect of cochlear stimulation, 272–273
human, auditory evoked responses in, 159–271
organization, 289
pattern of response, 290
point-to-point projection of cochlea, 272
representation, of frequency, 272–278, 290, 291
of spatial location of sound, 281–284
of temporal features of stimulus, 278–281

Cerebral cortex—*continued*
responses evoked in, 259–271
effect of anaesthesia, 273, 287, 289–290
in bats, 78, 79–80
role in sound localization, 219–225, 228, 232–233
tonotopic arrangement of frequencies, 273–278, 290, 291
V (vertex) potentials, 259–271
Chamaeleo, 19, 33
Characinidae, 6
Characteristic frequency, definition, 275
Chilonycteris, 69, 73, 74, 75, 81
Cholinesterase, in nerve fibres, 249
in papilla basilaris, 50
Clupea harengus (herring), auditory threshold, 7, 15
Clupeidae, 7
Cochlea. *See also specific structures* Basilar membrane, Tectorial membrane, Hair cell *etc.*
adrenergic innervation, 114, 117
afferent nerve fibres, 105, 120
basilar membrane. *See* Basilar membrane
degeneration of, 136, 137
and ion leakage, 141–142
effect of acetylcholine on microphonics, 175, 184
effect of age, 132–134
effect of industrial noise, 136
efferent inhibition in by olivo-cochlear bundle, 169–186
blocked by strychnine, 182, 183
hair cells and, 171–173
site of action, 169–175
transmitter substance, 175–178, 181
efferent nerve fibres in, 105, 120
activation of, 178
and endocochlear potential, 308
distribution of, 301
function of, 298–309
frequency units, 291
in bats, 73–77, 87
in birds, 46
in cat, 106, 107
in Ménière's syndrome, 140–141
in pitch discrimination, 9
length of, in birds, 47
nerve endings, 102
nerve fibres in, 52, 106–107
amounts, 292, 293
Cochlea, nerve fibres in—*continued*
distribution, 102, 104
in bats, 88
neurons, 115–116, 120
inhibition, 159–160
"normal" cell complement, 129, 140
number of hair cells in, 130–131
point-to-point projection on to cortex, 272
potentials evoked in, 117
in bats, 70, 71–72, 76, 80
recovery, 86
receptor, components, 89
origin of microphonics, 95, 296
peripheral innervation of, 89–125
ultrastructure, 89–125
responses in, in bats, 79
sensitivity to time-differences, 158
sharpening, 299, 300
spiral ligament, in bats, 74–75
stimulation of, effect on cortex, 272–273
structure, in bats, 73–74
hearing and, 126–142
study of, 139–140
by surface specimen technique, 126–127, 139
supporting cells, 89
in bats, 76, 86
synaptic activity in, 107, 116, 120, 123
transduction in, 120
uncrossed efferents in, 174–175
Cochlear microphonic, in bats, 72–73, 76–77, 86–87
in birds, 51, 52
origin of, 95
Cochlear nucleus, connexions from lemniscal column, 250
effect of olivary stimulation, 249
facilitation of discharge, 250, 251
inhibition in, 246
nerve fibres entering, 248, 255
Cockroach, vibration receptors, 297
Codfish (*Gadus morrhua*), hearing in, 6
Colliculus, echo sensitivity in, in bats, 80
potentials and body temperature, 80
response in, latency, 85
Columba, 43, 54, 55, 59–60, 174
Commissural pathways, role in sound localization, 216–219
Control mechanisms of auditory pathways, 246–258
Cortex. *See under* Cerebral cortex

Corti, arch of. *See* Arch of Corti
organ of. *See* Organ of Corti
Cortical representation, tonotopic, 272–278, 288, 289
Cottus scorpius, 6, 9, 10, 15, 168
Cyprinidae, 6, 189

Deafness, low-tone perceptive, 140
Deiters' cells, 89, 106
junction with hair cells, 141
Dimetrodon, 29, 35
Displacement, as auditory stimulus, 4–6, 13–14
Dogfish (*Scyliorhinus caniculus*), auditory threshold, 8
Dolphin, acoustic nerve fibres, 293, 294

Ear(s), *See also under individual regions*
air conducting, 18
evolution of, 27–29, 31–36
anatomy and function, in birds, 44–52
asymmetry of, 45, 62, 242–243
bone conducting, 19
evolution of, 20–27, 39
classification of, 18–19
efficiency, in birds, 43–44
evolution of, Anura, 30, 36
Archosauria, 29–30, 37
functional theory, 22–36
in reptiles, 27
lizards, 30
mammalian, 31–36, 39
premammalia, 29
text-book theory, 19–20
function of, 41
in birds, 41–63
in early amphibians, 19–20
in pelycosaurs, 28, 29–30
mobility of in bats, 68
performance according to behavioural studies, 42–44
time-differences between, 242
transitional forms, 19
Echolocation, ultrasonic. *See* Ultrasonic echolocation
Efferent auditory fibres. *See* Olivo-cochlear fibres
Electrotonus, concept of, 124
Endocochlear potentials, in birds, 51
fluctuations in, 304–309
Endolymph, mixing with perilymph, 142
Eogyrinus, hearing mechanism in, 19, 21, 22, 23, 27, 33
Eptesicus fuscus, 80
External ear, anatomy and function in birds, 44–46
in bats, 67–71

Facilitation, production of, 85, 246, 255
Falco tinnunculus, 54
Far-field, hearing in, 4–8, 188–189, 191
Fish. *See also under individual species*
effect of sound intensity, 16
hearing in, 4–17
locomotion of, methods of study, 189–193
near-field and far-field effects of sound sources, 4–6, 188–189
orientation by sound, 188–206
pitch discrimination in, 6, 8–10, 15, 16, 168
Forebrain, in bats, 77, 80
in birds, 56
Frequency. *See* Pitch
Frequency analysis, in fish, 9, 10, 16, 161, 164–165
in lizard, 161
Frequency modulated sounds,
responses to, 60–61, 278–280, 287
in bats, 79–80, 81

Gadus morrhua, 6
Gekko, 19, 30, 33
Gobius niger, 10
Goldfish (*Carassius auratus*), auditory threshold, 5, 7
maculae in, 14

Hair cells, 95, 109, 115
activity of, 125
and endocochlear potential, 308
degeneration of, 129, 132–134, 141
differential innervation of inner and outer, 106–112, 173
effect of acetylcholine, 175, 181
effect of *d*-tubocurarine, 177, 185
effect of industrial noise, 134
function, 13, 241
in inner ear of birds, 49
inner, 99
activity, 184
efferent innervation, 107, 116–117, 119, 123
and crossed efferents, 107, 173, 184, 185
and uncrossed efferents, 175

Hair cells, inner—*continued*
numbers of, 130
synaptic structure, 123
invaginations in, 123
junction with Deiters' cells, 141
mucopolysaccharides in, 99
outer, 98, 99
activity, 124, 184
afferent innervation, 105–106, 107, 109, 114, 116–117, 119, 120, 123
and crossed efferents, 105–106, 116, 169–174, 303
effect of damage to, 299
in foetus, 128–129
length of, 139
mitochondria in, 99, 300–301
numbers of, 130
role of, 141
synaptic activity, 123, 171–172, 174
postsynaptic events, 171
receptor poles, 99
site of origin of cochlear microphonics, 95
specialization of regions, 95, 99, 309
synaptic activity, 107, 120
transduction and, 13–14, 95, 120
width of area, 132
Hearing, absolute sensitivity of mechanism, 296–298
and auditory cortex, 278
and behavioural response in bats, 80–81
and degeneration of cochlea, 137
capacity in birds, 43–44
definition, 1
efficiency, correlated with nucleus laminaris in birds, 53–54
"extra-auditory", 60
in bats, 66–87
in fish, 4–17
experimental difficulties, 5
upper limits of, in birds, 55, 59–60
Hemispherectomy, effect on sound localization, 221
Hipposideros, 74, 75
Holbrookia, 19
Holocentrus ascencionis, 7
Hyomandibula, and middle-ear evolution, 19, 23, 24

Inferior colliculus, in sound localization, 216, 229, 231
Information transfer, 240, 295
Information transfer—*continued*
adrenergic innervation in, 114, 117
processing and ecological performance, **54**
coding, 144, 158
in birds, 52–57
mechanism, 144, 157–158
Inner ear, anatomy, in birds, 46–50
biochemistry and electrophysiology, in birds, 50–52
in bats, 88
Intensity, and sensory units, 15
and sound location in fish, 202–203
differentiation of, in birds, 44
effect on fish brain, 16
masking, 155
threshold, in bats, 78, 80, 84, 119
Interference, reduction of by bats, 78
Ion leaks, in cochlea, 141–142

Lacertilia (lizards), middle ear in, 30, 31
Lagena, as statoreceptor, 62–63
connexion with auditory nuclei, 63
function, 48
in birds, 46
in fish, 168
pitch discrimination and, 9, 168
Lagenar nerve fibres, course of, in birds, 48
Larynx, synchronization with stapedius muscle in bats, 73
Latency, of auditory responses, 262, 269
in bats, 78, 85–86
Lateralization of sound, 207–230
Lateral lemniscus and sound localization, 214, 229
centrifugal pathways, 250
Lateral line receptors, stimulation of, 4, 5, 12, 13, 17, 188
Lavia, 67
Lepomis gibbosus, 189, 191–193, 198
Lizards, evolution of ear in, 30, 31
frequency coding in, 161
Localization of sound, 62, 119, 210
auditory pathways, 212–216, 228, 231, 233
by fish, 188–206
by single ear, 242
cortical responses, 281–283
importance of memory, 202–203, 204
in bats. *See* Ultrasonic echolocation
in cats, 294, 295
in space, 207–233
methods of testing, 208–210

Localization of sound—*continued*
representation in cortex, 281–284
role of cerebellum, 216
role of cerebral cortex, 219–225, 228, 232–233
role of commissural pathways, 216–219
role of inferior colliculus, 216, 229, 231
role of medulla, 210–216
role of pinna, 234–243
role of tectum, 216
theory of mechanism, 225, 228–229

Macula, in goldfish, 14
in mammals, 14
Mauthner's cells, 190
Median nerve, stimulation of, 265, 271
Medulla, auditory centres, 53–54
nerve fibre distribution in, 52
neurophysiology of, in birds, 54–56
role in sound localization, 210–216
Membrana tympani, in birds, 46
Menhaden (*Breevoortia tyrannus*), response to sound, 198
Ménière's syndrome, cochlea in, 140–141
Microphonic potentials. *See* Cochlear microphonics
Middle ear, anatomy, in birds, 46
evolution of, 19–36
in bats, 71–73, 84
muscles, 255
contraction of, 84–85
in bats, 71, 84
in man, 84
transmission through, 85
Monaural sound localization, 210, 225, 228, 232
Monkey, nerve fibre patterns of activity, 144–168
organ of Corti, in, 128
Mormyridae, 7
Mucopolysaccharides, as mechano-electrical transducers, 99
Myelination, and origin of action potential, 124
Myotis lucifugus, 70, 71, 76, 77, 80, 81

Near-field, hearing in, 4–8, 188
Negaprion brevirostris, 7, 9
Noise, effect on hair cell population, 134
Noradrenaline, effect on cochlear nucleus, 250, 304
Nucleus angularis, in birds, 53
Nucleus laminaris, correlation with hearing efficiency, 53–53
information received by, 53
Nucleus magnocellularis, in birds, 53, 59

Off-response, 263–264, 270–271
Olivary complex, role in localization of sound, 210, 214–16, 229
stimulation of, 248, 249
Olivo-cochlear bundle, 246–248
See also Olivo-cochlear fibres
Olivo-cochlear fibres (auditory efferent fibres), acetylcholine as a transmitter substance, 116, 175–178, 246, 257
crossed, 169–174, 184
stimulation of, 178, 185, 257
transmitter substance, 175–178, 181, 184, 257
effect of *d*-tubocurarine, 176–177, 182
function of, 178–179, 246, 298–309
in cat, 105–106, 117, 124
inhibition of cochlea by, 169–186, 246
relation to (outer) hair cells, 106–107, 116, 169, 171–173, 304
uncrossed, 174–175
On-response, 263–264, 270–271
Organ of Corti, 92
afferent nerve fibres in, 122–123
degeneration of, 134–35, 136–137, 141
effect of *d*-tubocurarine on, 176–178, 182
effect of industrial noise on, 134–35, 136
foetal development, 128–129
in monkey, 128
length of, and hair cell number, 131
mass, 132
mechanism, 303–308
membrane thickenings, 122
nerve fibres in, 104–105, 106
nerve fibre distribution in, 107, 109, 112, 114, 115
number and distortion of hair cells, 130–132
reduction of distortion, 89, 91
sensory cells in, 128
degeneration of, 132–135
in foetus, 128
species differences, 128
structure, 92, 93
surface specimen technique for study of, 126–127
synaptic activity, 116
transmitter substance in, 175–178

Orientation through sound, in fishes, 188–206
  See also Localization of sound
Ostariophysi, 6, 8, 11
Owls, asymmetric ears, 45–46, 62
  external ear in, 45
  hearing efficiency, 54
  number of acoustic nerve fibres, 294
  sound localization by, 242

Paaw's cartilage in bats, 71
Pacinian corpuscle, 124
Papilla basilaris, 49, 50
Parakeets and parrots, medullary auditory function, 42, 54
Pelycosaurs, evolution of hearing in, 28, 29–30, 38, 39
Perilymph, flow of, 185
  mixing with endolymph, 142
  perfusion of, 182
Periodogram, and coefficient of synchronization, 149–150
Period time code, for frequency information, 144–157, 158, 161, 164–168
Phase-locking to stimulus cycle, 9, 12, 14, 146–151, 307
Phoxinus laevis, pitch discrimination in, 8, 16
  sound localization in, 205
Phyllostomus hastatus, 69, 76
Pigeon (*Columba*), crossed cochlear inhibition in, 174
  upper limit of hearing in, 55, 59–60
Pillar cells, 89, 106
Pinna, distortion of in bats, 70
  importance of size and shape, 70, 239, 241
  in bats, 67
  mobility of, in bats, 69
  role in localization of sound, 234–243
Pitch, amounts of nervous tissue devoted to, 291–292
  and sensory units, 15
  higher limits, 292, 293, 294
  range in humans, 66
  response to, limit, 166
    mechanism, 166
    in cats, 292
Pitch discrimination, 166, 279
  and basilar membrane, 167–168, 298–299, 300, 301–302
  and cochlear fibre number, 292–293
Pitch discrimination—*continued*
  and cortex, 272–278, 290, 291
  and periodicity, 9, 144–159, 165–166
  at brainstem level, 250–251
  in auditory pathways, 288
  in bats, 81, 86–87
  in birds, 45
  in cats, 250, 291–294
  in fish, 8–10, 15, 16, 161
  in humans, 66
  mechanism of (synchronization or frequency analysis), 9, 16, 144, 157–159, 161, 168
  role of efferent fibres (olivo-cochlear bundle), 178–179, 298–299
"Place" representation mechanism, 9, 161–162, 167–168, 273, 300
Plecotus, 67, 70, 78, 80
Power law, psychophysical, and vertex response, 262–263
Pteropus, 69, 74, 76
Pure tones, 60, 61

Ray (*Raja clavata*), labyrinthine neurons in, 9
Receptor(s), types of, 255
  ultrastructure, 89–125
Reptiles, evolution of ear in, 27
Responses, auditory,
  and localization of sound, 282–284
  cochlear, in bats, 78, 79
  effect of frequency, 279
  evoked in cortex, 259–271
    in bats, 79–80
  interspike intervals, distribution of, 145
  latency, 78, 85, 86, 146, 262, 264, 269
  mechanism, 119–120
  phase-locked, 9, 12, 14, 146–151, 159, 160, 161, 164, 307
    upper limit, 151
  probability of, 164, 165
  single unit, 78, 85, 86
  slow wave, 259–260
  species differences, 81
  synchronization coefficient, 149–150
  to two tones, 155–156, 158
Reticular formation, 258
Rhinolophus ferrumequinum, 68, 70, 73, 75, 81

Saccular microphonics, 6, 12–14
Sacculus, hair cells in, 13
  pitch discrimination and, 9, 168

Sargus annularis, 10
Sculpin (*Cottus scorpius*), 6, 9, 10, 15, 168
Scyliorhinus caniculus, 8
Sensory cells. *See* Hair cells
Shark, auditory threshold, 7, 8, 12
pitch discrimination in, 9
sound localization in, 203, 205
Shearing, as stimulus to hair cells, 13–14,91
Siluridae, 6
Song birds, 42
Sound. *See also* Auditory stimuli
direction of, and perception, 62
ground-conducted, 21, 23–25, 26, 27, 37
localization of. *See* Localization of sound
of own voice, 40
orientation by, in fishes, 188–206
production by imitation, 42
reflex orientation to, 225
transmission, in water and air, 5
underwater source of, effects, 4
experimental difficulties, 5
Space of Nuel, 76
Sphenodon, 19, 30
Spiral ligament, in cochlea, in bats, 74–75
Stapedius muscle, in bats, 71, 72
relation to larynx in bats, 73
Stapes, evolution of, 26, 30, 33, 34, 35, 36, 39
Strychnine, blocking efferent fibres, 182, 183
Sunfish (*Lepomis gibbosus*), locomotion, 189, 191–193, 198
Swim bladder, role in sound perception, 5, 6–8, 11–12

Tactile stimuli and response, 265, 270
Tadarida, 67, 70, 72
Tectorial membrane, 93, 95
Tectum, role in sound location, 216
Tegmentum vasculosum, in birds, 48–49
Temporal discrimination, and olivo-cochlear bundle, 300
Thigmotaxis, 201
Time-differences, measurement by nervous system, 167, 212, 233, 241–243
Tonality, perception of, 60
Tones, two, response to, 155–156, 158
Tonotopicity, in auditory cortex, 272–278, 288, 289
in cochlea, 50, 300–302
Torus semicircularis, in birds, 56
Tragus, in bats, 67
Transduction, and hair cells, 13–14, 95, 120
Transmitter substance, of cochlear efferent inhibition, 175–178, 182–184
effect of magnesium, 181–182
in organ of Corti, 175–178
role of calcium, 170, 181
Trapezoid body, role in localization of sound, 212, 228
d-Tubocurarine, effect on organ of Corti, 176–178, 182
Tuning curves, 240
in bats, 78
sharpening process, 240, 299, 300, 303
Tympanic lamina cells, 92
Tympanic membrane, in bats, 71
Tyto alba, 45, 54, 294

Ultrasonic echolocation, in bats, 66, 68
difficulties, 87–88
ear mobility and, 68–69
mechanism of, 72–73, 78–79
type of signal employed, 79
Ultrasonic stimulation, effects on bats, 77–78
Urodela, middle-ear evolution in, 30

Vertebrates, evolution of auditory conducting apparatus, 18–40
V(vertex) potentials, in human, 260–268, 269–271
Vestibulo-tympanic mechanism, evolution of, 18, 19, 20, 21
Volley theory of pitch discrimination, 9, 10, 144, 161

Weberian ossicles, 6, 11

Author and subject indexes prepared by Mr. William Hill.

Printed by Spottiswoode, Ballantyne & Co. Ltd., London and Colchester